Sulfur in the Environment

BOOKS IN SOILS, PLANTS, AND THE ENVIRONMENT

Soil Biochemistry, Volume 1, edited by A. D. McLaren and G. H. Peterson
Soil Biochemistry, Volume 2, edited by A. D. McLaren and J. Skujiņš
Soil Biochemistry, Volume 3, edited by E. A. Paul and A. D. McLaren
Soil Biochemistry, Volume 4, edited by E. A. Paul and A. D. McLaren
Soil Biochemistry, Volume 5, edited by E. A. Paul and J. N. Ladd
Soil Biochemistry, Volume 6, edited by Jean-Marc Bollag and G. Stotzky
Soil Biochemistry, Volume 7, edited by G. Stotzky and Jean-Marc Bollag
Soil Biochemistry, Volume 8, edited by Jean-Marc Bollag and G. Stotzky
Soil Biochemistry, Volume 9, edited by G. Stotzky and Jean-Marc Bollag

Organic Chemicals in the Soil Environment, Volumes 1 and 2, edited by C. A. I. Goring and J. W. Hamaker
Humic Substances in the Environment, M. Schnitzer and S. U. Khan
Microbial Life in the Soil: An Introduction, T. Hattori
Principles of Soil Chemistry, Kim H. Tan
Soil Analysis: Instrumental Techniques and Related Procedures, edited by Keith A. Smith
Soil Reclamation Processes: Microbiological Analyses and Applications, edited by Robert L. Tate III and Donald A. Klein
Symbiotic Nitrogen Fixation Technology, edited by Gerald H. Elkan
SoilWater Interactions: Mechanisms and Applications, Shingo Iwata and Toshio Tabuchi with Benno P. Warkentin
Soil Analysis: Modern Instrumental Techniques, Second Edition, edited by Keith A. Smith
Soil Analysis: Physical Methods, edited by Keith A. Smith and Chris E. Mullins
Growth and Mineral Nutrition of Field Crops, N. K. Fageria, V. C. Baligar, and Charles Allan Jones
Semiarid Lands and Deserts: Soil Resource and Reclamation, edited by J. Skujiņš
Plant Roots: The Hidden Half, edited by Yoav Waisel, Amram Eshel, and Uzi Kafkafi
Plant Biochemical Regulators, edited by Harold W. Gausman
Maximizing Crop Yields, N. K. Fageria

Transgenic Plants: Fundamentals and Applications, edited by Andrew Hiatt
Soil Microbial Ecology: Applications in Agricultural and Environmental Management, edited by F. Blaine Metting, Jr.
Principles of Soil Chemistry: Second Edition, Kim H. Tan
Water Flow in Soils, edited by Tsuyoshi Miyazaki
Handbook of Plant and Crop Stress, edited by Mohammad Pessarakli
Genetic Improvement of Field Crops, edited by Gustavo A. Slafer
Agricultural Field Experiments: Design and Analysis, Roger G. Petersen
Environmental Soil Science, Kim H. Tan
Mechanisms of Plant Growth and Improved Productivity: Modern Approaches, edited by Amarjit S. Basra
Selenium in the Environment, edited by W. T. Frankenberger, Jr., and Sally Benson
Plant–Environment Interactions, edited by Robert E. Wilkinson
Handbook of Plant and Crop Physiology, edited by Mohammad Pessarakli
Handbook of Phytoalexin Metabolism and Action, edited by M. Daniel and R. P. Purkayastha
Soil–Water Interactions: Mechanisms and Applications, Second Edition, Revised and Expanded, Shingo Iwata, Toshio Tabuchi, and Benno P. Warkentin
Stored-Grain Ecosystems, edited by Digvir S. Jayas, Noel D. G. White, and William E. Muir
Agrochemicals from Natural Products, edited by C. R. A. Godfrey
Seed Development and Germination, edited by Jaime Kigel and Gad Galili
Nitrogen Fertilization in the Environment, edited by Peter Edward Bacon
Phytohormones in Soils: Microbial Production and Function, William T. Frankenberger, Jr., and Muhammad Arshad
Handbook of Weed Management Systems, edited by Albert E. Smith
Soil Sampling, Preparation, and Analysis, Kim H. Tan
Soil Erosion, Conservation, and Rehabilitation, edited by Menachem Agassi
Plant Roots: The Hidden Half, Second Edition, Revised and Expanded, edited by Yoav Waisel, Amram Eshel, and Uzi Kafkafi
Photoassimilate Distribution in Plants and Crops: Source–Sink Relationships, edited by Eli Zamski and Arthur A. Schaffer
Mass Spectrometry of Soils, edited by Thomas W. Boutton and Shinichi Yamasaki
Handbook of Photosynthesis, edited by Mohammad Pessarakli
Chemical and Isotopic Groundwater Hydrology: The Applied Approach, Second Edition, Revised and Expanded, Emanuel Mazor
Fauna in Soil Ecosystems: Recycling Processes, Nutrient Fluxes, and Agricultural Production, edited by Gero Benckiser
Soil and Plant Analysis in Sustainable Agriculture and Environment, edited by Teresa Hood and J. Benton Jones, Jr.
Seeds Handbook: Biology, Production, Processing, and Storage: B. B. Desai, P. M. Kotecha, and D. K. Salunkhe

Modern Soil Microbiology, edited by J. D. van Elsas, J. T. Trevors, and E. M. H. Wellington
Growth and Mineral Nutrition of Field Crops: Second Edition, N. K. Fageria, V. C. Baligar, and Charles Allan Jones
Fungal Pathogenesis in Plants and Crops: Molecular Biology and Host Defense Mechanisms, P. Vidhyasekaran
Plant Pathogen Detection and Disease Diagnosis, P. Narayanasamy
Agricultural Systems Modeling and Simulation, edited by Robert M. Peart and R. Bruce Curry
Agricultural Biotechnology, edited by Arie Altman
Plant–Microbe Interactions and Biological Control, edited by Greg J. Boland and L. David Kuykendall
Handbook of Soil Conditioners: Substances That Enhance the Physical Properties of Soil, edited by Arthur Wallace and Richard E. Terry
Environmental Chemistry of Selenium, edited by William T. Frankenberger, Jr., and Richard A. Engberg
Principles of Soil Chemistry: Third Edition, Revised and Expanded, Kim H. Tan
Sulfur in the Environment, edited by Douglas G. Maynard
Soil–Machine Interactions: A Finite Element Perspective, edited by Jie Shen and Radhey Lal Kushwaha

Additional Volumes in Preparation

Mycotoxins in Agriculture and Food Safety, edited by K. K. Sinha and Deepak Bhatnagar

Sulfur in the Environment

Edited by

Douglas G. Maynard

Canadian Forest Service, Pacific Forestry Centre, Victoria, British Columbia, Canada

Marcel Dekker, Inc. New York • Basel • Hong Kong

Library of Congress Cataloging-in-Publication Data

Sulfur in the environment / edited by Douglas G. Maynard.
p. cm. — (Books in soils, plants, and the environment ; v.67)
Includes bibliographical references and index.
ISBN 0-8247-8992-X (alk. paper)
1. Sulfur—Environmental aspects. I. Maynard, Douglas G. II. Series.
QH545.S9S88 1998
572'.554—dc21 97-52812
CIP

The publisher offers discounts on this book when ordered in bulk quantities. For more information, write to Special Sales/Professional Marketing at the address below.

This book is printed on acid-free paper.

MARCEL DEKKER, INC.
270 Madison Avenue, New York, New York 10016
http://www.dekker.com

Current printing (last digit):
10 9 8 7 6 5 4 3 2 1

PRINTED IN THE UNITED STATES OF AMERICA

Preface

Sulfur (S) exists in the environment in a number of oxidation states and is extremely difficult to measure in environmental material. Deficiencies of S and possible links of S emissions to ecological impact were identified prior to the twentieth century; however, it was not until the late 1960s and early 1970s (continuing through the 1980s) that intensive research on S was carried out. Despite considerable research on S processes in terrestrial ecosystems, there are still large gaps in our understanding of the S processes in these ecosystems. This book focuses on S research in terrestrial temperate ecosystems during the past 10 to 15 years and suggests where further research may be required.

The first chapter (Maynard) provides a general overview on S in temperate terrestrial ecosystems and sets the stage for the more detailed discussions on S in the environment. Chapters 2 through 5 focus on agroecosystems in North America and New Zealand and Australia. Janzen and Ellert (Chapter 2) summarize the wealth of information on S in the temperate agroecosystems. This chapter reflects the more recent emphasis in agricultural research on the links between agroecosystems and other ecosystems, particularly the atmosphere. In general the discussion focuses on systems low in S. Chapter 3 (Saggar et al.) focuses on a little-known yet increasingly important aspect of S research: soil–plant–animal interactions. The S cycle in these systems is complicated by the introduction of animals that cycle S through their system, resulting in an uneven input of S back to the soil through excreta. Additional pathways of S losses also occur through removal of animal products. The following chapter (Chapter 4, Bozwell and Gregg) continues the discussion on the soil–plant–animal ecosystem as it relates to the S requirements and the S fertilization required to sustain these pasture systems. The final chapter on S in agroecosystems

(Chapter 5, Watkinson and Bolan) looks specifically at the oxidation processes of elemental S (S°). Elemental S, one of the cheapest and most easily attainable fertilizer S forms, is becoming increasingly important, particularly in perennial crop production systems such as pastures.

The second part of the book (Chapters 6 to 10) focuses on S in forested ecosystems. One difficult issue in understanding the dynamics of forest ecosystem is that of spatial scale. As human exploitation of forested lands continues and greater proportions of the forested land become affected, it is increasingly important to look at the dynamics of S at various scales. Chapters 6 (Hogan et al.) and 9 (Arp and Meng) examine the role of S at the cellular level and integrate it to the individual tree level. Chapter 6 discusses the role of excess S at a cellular level and how it affects the overall tree biology. Arp and Meng (Chapter 9) model the impact of low levels of SO_2 on net photosynthesis from a cellular level up to a canopy level.

The seventh and eighth chapters assess S in high S input systems (eastern North America, Johnson and Mitchell) and low forest ecosystems (Southern Hemisphere, Lambert and Turner) at the stand and regional levels. Johnson and Mitchell (Chapter 7) use a dynamic model approach to assess the implications of reduced S deposition for outputs from these forest ecosystems. Lambert and Turner (Chapter 8) discuss the role of S in the temperate and subtropical forest ecosystems of the Southern Hemisphere. These forests differ from the temperate forests of North America because there is a relatively low atmospheric input of S (currently) and the native trees are adapted to low sulfur conditions. These systems maintain themselves primarily through the cycling and conservation of nutrients.

The final chapter of the book (Oja and Arp) assesses the impact of acid deposition on soils and forest ecosystems using a steady-state mass balance model. The basis of the model is to determine the potential impact to the ecosystem of current and predicted future rates of acidic deposition (including both S and N). This approach has been widely used in Europe and is currently being applied to areas in eastern Canada. In the United States, there is still concern about the acceptability of this approach. This chapter demonstrates the use of a steady-state mass balance model to assess the use of critical loads in determining the risk to the ecosystem from acidic deposition.

This book is not all-inclusive but focuses on S in temperate agricultural and forest ecosystems with emphasis on recent research trends. New technologies in S analysis have increased our ability to measure and quantify different forms of S. The biggest influence on current research, however, is the integration of the experimental findings into models of various spatial scales from the cellular to the landscape level.

Douglas G. Maynard

Contents

Contributors

Paul A. Arp, Ph.D. Professor, Faculty of Forestry and Environmental Management, University of New Brunswick, Fredericton, New Brunswick, Canada

Nanthi S. Bolan, Ph.D. Senior Lecturer, Department of Soil Science, Massey University, Palmerston North, New Zealand

Colin C. Boswell, Ph.D. Senior Scientist, Department of Agronomy, New Zealand Pastoral Agricultural Research Institute, Invermay Agricultural Centre, Mosgiel, New Zealand

Benjamin H. Ellert, Ph.D. Research Scientist, Land Resource Sciences, Agriculture and Agri-Food Canada, Lethbridge, Alberta, Canada

Siegfried Fink, Ph.D. University Professor, Institute for Forest Botany and Tree Physiology, Albert-Ludwigs University, Freiburg, Germany

Paul E. H. Gregg, Ph.D. Associate Professor, Department of Soil Science, Massey University, Palmerston North, New Zealand

Michael J. Hedley, Ph.D. Associate Professor, Fertilizer and Lime Research Centre, Massey University, Palmerston North, New Zealand

Gary D. Hogan, Ph.D. Research Scientist, Natural Resources Canada, Canadian Forest Service, Great Lakes Forestry Centre, Sault Ste. Marie, Ontario, Canada

H. Henry Janzen, Ph.D. Research Scientist, Land Resource Sciences, Agriculture and Agri-Food Canada, Lethbridge, Alberta, Canada

Dale W. Johnson, Ph.D. Research Professor, Biological Sciences Center, Desert Research Institute, Reno, Nevada

Marcia J. Lambert, B.Sc.(Hons), M.Sc.(Hons), ASTC Research Scientist and Director, FORSCI Pty. Ltd., Eastwood, New South Wales, Australia

Douglas G. Maynard, B.Sc.(Agric), M.Sc., Ph.D. Research Scientist, Natural Resources Canada, Canadian Forest Service, Pacific Forestry Centre, Victoria, British Columbia, Canada

Fan-Rui Meng Professor, Faculty of Forestry and Environmental Management, University of New Brunswick, Fredericton, New Brunswick, Canada

Myron J. Mitchell, Ph.D. Professor, Faculty of Environmental and Forest Biology, College of Environmental Science and Forestry, State University of New York, Syracuse, New York

Tõnu Oja, Ph.D. Professor, Institute of Geography, University of Tartu, Tartu, Estonia

Sathien Phimsarn, Ph.D.* Fertilizer and Lime Research Centre, Massey University, Palmerston North, New Zealand

Heinz Rennenberg, Ph.D. Professor, Institute for Forest Botany and Tree Physiology, Albert-Ludwigs University, Freiburg, Germany

Surinder Saggar, B.Sc.(Agric.) with Hons., M.Sc., Ph.D. Senior Scientist, Environmental Quality Group, Landcare Research NZ Ltd., Palmerston North, New Zealand

John Turner, B.Sc.(For.), Ph.D. Research Scientist and Director, FORSCI Pty. Ltd., Eastwood, New South Wales, Australia

John H. Watkinson, B.Sc., Ph.D., F.N.Z.I.A.S., F.N.Z.S.S.S. Scientist, Department of Soils and Fertilizers, AgResearch, Ruakura Research Centre, Hamilton, New Zealand

**Current affiliation*: Ministry of Agriculture and Cooperatives, Chatucchak, Bangkok, Thailand

1
Introduction

Douglas G. Maynard
Pacific Forestry Centre, Victoria, British Columbia, Canada

Considerable advances have been made in the past 10 to 15 years toward an understanding of the sulfur (S) cycle in terrestrial ecosystems (e.g., see reviews of Tabatabai, 1986; Howarth et al., 1992; Robarge and Johnson, 1992). During this period, the emphasis of S research has evolved, reflecting changes in our understanding of S dynamics and interaction with other elements, new technologies, the changing rates of atmospheric S deposition in various regions, and the social and political realities of the 1990s. The purpose of this chapter is to provide an overview of S research during the past 10 to 15 years in temperate terrestrial ecosystems as a backdrop to the other chapters of this book.

Sulfur has often been considered a secondary nutrient with respect to the nutrient requirements of plants. The importance of S, however, varies considerably throughout the world depending on the plant species being grown, its S requirements, and the S status of the predominant soil types. Inputs of S either as by-products of chemicals and other fertilizers (e.g., superphosphates) or as atmospheric deposition from natural and anthropogenic emissions are also factors.

Historically, one of the main reasons so little work has been done on S, relative to other plant nutrients, is the difficulty in measuring S in soils (Tillman, 1988). The development of new methods, including ion chromatography (Dick and Tabatabai, 1979; Nieto and Frankenberger, 1985) and inductively coupled plasma atomic emission spectroscopy (ICP-AES)(McQuaker et al., 1979; Munter and Grande, 1980) has spurred further research. The various new methodologies used for determining plant-available S have been reviewed elsewhere (Jones, 1986; Tabatabai, 1992).

Aside from the difficulties associated with the analysis of S in biological material, extractants are often poorly correlated with plant response partially because of the S mineralized from organic matter. Recent methods have been developed that extract and measure that portion of the organic S thought to be directly related to the readily mineralizable organic S (Blair et al., 1993a; Zhao and McGrath, 1994; Ch. 3). Other factors, such as uptake of S from various soil depths, and potential inputs and losses from the system continue to make the prediction of S availability difficult (Zhao and McGrath, 1994).

Sulfur deficiencies are generally alleviated by additions of S fertilizers in agricultural systems (Ch. 4). The most widely used forms are sulfates, although elemental S (S°) is becoming more common (Ch. 4). Elemental S, which is widely available and has a high S analysis, must be oxidized to SO_4^{2-} to be available for plant uptake (Ch. 5). Thus, the oxidation rate of various forms of S°, which is affected by factors such as particle size, is important is determining the timing and amount of application. The mechanism of S° oxidation and related factors have been modeled to improve the increasingly widespread use of S° as a fertilizer (Blair et al., 1993b; Ch. 5).

The high proportion of organic S in soils and the poor correlation of inorganic SO_4^{2-}-S with plant growth led to considerable studies on organic S and the mineralization-immobilization processes. An important component of this work involved the identification of biologically significant fractions of organic S. Most of the initial work was carried out on agricultural soils (e.g., Freney et al., 1969, 1975). Organic S fractionation techniques similar to those used in agroecosystems were later applied in the study of S dynamics in forested ecosystems. (e.g., David et al., 1982; Fitzgerald et al., 1983). Thorough reviews on the various fractionation chemical and physical fractionation schemes have been published (Freney, 1986; Mitchell et al., 1992; Schoenau and Germida, 1992).

Since the mid-1980s, there has been little change in soil organic S fractionation techniques (Schoneau and Germida, 1992). The separation of organic S into C-bonded S and ester sulfate chemical fractions remains the most common technique used in studies on organic S processes in soils. Sulfur and N directly bonded to C would mineralize as a result of C oxidation to provide energy. McGill and Cole (1981) hypothesized that any element associated with organic matter for which there are specific degradation mechanisms may be mobilized independently of other elements. Thus, the S associated with the HI-reducible S fraction (ester sulfates) would be mineralized by enzymes in response to the need for S (Fitzgerald, 1976). Many studies have found evidence to support this process of S mineralization, but other studies have found that it did not hold up in all soils (see Ch. 2).

The relationship of S to carbon (C), nitrogen (N), and phosphorus (P) and the importance of these ratios in organic material in controlling the mineralization of S has also been widely studied (e.g., Bolin and Cook, 1983). It was originally thought, because organic S and N are closely linked, these two elements would mineralize in the same ratio as they are found in soil organic matter (Walker, 1957; White, 1959). As hypothesized by McGill and Cole (1981) and demonstrated in numerous incubation studies (e.g., Maynard et al., 1983), however, the rate of S mineralization often does not parallel that of N.

The attempts to identify and quantify a labile, easily mineralizable organic S fraction shifted to studies of microbial S and sulfatase enzyme activity. Sulfatase enzymes (arlysulfatases) have been detected in many soils and changes in the activity and kinetics of arlysulfatases under different land uses have been measured (Farrell et al., 1994). Their role in S mineralization processes has been proven, but activity is often poorly correlated to rates of S mineralization (Ch. 2).

Methods have been developed for estimating microbial S (Saggar et al., 1981; Strick and Nakas, 1984; Wu et al., 1994; Banerjee and Chapman, 1996). This fraction, which accounts for $<3\%$ of organic S in soil, has been considered important in the short-term turnover of organic S; however, the processes involved and its role in relation to plant availability have not been fully elucidated.

In agricultural systems, the emphasis has been on soluble SO_4^{2-}-S and the identification of labile organic S fractions. In most temperate soils, adsorbed SO_4^{2-} is negligible (Ch. 2), but adsorbed SO_4^{2-} is important in many tropical agricultural soils (Fox and Blair, 1986). The mechanisms of SO_4^{2-} adsorption and the important role of adsorbed SO_4^{2-} in S retention processes in areas of high S deposition (primarily forested ecosystems) are presented elsewhere in this volume (Ch. 7).

As S inputs from atmospheric deposition and fertilizer sources such as ammonium sulfate and superphosphates decrease, S deficiencies have become more common in grassland and arable crops in Europe (Zhao and McGrath, 1994) and elsewhere. With these inputs diminished, and continued losses of S through leaching and removal in animal products, productivity of grazed pastures in New Zealand and Australia may not be sustainable (Ch. 3; Ch. 4). Consequently, deficiencies of S may become increasingly prevalent. Janzen and Ellert (Ch. 2) indicate that in many areas, S deposition currently equals or exceeds S removed in products. Even with decreases in S deposition, they suggest that the S removed can be easily replaced through additions of small amounts (15–25 kg S ha^{-1} yr^{-1}). We still, however, require information on S in soils to more wisely use the fertilizer resource and determine areas that may be susceptible to S deficiencies.

In the late 1960s and early 1970s, S from industrial emissions was considered the primary source of acidic deposition. Concern about S emissions led to a large body of research on S in forested and aquatic ecosystems (e.g., Barnard and Lucier, 1991). In the past 15 years, S emissions have declined in both North America (17%) and Europe (39%) (Rodhe et al., 1995). At the same time, nitrogen (N) deposition has been increasing and N saturation of forest ecosystems is potentially a more significant concern (EPA 1995).

Acidic deposition, as S and N compounds, can affect forest ecosystems directly or indirectly. Direct effects include the adsorption of gaseous S by the plant (Ch. 6). Sulfur deposition may also affect vegetation indirectly, by making it more susceptible to biotic or abiotic agents. The potential effects of long-term chronic acidic deposition on soil processes include leaching of cations, increased solubility of potentially toxic elements such as aluminum, nutrient imbalances, and impacts on soil microbial populations (Foster, 1989; Binkley et al., 1989). Accelerated leaching of elements as a result of soil acidification or N and S saturation can lead to excess inputs into aquatic ecosystems (EPA, 1995).

The relative importance of SO_2 as a pollutant in Europe and eastern North America has diminished over recent years because of reduced S emissions and the increasing recognition of the role of N compounds and ozone (WHO, 1994). More is now known about how gaseous S at very low concentrations affects plants (Ch. 6). Mathematical models have been developed that describe the effect of low rates of SO_2 exposure on net photosynthesis (Arp and Meng, Ch. 9). Such models will continue to be important in understanding the role of acid fog and mists on forests (Cape, 1993).

Research on acid deposition has shifted focus to evaluate the effects of diminishing S deposition and the interactions of excess S with N and various other pollutants (Taylor et al., 1994). In many areas of Europe and North America, S deposition continues to decrease; however, even at the reduced deposition levels, some ecosystems remain at high risk. Attempts to implement emission control strategies led to the development of the critical and target load concept (Bull, 1991). The use of the term *critical load* first arose in Canada in the late 1970s, but it was at the Stockholm Conference in 1982 that the concept was first defined (as threshold concentrations) (Bull, 1992).

The definition of critical loads proposed by the UN/ECE (1988) was "A quantitative estimate of an exposure to one or more pollutants below which significant harmful effects on specified sensitive elements of the environment do not occur according to our present knowledge" (Brodin and Kuylenstierna, 1992). In the 1990s considerable effort has been devoted to the development of models that estimate critical loads on a large scale.

Critical loads can be calculated at three complexity levels (Brodin and Kuylenstierna, 1992). The empirical approach is the least complex and is based on existing soils information and acid deposition levels (e.g., Hornung et al., 1995). The second approach involves the derivation of critical loads with the use of a steady-state mass balance calculation of input and output data based on a steady-state model (see Ch. 10). This is currently the approach followed in Europe to apply the critical load concept on a large scale with a lack of adequate input data (Posch et al., 1995). A similar model has been applied to upland forests in southern Ontario (Arp et al., 1996). The most complex approach uses dynamic models such as the Nutrient Cycling Model (NuCM; Lui et al., 1991) and requires detailed information on rates of acid-forming and acid-consuming processes within the ecosystem (Ch. 7). Dynamic models include time-dependent processes, in contrast to the steady-state mass balance models that do not consider time scale of acidification and recovery (Warfvinge et al., 1992). This model has been used to assess the possible ecological implications under reducing S (and N) deposition rates in forest ecosystems (Johnson et al., 1995; Johnson et al., 1996; Ch. 7).

The fifth International Conference on Acid Deposition ("Acid Reign'95?") produced an overview of the scientific observations about acidic deposition that are known with confidence (Rodhe et al., 1995). Included in this category were the following (1) Sulfur deposition has increased concentrations of S, particularly adsorbed SO_4^{2-}, in the soils of certain regions of Europe and eastern North America. This has led to decreases of exchangeable bases, leaching of base cations and nutrient deficiencies, particularly magnesium (Mg) and potassium (K). (2) In parts of northern Europe, soil pH has decreased up to 1.0 unit during the past 30 to 60 years, partly as a result of acidic deposition. (3) Acidification has increased the availability of metals [e.g., aluminum (Al)] in soils. The consequences of increased Al are not entirely understood but increased solution concentrations of Al or the calcium(Ca)/Al molar ratio are thought to indicate increased risk to trees from acidification (Cronan and Grigal, 1995).

As S deposition to these areas decreases, adsorbed SO_4^{2-} may be released to drainage waters delaying the recovery of aquatic ecosystems.

Other potential concerns related to the declining S emissions are the extent and rate at which these ecosystems will recover. Modeling simulations (NuCM) indicated changes to the Ca/Al ratio (for a discussion on Ca/Al ratios, see Ch. 10) that were not as sensitive in the short-term to reduced S deposition as were solution values of Al^{3+} (Johnson et al., 1996). They hypothesized that if plant roots were sensitive strictly to the Ca/Al ratio, then reduced S deposition would have less benefit in the short term than if roots were affected by absolute values of Al^{3+}.

In forested systems, much of the focus in the past has been on excessive S deposition; however, in the absence of S inputs, the sustainability of the forest is almost entirely dependent on nutrient cycling (including S) within the system (Nys et al., 1990). In forested ecosystems with minimal atmospheric inputs, such as in Australia, S deficiencies have been observed in some plantations (see Ch. 8).

Whole-tree harvesting and shortened rotations or intensive plantation silviculture may potentially lead to S deficiencies. For example, there is increased harvesting activity in the boreal forests, which account for 25% of the world's forested lands. Many of the Gray Luvisolic soils cleared for agriculture began showing S deficiencies after two or three decades of continuous cropping (Beaton and Soper, 1986). Similar soil types are the predominant soils of much of the boreal forest currently being harvested for the first time. Thus, depending on the rate of S removal by harvesting and rates of S inputs (either through atmospheric deposition or release of S in available forms), deficiencies may occur.

In areas of S deposition, the release of S from the soil system following harvest may potentially be a problem. If the trees and understory vegetation retain S within the ecosystem, harvesting (depending on the harvest system used) may result in the loss of the vegetation sink. The S would be susceptible to leaching (particularly in soils with little SO_4^{2-} adsorption capacity) with the potential for leaching of base cations or other indirect effects. This could also have implications to aquatic ecosystems if significant leaching occurred.

Much was written in the 1970s and 1980s about the role of S as a nutrient and as a toxic compound affecting agricultural and forest ecosystems. Sulfur research during the past 10 to 15 years has changed in several important directions; however, the organic matter fractionation schemes and broad classification of organic S into C-bonded S and HI-reducible S (ester sulfates) still remain the fundamental tools for study of soil S. Analytical techniques have improved to allow further differentiation of organic and inorganic S. Models to predict S adsorption-desorption processes under different S deposition scenarios, critical loads for acidic deposition and effects of low-level S deposition on processes such as photosynthesis have led to increased knowledge of S processes in our terrestrial ecosystems.

Increasing demand for food and fiber production from a growing population has resulted in a need to continually increase productivity on agricultural and forested lands. Thus, understanding the dynamics of S, its relationship with other nutrients and pollutants, and the role of S in primary productivity is essential. The need for research into the processes of S transformations in terrestrial ecosystems will continue as these demands and stresses on our natural resources increase.

REFERENCES

Arp, P. A., Oja, T., and Marsh, M. 1996. Calculating critical S and N loads and current exceedances for upland forests in southern Ontario. Can. J. For. Res. 26: 696–709

Banerjee, M. R., and Chapman, S. J. 1996. The significance of microbial biomass sulphur in soil. Biol Fert. Soils 22: 116–125.

Barnard, J. E., and Lucier, A. A. 1991. Report 16: Changes in forest health and productivity in the United States and Canada *in* P. M. Irving (ed.), Acidic deposition: State of science and technology, Vol. III, terrestrial, materials, health and visibility effects. The U. S. National Acid Precipitation Assessment Program, Washington, DC.

Beaton, J. D., and Soper, R. J. 1986. Plant response to sulfur in western Canada. Pages 375–403 *in* M. A. Tabatabai (ed.), Sulfur in Agriculture, Agronomy Monograph No. 27, American Society of Agronomy, Madison, WI.

Binkley, D., Driscoll, C. T., Allen, H. L., Schoeneberger, P., and McAvoy, D. 1989. Acidic deposition and forest soils: Context and case studies in the southeastern United States. Springer-Verlag, New York, N.Y.

Blair, G. J., Lefroy, R. D. B., Chinoim, N., and Anderson, G. C. 1993a. Sulfur soil testing. Plant and Soil 155/156: 383–386.

Blair, G. J., Lefroy, R. D. B., Dana, M., and Anderson, G. C. 1993b. Modelling of sulfur oxidation from elemental sulfur. Plant and Soil 155/156: 379–382.

Bolin, B., and Cook, R. B. (eds.) 1983. The major biogeochemical cycles and their interactions. SCOPE 21, John Wiley & Sons, Chichester, England.

Brodin, Y. W., and Kuylenstierna, J. C. I. 1992. Acidification and critical loads in Nordic Countries: a background. Ambio 21: 332–338.

Bull, K. R. 1991. The critical loads/levels approach to gaseous pollutant emission control. Environ. Pollut. 69:105–23.

Bull, K. R. 1992. An introduction to critical loads. Environ. Pollut. 77: 173–176.

Cape J. N. 1993. Direct damage to vegetation caused by acid rain and polluted cloud: Definition of critical levels for forest trees. Environ. Pollut. 82:167–80.

Cronan, C. S., and Grigal, D. F. 1995. Use of calcium/aluminum ratios as indicators of stress in forest ecosystems. J. Environ. Qual. 24: 209–226.

David, M. B., Mitchell, M. J., and Nakas, J. P. 1982. Organic and inorganic sulfur constituents of a forest soil and their relationship to microbial activity. Soil Sci. Soc. Am. J. 46: 847–852.

Dick, W. A., and Tabatabai, M. A. 1979. Ion chromatographic determination of sulfate and nitrate in soils. Soil Sci. Soc. Am. J. 43:899–904.

Environmental Protection Agency (EPA). 1995. Acid deposition standard feasibility study report to congress. United States Environmental Protection Agency, Office of Air and Radiation, Acid Rain Division, Washington, DC. EPA/430-R-95-001a.

Farrell, R. E., Gupta, V. S. R., and Germida, J. J. 1994. Effects of cultivation on the activity and kinetics of arylsufatase in Saskatchewan soils. Soil Biol. Biochem. 26: 1033–1040.

Fitzgerald, J. W. 1976. Sulfate ester formation and hydrolysis: a potentially important yet often ignored aspect of sulfur cycle of aerobic soils. Bacteriol. Rev. 40: 698–721.

Fitzgerald, J. W., Ash, J. T., Strickland, T. C., and Swank, W. T. 1983. Formation of organic sulfur in forest soils: a biologically mediated process. Can. J. For. Res. 13: 1077–1082.

Foster N. W. 1989. Acidic deposition: what is fact, what is speculation, what is needed? Water Air Soil Pollut. 48:299–306.

Fox, R. L., and Blair, G. J. 1986. Plant response to sulfur in tropical soils. Pages 405–434 *in* M. A. Tabatabai (ed.), Sulfur in Agriculture, Agronomy Monograph No. 27, American Society of Agronomy, Madison, WI.

Freney, J. R. 1986. Forms and reactions of organic sulfur compounds in soils. Pages 207–232 *in* M. A. Tabatabai (ed.), Sulfur in Agriculture, Agronomy Monograph No. 27, American Society of Agronomy, Madison, WI.

Freney, J. R., Melville, G. E., and Williams, C. H. 1969. Extraction, chemical nature and properties of soil organic sulphur. J. Sci. Food Agric. 20: 440–445.

Freney, J. R., Melville, G. E., and Williams, C. H. 1975. Soil organic matter fractions as sources of plant-available sulphur. Soil Biol. Biochem. 7: 217–221.

Hornung, M., Bull, K. R., Cresser, M., Hall, J., Langan, S. J., Loveland, P., and Smith, C. 1995. An empirical map of critical loads of acidity for soils in Great Britain. Environ. Pollut. 90: 301–310.

Howarth, R. W., Stewart, J. W. B., and Ivanov, M. V. 1992. Sulphur cycling on the continents: Wetlands, terrestrial ecosystems, and associated water bodies. SCOPE 48, John Wiley & Sons, Chichester, West Sussex, England.

Johnson, D. W., Binkley, D., and Conklin, P. 1995. Simulated effects of atmospheric deposition, harvesting, and species change on nutrient cycling in a loblolly pine forest. For. Ecol. Mange. 76: 29–45.

Johnson, D. W., Susfalk, R. B., and Brewer, P. F. 1996. Simulated response of red spruce forest soils to reduced sulfur and nitrogen deposition. J. Environ. Qual. 25: 1300–1309.

Jones, M. B. 1986. Sulfur availability indexes. Pages 549–566 *in* M. A. Tabatabai (ed.), Sulfur in Agriculture, Agronomy Monograph No. 27, American Society of Agronomy, Madison, WI.

Lui, S., Munson, R., Johnson, D. W., Gherini, S., Summers, K., Hudson, R., Wilkinson, K., and Pitelka. 1991. Application of a nutrient cycling model (NuCM) to northern mixed hardwood and southern coniferous forest. Tree Physiol. 9: 173–182.

Maynard, D. G., Stewart, J. W. B., and Bettany, J. R. 1983. Sulfur and nitrogen mineralization in soils compared using two incubation techniques. Soil Biol. Biochem. 15: 251–256.

McGill, W.B., and Cole, C. V. 1981. Comparative aspects of cycling of organic C, N, S and P through soil organic matter. Geoderma 26: 267–286.

McQuaker, N. R., Brown, D. F., and Kluckner, P. D. 1979. Digestion of environmental materials for analysis by inductively coupled plasma-atomic emission spectrometry. Anal. Chem. 51: 1082–1084.

Mitchell, M. J., David, M. B., and R. B. Harrison. 1992. Sulphur dynamics in forest ecosystems. Pages 215–254 *in* R. W. Howarth et al. (eds.), Sulphur cycling on the continents: Wetlands, terrestrial ecosystems, and associated water bodies. SCOPE 48, John Wiley & Sons, Chichester, West Sussex, England.

Munter, R. C., and Grande, R. A. 1980. Plant tissue and soil extract analysis by ICP-atomic emission spectrometry. Pages 653–672 *in* R. M. Barnes (ed.), Developments in atomic plasma spectrochemical analysis. Proceedings of the International Winter Conference, 7–11 January 1980, San Juan, Puerto Rico. Heyden, London, England.

Nieto, K. F., and Frankenberger, W. T. 1985. Single column ion chromatography: I. Analysis of inorganic anions in soils. Soil Sci. Soc. Am. J. 49: 587–592.

Nys, C., Stevens, P., and J. Ranger. 1990. Sulphur nutrition of forests examined using a sulphur budget approach. Pages 356–372 *in* A. F. Harrison et al. (eds.), Nutrient cycling in terrestrial ecosystems. Field methods, application and interpretation. Elsevier Applied Science, London, England.

Posch, M., de Smet, P. A. M., Hettelingh, J.-P., and Downing, R. J. (eds.) 1995. Calculation and mapping of critical thresholds in Europe: Status report 1995. Coordination Center for Effects, National Institute of Public Health and the Environment, RIVM Report No. 259101004, Bilthoven, the Netherlands.

Robarge, W. P., and Johnson, D. W. 1992. The effects of acidic deposition on forested soils. Adv. Agron. 47:1–83.

Rodhe, H., Grennfelt, P., Wisniewski, J., Ågren, C., Bengtsson, G., Johansson, K., Kauppi, P., Kucera, V., Rasmussen, L., Rosseland, B., Schotte, L., and Selldén, G. 1995. Acid reign '95?—Conference summary statement. Water, Air, and Soil Pollut. 85: 1–14.

Saggar, S. K., Bettany, J. R., and Stewart, J. W. B. 1981. Measurement of microbial sulphur in soil. Soil Biol. Biochem. 13: 493–498.

Schoenau, J. J., and Germida, J. J. 1992. Sulphur cycling in upland agricultural systems. Pages 261–277 *in* R. W. Howarth et al. (eds.), Sulphur cycling on the continents: Wetlands, terrestrial ecosystems, and associated water bodies. SCOPE 48, John Wiley & Sons, Chichester, West Sussex, England.

Strick, E., and Nakas, J. P. 1984. Calibration of a microbial sulfur technique for use in forest soils. Soil Biol. Biochem. 16: 289–292.

Tabatabai, M. A. (ed.) 1986. Sulfur in agriculture. Agronomy Monograph No. 27, American Society of Agronomy, Madison, WI. 668 pp.

Tabatabai, M. A. 1992. Methods of measurement of sulphur in soils, plant materials and waters. Appendix, Pages 307–344 *in* R. W. Howarth et al. (eds.), Sulphur cycling on the continents: Wetlands, terrestrial ecosystems, and associated water bodies. SCOPE 48, John Wiley & Sons, Chichester, West Sussex, England.

Taylor, G. E. Jr., Johnson, D. W., and Anderson, C. P. 1994. Air pollution and forest ecosystems: A regional to global perspective. Ecol. Appl. 4: 662–689.

Tillman, R. W. 1988. A review of sulphur chemistry in soils. Pages 4–8 *in* R. E. White and L. D. Currie (eds.), Proceedings of the workshop towards the more efficient use of soil and fertilizer sulphur, Massey University, Palmerston North, 17–18 February 1988. Occasional Report No. 2, Fertilizer & Lime Research Centre, Massey University, Palmerston North, New Zealand.

Walker, T. W. 1957. The sulphur cycle in grassland soils. J. Brit. Grassland Soc. 12: 10–18.

Warfvinge, P., Holmberg, M., Posch, M., and Wright, R. F. 1992. The use of dynamic models to set target loads. Ambio 21: 369–376.

White, J. G. 1959. Mineralization of nitrogen and sulphur in sulphur-deficient soils. New Zealand J. Agric. Res. 2: 255–258.

World Health Organization (WHO) 1994. Updating and revision of the air quality guidelines for Europe. Report of the WHO Working Group on Ecotoxic Effects. Les Diablerets, Switzerland, September 21–23, 1994. Report EUR/ICP/CEH 230/B 04295. (English only, unedited.)

Wu, J., O'Donnell A. G., and Syers, J. K. 1994. Fumigation-extraction method for the measurement of soil microbial biomass-S. Soil Biol. Biochem. 26: 117–125.

Zhao, F., and McGrath, S. P. 1994. Extractable sulphate and organic sulphur in soils and their availability to plants. Plant and Soil 164:243–50.

2
Sulfur Dynamics in Cultivated, Temperate Agroecosystems

H. Henry Janzen and Benjamin H. Ellert
Agriculture and Agri-Food Canada, Lethbridge, Alberta, Canada

I. INTRODUCTION

Sulfur(S) is highly reactive, existing in six oxidation states and moving freely among the lithosphere, hydrosphere, and atmosphere. Within the biosphere, it is found in numerous chemical combinations and serves multiple biochemical functions. As a result, the pathways of S in any ecosystem are complex and intertwined with those of other elements.

The transformations of S in agroecosystems first came to prominence with the recognition, several decades ago, of widespread S deficiency in soils throughout the world (Beaton and Soper, 1986; Bettany et al., 1983; Tisdale et al., 1986). The effects of these deficiencies on crop yields spawned research into processes governing the S fertility of soils. This work focused primarily on dynamics among S pools in soil, principally the release of plant-available sulfate from organic matter and insoluble inorganic S compounds. At the same time, another avenue of research sought to determine the fate of S fertilizers applied to the soil. The overriding objective of much of this work was the development of agronomic practices that minimized yield depression induced by S deficiency.

More recently, the research emphasis has shifted to a broader perspective, with the recognition that the S cycle within an agroecosystem is not closed, but is linked to other ecosystems via the atmosphere (Ivanov, 1983). The focus of S research has therefore expanded beyond processes within ecosystems to the movements among ecosystems via atmospheric fluxes.

Agriculture has claimed a large share of productive terrestrial environments across the globe (Elliott and Cole, 1989; Stewart et al., 1990). As a result, management practices imposed on agricultural lands affect S dynamics, not only within agroecosystems, but also among other terrestrial biomes.

Our objective is to review the S dynamics within agroecosystems and their interaction with S cycles in the surrounding environment. We focus primarily on arable agricultural systems in temperate regions and, where possible, rely on studies reported within the last decade.

II. S POOLS IN AGROECOSYSTEMS

The forms of S in the atmosphere, biota, and soil have been thoroughly described in earlier reviews (Duke and Reisenauer, 1986; Freney, 1986; Ryaboshapko, 1983). We present only a brief description of these various forms as background to subsequent discussion of S dynamics.

A. Atmosphere

Like the cycles of nitrogen and carbon, the S cycle has a significant atmospheric component that results in the spatial and chemical redistribution of S. Most of the volatile S compounds are quite unstable, however, so that atmospheric S concentrations are very low relative to those of carbon and nitrogen.

The atmosphere contains a range of S compounds including hydrogen sulfide (H_2S), carbon disulfide (CS_2), dimethyl sulfide (CH_3SCH_3), carbonyl sulfide (COS), S dioxide (SO_2), and sulfate (SO_4) aerosols (Noggle et al., 1986; Ryaboshapko, 1983; Trudinger, 1986). Natural and anthropogenic sources account for approximately equivalent portions of this S.

Of the biogenic S fluxes to the atmosphere, CH_3SCH_3 accounts for about 75%, H_2S for 15%, and CS_2 + COS for 10% (Kelly and Smith, 1989). Hydrogen sulfide and CH_3SCH_3 are rapidly oxidized to SO_2 and SO_4, and their residence time in the atmosphere is short (Trudinger, 1986).

Atmospheric SO_2 is largely of anthropogenic origin, derived primarily from the combustion of fossil fuels and smelting of ores (Trudinger, 1986). As a result, SO_2 concentrations vary widely among geographical regions; near point sources they may be as high as several milligrams S m^{-3} whereas in remote regions, concentrations are often as low as 0.1 $\mu g\ m^{-3}$

or less (Ryaboshapko, 1983). Sulfur dioxide is removed from the atmosphere by oxidation to SO_4, or by wet and dry deposition.

The atmosphere also contains sulfate derived from ocean sprays, oxidation of atmospheric SO_2, volcanic activity, and aeolian weathering of soils (Ryaboshapko, 1983).

B. Plant Biomass

The multiple oxidation states and the diversity of its compounds give S a unique versatility in plant metabolism (Clarkson and Hanson, 1980; Duke and Reisenauer, 1986; Schmidt and Jager, 1992; Thompson et al., 1986). Its primary nutritional role is in the synthesis of the amino acids cysteine, cystine, and methionine, but S also occurs in other essential biochemicals such as cofactors, iron-S proteins, thioredoxins, and sulfolipids. In addition, S occurs in various compounds of uncertain function, including various S-substituted cysteines and a wide range of sulfate esters, glucosinolates, polysaccharide sulfates, and lipid sulfates.

The primary metabolic fate of assimilated S is the synthesis of protein. Unlike nitrogen and other nutrients, however, S is often taken up in excess of plant requirements and stored as various ester sulfates or as free sulfate (Mayland and Robbins, 1994; Rennenberg, 1984). In a pot study, for example, soluble S accounted for 67 to 94% of the S in canola (*Brassica napus* L.) residue (Roppel, 1991). This S occurred, apparently, in the form of glucosinolates or inorganic sulfate.

The amount of S stored in plant biomass in a given agroecosystem is determined by crop type, nutritional S status, and crop yield. Reported values of crop S content range from a few kilograms S ha^{-1} to almost 100 kg ha^{-1}, though S content in temperate regions is usually less than 50 kg S ha^{-1} (Duke and Reisenauer, 1986; Tabatabai, 1984).

C. Soil

The S concentration in the plow layer of agricultural soils usually ranges from 50 to 1000 mg S kg^{-1} soil, though appreciably higher concentrations occur in organic and saline soils (Bailey, 1985; Freney, 1986; Freney and Williams, 1983). Most of the S in the surface horizon is usually present in organic combinations (typically >95%), with the remainder occurring as inorganic sulfate. The inorganic sulfate may exist in the form of soluble sulfate salts, sparingly soluble salts (e.g., gypsum), or in co-precipitates with $CaCO_3$. In weathered and acidic soils, significant amounts of sulfate may also be adsorbed to soil colloids. Although some reduced S (e.g., elemental S, thiosulfate, or sulfide) may be present (Zhong and Morra, 1994), these compounds are transitory in aerobic soils and their concentrations are usually negligible.

The chemical nature of organic S has not been fully characterized (Freney, 1986). A significant proportion (up to 30%) occurs as amino acids bound to soil colloids. In addition, small amounts of S may occur in free amino acids, sulfated polysaccharides, and sulfolipids. The majority of the organic S in soils, however, is present in a diversity of complex organic molecules.

The heterogeneous, undefined nature of organic S in soils has led many researchers to fractionate the S into broad, functionally distinct groups. The most widely used separation is based on the nature of the chemical bond and delineates two fractions: (1) carbon-bonded S, which includes all compounds in which S is covalently bonded to carbon, and (2) HI-reducible S (HI—S), defined as S that can be reduced by hydriodic acid to H_2S, including ester sulfates (—C—O—S—), sulfamates (—C—N—S), and the second S in S—sulfocysteine (—C—S—S—)(Freney, 1986). HI—S (organic sulfate), which typically accounts for approximately half of soil organic S, is often considered to be a relatively labile fraction, associated largely with low molecular weight materials (Biederbeck, 1978; McLaren et al., 1985; Saggar et al., 1984).

Some researchers have used physical fractionation techniques to characterize organic S. For example, Anderson et al. (1981) observed that more than 70% of the total S and more than 80% of the HI-S occurred in the clay-sized fractions, and proposed that they constituted a storehouse of S, part of which was potentially labile. More recently, Eriksen et al. (1995a, b) separated organic S into three fractions: non-protected S, protected S, and insoluble S. They concluded that the soils they studied contained a large pool of passive organic S (52 to 82% of total organic S), including S protected within aggregates and insoluble S, which probably contributed very little to S mineralization.

A relatively recent advance has been the fractionation of organic S into biotic and abiotic components (Castellano and Dick, 1991b; Gupta et al., 1988; Saggar et al., 1981). Although it typically accounts for less than 5% of organic S (Chapman, 1987a; Gupta and Germida, 1988), the microbial biomass represents a highly labile fraction of soil that may be responsible for much of the S turnover in some soils. The S in microbial biomass usually exists chiefly in the form of proteins, but various organisms, most notably fungi, can store excess S in the form of organic sulfates (Saggar et al., 1981). As a result, the nature of the microbial population as well as the S status of its immediate environment may influence the potential turnover of microbial S. Unlike many other fractionation schemes, measurement of microbial biomass delineates a functionally distinct portion of organic S. Estimates of biomass S remain somewhat qualitative, however, dependent on methods employed (Banerjee et al., 1993) and subject to artifacts [e.g., presence of plant roots (Perrott and Sarathchandra, 1990)].

An alternative method of characterizing organic matter involves the use of kinetic approaches to separate fractions of organic S on the basis of turnover rates in soil. The identification of an 'active fraction' by this approach has not been highly successful, however, perhaps because of the heterogeneity of S substrates in soil organic matter (Ellert and Bettany, 1988; Ghani et al., 1991).

III. SULFUR FLUXES IN AGROECOSYSTEMS

Sulfur fluxes occur both among the major pools (atmosphere, soil, and plant biomass) and among fractions within each of these pools (Fig. 1). We will consider the dominant processes among soil fractions and the transfer of S among soil, atmosphere, and plant biomass pools.

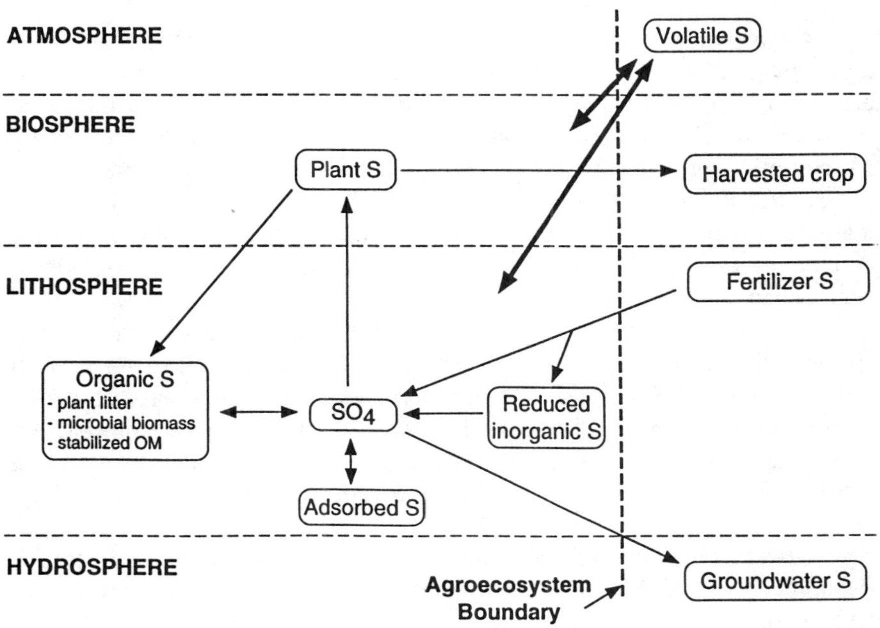

Figure 1 Simplified overview of the main pools and fluxes of sulfur in cultivated agroecosystems.

A. Sulfur Dynamics in Soil

1. Mineralization—Immobilization

Sulfur mineralization refers to the decomposition of organic S to inorganic sulfate. More research attention has been devoted to this process than to any other S flux in agroecosystems, probably because of its importance to S nutrition of crops.

The substrates for S mineralization in agroecosystems include both indigenous organic matter and fresh detritus (predominantly crop residues and microbial cells). Most studies of S mineralization have focused on the former fraction; crop residues, for example, are routinely removed from soils prior to the establishment of laboratory incubation experiments. These studies have shown that S mineralization is strongly influenced by soil factors that govern the rate of microbial activity. For example, mineralization rates are strongly related to soil temperature (Ellert and Bettany, 1992; MacDonald et al., 1995; McLaren et al., 1988; Pirela and Tabatabai, 1988). The amount of mineralization also appears to be influenced by sulfate concentration in the soil solution. Thus S mineralization is often higher in intermittently leached incubations than in closed systems where sulfate is allowed to accumulate (Maynard et al., 1983; McLaren et al., 1988; Tan et al., 1994).

Crop residues are an important source of mineralizable S in many agroecosystems. Several recent studies suggest that the pattern of S mineralization from residues is very different from that of N or other nutrients. In many cases, S mineralization exhibits a rapid flush of sulfate shortly after application of residues, followed by a very slow, linear phase of S release (Chapman, 1987b; Roppel, 1991; Till et al., 1982). The initial release of S may be attributed to the solubilization of sulfate present in the residue in inorganic form or the hydrolysis of organic sulfates during senescence.

The rate and amount of S release from residues varies among crop species. Tabatabai and Chae (1991) observed an initial release of S from alfalfa residue, but a significant immobilization in treatments with added corn or soybean residue. Because S deficiency rarely limits decomposition and because of the heterogeneity of S compounds, S mineralization from residues is not closely related to their C:S ratios (Coughenour et al., 1980; Freney, 1986).

Microbial biomass is another important source of mineralizable S (Chapman, 1987a; Smith and Paul, 1990). For example, after drying of soil, there may be a flush of soluble sulfate, apparently from the decomposition of microbial cells (Ghani et al., 1990; Gupta and Germida, 1989b). The mineralization of microbial biomass S may be accelerated by amoebal grazing (Gupta and Germida, 1989a). Because much of the microbial S may be derived from crop residues (Chapman, 1987b), the microbial biomass serves as an intermediate between residues and inorganic sulfate.

Considerable effort has been devoted to characterizing the biochemical source of mineralized S. The conventional view is that ester sulfate represents the labile organic S fraction and therefore accounts for much of the S mineralized. Some studies support this view (e.g., McLaren et al., 1985) but others report that ester sulfate (HI-reducible S) may be no more important than C-bonded S as a source of mineralizable S (Castellano and Dick, 1991a). For example, Ghani et al. (1991) observed that C-bonded S was a better predictor than HI-S of S mineralization; indeed, they observed an increase in HI-S in some soils during mineralization. The classification of S on the basis of bond classes may not have enough sensitivity to distinguish between labile and refractory fractions (Castellano and Dick, 1991a; Eriksen et al., 1995d; Lou and Warman, 1992; Schoenau and Germida, 1992). For example, S in amino acids is highly labile but would fall into the C-bonded category. Furthermore, much of the ester sulfate, while intrinsically labile, may be inaccessible to enzymes (Lou and Warman, 1994).

McGill and Cole (1981) proposed that mineralization of organic S occurs by two distinct mechanisms: (1) biological mineralization: the release of C-bonded S as a secondary result of C oxidation and (2) biochemical mineralization: the enzymatic hydrolysis of ester sulfate. The latter process is subject to end-product control, and any factor that depletes soluble sulfate should therefore promote biochemical mineralization. This hypothesis is supported by observations that continual removal of sulfate by leaching or plant uptake stimulates mineralization (Maynard et al., 1983, 1984, 1985). Other studies, however, have furnished results, such as the lack of HI-S mineralization in the presence of applied C substrate (Ghani et al., 1992), that do not conform to the model. Thus, while this model provides a unifying framework, it may not fully account for the complexity of S mineralization processes.

The activity of sulfatase, an enzyme that catalyzes ester sulfate hydrolysis, has been suggested as a possible indicator of potential S mineralization (Gupta et al., 1993). The relationship between S mineralization and sulfatase activity may be weak, however (Haynes and Williams, 1992), apparently because of several factors: (1) sulfate accumulation is the result of *net* mineralization (rather than solely of S release), (2) there are a variety of ester sulfates in soil, each hydrolyzed by specific enzymes, and (3) only a portion of mineralized S is derived from ester sulfate. Furthermore, sulfatase activity in soil is usually not a limiting factor in S release from ester sulfates (Ganeshamurthy and Nielsen, 1990).

The rate of dimethyl sulfoxide (DMSO) reduction to dimethyl sulfide has been suggested as another indicator of S mineralization (Sparling and Searle, 1993). Although this technique may be a useful assay of microbial activity, it was not very reliable in predicting S mineralization.

Immobilization, the assimilatory reduction of sulfate by microorganisms, often occurs concurrently with mineralization (Maynard et al., 1985; McLaren et al., 1985). The amount of immobilization typically reflects the microbial demands for S; thus the stimulation of microbial populations by addition of C substrate enhances immobilization. For example, O'Donnell et al. (1994) observed that 21 to 34% of added $^{35}SO_4$ was immobilized in glucose-amended soil within 3 days of S application, compared to only 0.5 to 13% without addition of glucose. Application of residues with insufficient S concentration to satisfy microbial S demands, therefore, often results in net immobilization (Tabatabai and Chae, 1991). Other events that stimulate microbial growth may similarly result in immobilization; Gupta and Germida (1989b) observed appreciable immobilization following rewetting of an air-dried soil. As well, addition of sulfate may also favor immobilization. For example, Lou and Warman (1992) observed the formation of large amounts of ester sulfate upon addition of inorganic sulfate to soil. This newly formed ester sulfate was more labile than indigenous ester sulfate, apparently because the latter is protected from enzyme activity by humic polymers (Lou and Warman, 1994).

2. Adsorption—Desorption

Adsorbed sulfate is an important source of plant-available S in soils containing high amounts of 1:1 clays and hydrous oxides of iron (Fe) and aluminum (Al) (Bohn et al., 1986; Courchesne, 1992; Pasricha and Fox, 1993). The net dynamics of this sulfate are influenced by a number of factors, including anion exchange capacity, solution sulfate concentration, pH, competition from other anions (notably phosphate), calcium addition, and moisture content (Bolan et al., 1986; Comfort et al., 1991; Curtin and Syers, 1990). In many temperate soils, however, adsorbed sulfate plays only a minor role in S dynamics (e.g., Bettany et al., 1983).

3. Oxidation–Reduction

Virtually all soils exhibit significant ability to transform reduced forms of S to sulfate via oxidation reactions (Germida and Janzen, 1993). This oxidation is largely mediated by soil microorganisms. Numerous microbial species, both heterotrophic and autotrophic, are capable of oxidizing S, although the dominant organisms may vary among ecosystems. The ubiquity of S-oxidizing organisms in soils probably occurs because the mineralization of C-bonded S involves inorganic, reduced S intermediates and end-products such as elemental S, thiosulfate, and hydrogen sulfide (Morra and Dick, 1985; Zhong and Morra, 1994).

An understanding of S oxidation processes in agricultural soils has been derived largely from studies of the transformations of fertilizers containing reduced S forms (Wainwright, 1984). Because oxidation is predominantly mediated by soil microorganisms, its rate is strongly influenced by a host of variables that affect biological activities, including moisture, temperature, aeration, soil organic matter content, and pH (Germida and Janzen, 1993). As well oxidation is favored by S characteristics that increase accessibility of the substrate to microbial activity; this oxidation is accelerated when the S is finely divided (higher surface area) and by mixing the S particles into the soil. Most of the S oxidized is usually released into the soil as sulfate, though some immobilization has been observed during the initial stages of oxidation (He et al., 1994).

In anaerobic conditions, sulfate can be reduced to sulfide by the activity of bacteria that are strict anaerobes, primarily *Desulfovibrio* and *Desulfotomaculum* species (Freney and Williams, 1983). This process generally occurs only in waterlogged soils and is therefore probably not important in most arable agroecosystems (Konopka et al., 1986).

4. Solubilization–Precipitation

In arid soils, sulfate in the soil solution may be in equilibrium with gypsum, which is sufficiently soluble to contribute to soil S fertility (Bettany et al., 1983). Other sulfate salts present in soil, such as barium sulfate, strontium sulfate, and sulfate co-precipitated with carbonates, do not have sufficient solubility to influence significantly the activity of sulfate in the soil solution.

B. Exchange of S Between Soil and Plant Biomass

Plants acquire most of their nutritional S by absorption and assimilation of sulfate. Except for SO_2 absorbed by shoots near S-emitting industrial sources, most of the sulfate required by plants is absorbed by the roots and reduced in the photosynthesizing leaves (Cram, 1990). The uptake of S is increased by mycorrhizal infection, although effects on S nutrition are probably minimal because of the high mobility of sulfate in the soil solution (Rhodes and Gerdemann, 1978a,b).

Aside from litter inputs, plants may also return S to the soil via root exudation and related processes. Few studies have attempted to quantify rhizodeposition of S, but reports of substantial exudation of organic material (Lynch and Whipps, 1990) and the presence of amino acids in these exudates (Rovira, 1969) suggest that some S may enter the soil via this route.

C. Exchange of S Between Plant and Atmosphere

Plants may release measurable amounts of volatile S into the atmosphere (Haines et al., 1989; Rennenberg, 1984). The main forms of S released appear to be H_2S and dimethyl disulfide although release of other volatile S compounds, such as carbonyl sulfide, mercaptans, carbon disulfide, and SO_2, has also been reported (Aneja, 1990; Hartel and Haines, 1992; Rennenberg et al., 1990).

Rates of S emission depend on plant species and environmental conditions. In comparisons among species, for example, highest rates of volatile S emission are often observed in members of the Cruciferae and Liliaceae families (Grundon, 1980a; Kanda and Tsuruta, 1995). High concentrations of S in foliage, whether from exposure to SO_2 or from high uptake from soil, favors S emission (Rennenberg, 1984; Schoenau and Bettany, 1989; Winner et al., 1981). Indeed, emission of H_2S and other volatile S compounds has been proposed as a mechanism for removing excess inorganic S in plant tissues (Ernst, 1990; Rennenberg, 1984; Rennenberg, 1989; Winner et al., 1981). Other factors which affect rates of S emission from plants include temperature, light intensity, plant age, and plant injury (Grundon and Asher, 1986; Trust and Fry, 1992; Wilson et al., 1978; Winner et al., 1981).

Current estimates of volatile S emission from agricultural crops are somewhat tenuous because most measurements to date have been performed in controlled conditions with relatively few species (Ernst, 1990). Extrapolation of these findings to the field scale suggests that S emission from plants in agroecosystems may range from about 0.1 to 3 kg S ha^{-1} yr^{-1} (Aneja, 1990; Grundon, 1980b). In a recent field study with wheat and corn, however, Kanda et al. (1995) estimated much lower emission rates from plants, ranging from 0.025 to 0.073 kg S ha^{-1}.

Emission of gaseous S from decomposing plant litter may be higher than that from living tissues. For example, Aneja and Cooper (1989) cite data suggesting that S emission rates from decomposing leaves may be 10 to 100 times that from living leaves.

Some volatile S may also be generated by the burning of crop residues. About half of the S content in plant residues are believed to be released to the atmosphere upon burning (Aneja, 1990; Lefroy et al., 1994). The primary species of S released by burning is probably SO_2, though some S may also be released as COS, H_2S, and sulfate aerosol.

Crop plants may also be important sinks for atmospheric S, particularly in areas with high concentrations of atmospheric SO_2 (De Kok, 1990). In conditions of soil S deficiency, plants can derive a large proportion of their nutritional requirements for S by absorption of SO_2 (Faller, 1970; Noggle et al., 1986) and exposure to low SO_2 concentrations can increase plant growth and yield (Huang and Murray, 1993; McLeod, 1992). At high concentrations, SO_2 can reduce plant growth despite enhanced S nutrition (e.g., Clapperton and Reid, 1994; Murray et al., 1994). Crop plants may also be a sink for COS (Kelly and Baker, 1990) and H_2S (De Kok, 1990; Poortinga and De Kok, 1995).

Much of the SO_2 absorbed by plant foliage quickly enters the soil in litter. In grassland, for example, much of the SO_2 absorbed by live vegetation in a given year is returned to the soil by the next year (Milchunas and Lauenroth, 1984).

D. Exchange of S Between Soil and Atmosphere

Like plants, soils serve both as a source and a sink for atmospheric S. During decomposition of organic materials in soil, various reduced S volatiles, including H_2S, COS, DMS, and CS_2, may be released into the atmosphere (Aneja, 1990; Staubes et al., 1989). Although highest emission rates are usually observed from waterlogged soils (Noggle et al., 1986), volatile S loss may also be observed from aerobic soils. For example, in a review of 27 inland U.S. soils described by Aneja (1990), emission rates ranged from 0.1 to about 3 kg S ha^{-1} yr^{-1}, with a mean rate of $\leq$ 0.2 kg S ha^{-1} yr^{-1}. Factors that favor the emission of volatile S compounds include favorable soil moisture, high temperature, and high soil organic matter content (Staubes et al., 1989).

Although emission of volatile S may be significant, most agricultural soils probably derive a net gain from exchanges with the atmosphere. Much of the atmospheric input occurs as sulfate, scavenged from atmospheric SO_2, and dissolved in precipitation (wet deposition). Precipitation may also contain methane sulfonate, a product of photochemical oxidation of dimethyl sulfide in the atmosphere (Kelly and Baker, 1990). Addition of S in the form of precipitation has been reported to range from 1 to 168 kg S ha^{-1} yr^{-1} (Olson and Rehm, 1986), depending largely on proximity to emission sources of SO_2. In most rural, inland regions, S inputs in precipitation are less than 10 kg S ha^{-1} yr^{-1} (Olson and Rehm, 1986; Tabatabai, 1984; Terman, 1978).

The amount of sulfate deposition in precipitation appears to be diminishing over time in response to reduced SO_2 emissions. For example, annual inputs of S in precipitation measured from 1985 to 1987 in Wisconsin by Andraski and Bundy (1990) were about 42% less than those measured about 16 years earlier.

Soils may also gain atmospheric S through dry deposition. Of the various processes of dry deposition, sorption of SO_2 may be the most important, especially in proximity to emission sites (Nyborg et al., 1991). Upon contact with soil, SO_2 is transformed to sulfite and subsequently oxidized to sulfate (Rasmuson et al., 1990). Alternatively, SO_2 may absorbed by reaction with organic matter or $CaCO_3$ (Cihacek and Bremner, 1992). Soil properties that affect the amount of SO_2 absorbed include pH, clay content, moisture content, $CaCO_3$ concentration, ion exchange capacity, organic matter content, soil structure, and sulfate adsorption capacity (Cihacek and Bremer, 1988; Rasmuson et al., 1990; Ryaboshapko, 1983). Soils may also receive inputs of sulfate aerosol, produced by the oxidation of SO_2 in the atmosphere (Ryaboshapko, 1983; Wesely et al., 1985; Young et al., 1988), though these may be much smaller than those of dry deposition of SO_2. Bristow and Garwood (1984) reported dry deposition of sulfate of less than 1 kg S ha^{-1} yr^{-1} compared with SO_2 deposition of 19 to 40 kg S ha^{-1} yr^{-1} at a grassland site in Great Britain. Soils also have the capacity for absorption of H_2S (Cihacek and Bremner, 1990; Cihacek and Bremner, 1993).

At some sites, S inputs from dry deposition may be higher than those from precipitation. Using a S balance technique, Morra and Dick (1988) estimated dry deposition rates ranging from 14 to 39 kg ha^{-1} yr^{-1} at a site in Ohio and cited other estimates ranging form 0.8 to 50 kg ha^{-1} yr^{-1}.

Based on reported estimates of wet and dry S deposition, rural, inland agroecosystems may gain approximately 5 to 15 kg S ha^{-1} yr^{-1} from atmospheric sources. In areas with industrial activity, inputs may be substantially higher. As a result, the atmosphere is an important source of S for agricultural production and, in some cases, inputs from this source may compensate almost entirely for S losses and S removed in harvested products.

With stricter controls on industrial S emissions, the inputs of atmospheric S into agricultural ecosystems will likely diminish with time. In industrialized areas, change may eventually induce widespread deficiencies in agricultural soils (McGrath and Zhao, 1995, 1996).

E. Exchange of S Between Soil and the Hydrosphere

Sulfate is highly soluble and therefore susceptible to leaching and surface runoff. The magnitude of leaching losses varies widely, however, depending on leaching intensity, concentrations of sulfate in the soil profile, and soil characteristics. Studies reviewed by (Tabatabai (1984) indicated leaching losses ranging from negligible amounts to as high as 320 kg S ha^{-1} yr^{-1}. In general, however, leaching losses are minimal except in cases where S inputs from fertilizer or other sources exceed the assimilatory capacity of plants and microorganisms (Nguyen and Goh, 1992).

In soils subjected to high leaching intensity, sulfate may be carried out of the soil profile and into the groundwater by percolating water. In semi-arid conditions, however, sulfate leached from the surface horizons may accumulate at lower depths as gypsum or other sulfate salts (Bettany et al., 1983). This sulfate can be accessed by crop plants and is often sufficient to sustain crop requirements (Bole and Pittman, 1984; Hue and Cope, 1987; Sawyer and Ebelhar, 1995; Zhao and McGrath, 1994).

Although most studies of S leaching have focused on movement of inorganic sulfate, there is growing evidence for the translocation of S-containing, low molecular weight organic compounds in soil profiles (Roberts et al., 1989; Schoenau and Bettany, 1987). Using ^{34}S techniques, Chae and Krouse (1986) demonstrated appreciable leaching of organic S (primarily C-bonded S) down the profile in an irrigated site with a history of heavy manure applications. In semi-arid conditions, any leached organic S may be retained in the soil profile but under high leaching intensity, it may represent a significant export of S from the ecosystem.

Translocation of soluble S in the soil profile typically occurs downward, but evaporative forces may result in the upward movement of subsoil sulfate into the rooting zone (Castellano and Dick, 1990).

F. Agricultural Inputs and Removals

In all arable agroecosystems, significant amounts of S are removed in the form of plant or animal products. In many cases, however, much of the S assimilated by the crop is retained in the system as crop residues or animal wastes. Removal or burning of these residues appreciably increases the loss of S from the agroecosystem.

Soil erosion represents another possible loss of S from agroecosystems, although its greatest effect may be the redistribution of S rather than its export from the system. In some arid environments, wind erosion can lift significant amounts of S into the atmosphere as dust particles (Brimblecombe et al., 1989).

The primary S input in agroecosystems is in S-containing fertilizers. Historically, high amounts of S have been applied as incidental constituents of P fertilizers. With growing recognition of S deficiency and the development of higher analysis P fertilizers, there has been a shift to intentional application of S fertilizers. These fertilizers contain a variety of S forms, including sulfate, thiosulfate, elemental S, sulfide, sulfite, and sulfur dioxide (Germida and Janzen, 1993). Rates of application to arable croplands are typically in the range of 15 to 30 kg S ha^{-1} (e.g., Beaton and Soper, 1986; Mahler et al., 1993).

In localized areas, some S may be applied to soils in the form of nutritive amendments such as animal manure, composts, or sewage sludge. Analysis of 4 sewage sludges by Tabatabai and Chae (1991) indicated total S concentrations ranging from 6.6 to 22.6 g S kg^{-1}. A 50 Mg ha^{-1} application of these materials would therefore supply 330 to 1130 kg S ha^{-1}. A series of animal manures analyzed in the same study showed appreciably lower concentrations of S, ranging from 1.5 to 2.0 g S kg^{-1}.

IV. MANAGEMENT EFFECTS ON S DYNAMICS IN AGROECOSYSTEMS

A. Conversion to Arable Agriculture

Cultivation of soils for agricultural production usually results in loss of organic matter. This loss of organic matter is almost always associated with a concomitant loss of S. For example, Ellert and Bettany (1992) observed appreciably higher organic and mineralizable S concentrations in a recently cleared forest soil than in a comparable soil that had been under long-term fallow-wheat cropping. Similarly, intensive vegetable production has also been observed to result in loss of organic S, relative to that in adjacent, comparatively undisturbed soil (Olson and Lowe, 1990). Amounts of S loss vary depending on soil type, cropping practices, and duration of cultivation, but total S concentrations in cultivated soils are often 20 to 40% lower than those in uncultivated soils (Table 1).

Losses of S upon cultivation are usually lower than those of C and N (Bettany et al., 1980), possibly because S is more resistant to decomposition. Alternatively, the higher retention of S relative to C or N may reflect the lower susceptibility of S to loss from the ecosystem. Whereas both C and N are readily lost in gaseous form, any S losses occur primarily via leaching. Even then, the leaching potential of S is generally considered to be lower than that of N. The S accumulating from the decomposition of organic matter may be stored as gypsum in semi-arid soils, or may be assimilated by microorganisms, which can store excess S in various ester sulfate forms (Stewart and Cole, 1989). This latter explanation is partially supported by the observation that the relative proportion of HI-reducible S generally increases upon cultivation (Freney and Williams, 1983). For example, Bettany et al. (1980) found that HI-S, expressed as a proportion of total S, increased from 51% to 56% following cultivation.

B. Crop Sequence

The selection of crop species and rotations can influence soil S dynamics in various ways, including differential nutrient extraction, amount and composition of residue input, and degree of soil disturbance. Often the most extreme differences in soil S are observed between soils under arable cropping and those under perennial pastures. Soils under pasture typically have much higher total S concentrations than arable soils (Haynes and Williams, 1992; Keer et al., 1990; McLaren and Swift, 1977). Microbial biomass S in pasture soils is also often higher than that in arable soil (Ghani et al., 1990; Wu et al., 1994).

Few studies have compared S dynamics among different rotations of annual crops. A comparison of two cereal cropping systems suggested that S deficiency was more intense under continuous cropping than under a fallow-wheat system (Rasmussen and Douglas, 1992). Another study, however, reported slightly higher accumulation of sulfate in a continuous-cropping system, than in a rotation with periodic fallow (Grant and Lafond, 1994). A comparison of soil cropped to winter rape with that left uncropped demonstrated few differences in concentrations of total, ester, C-bonded or sulfate S, but arylsulfatase activity was increased by the presence of plants (Castellano and Dick, 1990). As well, microbial biomass S may be higher in cropped than in uncropped soil (Ghani et al., 1990). These studies provide some evidence of significant effects of crop sequence on S transformations, though the results are not yet conclusive

Additional understanding of S dynamics in various crop sequences can be inferred from studies describing S mineralization from crop residues. Laboratory incubation studies have shown large differences in S availability among residues. Plant litter from cruciferous crops like rape provide a large and rapid release of available S, apparently because of the accumulation of inorganic and organic sulfates in these residues (Roppel, 1991; Wu et al., 1993). Residues from other crops may release only minimal amounts of available S and in many cases (e.g. cereal residues), result in the net immobilization of sulfate (Tabatabai and Chae, 1991; Wu et al., 1993). In contrast to release of nutrients like nitrogen, S release typically occurs very rapidly; for example, Wu et al. (1993) observed that more than 80% of S in rape residue was released as sulfate or assimilated into microbial biomass within 5 days of application. As a result, S-rich residues can be considered an important source of plant-available S, and the choice

Table 1 Impacts of Management on Total S Concentration in Surface Soils of Selected Studies

Location: soil type (reference)	Management history	No. of soils	Years managed	Total S (mg S kg^{-1})
Sask., Canada: Black Chernozem (Bettany and Stewart, 1983)	Native	1	0	410
	Cultivated	1	5	400
	Cultivated	1	60	330
	Cultivated	1	90	280
Berkshire, England: wooded soil (Cowling and Jones, 1970)	Beech forest	1	0	407
	Cultivated	1	100+	327
New Zealand: yellow-gray earths (Sears et al., 1965)	Pasture	3	0	340
	Cultivated	3	3	300
New South Wales: red-brown earths (Williams and Lipsett, 1961)	Uncultivated	55	0	128
	Cultivated	55	60	92
Iowa, USA: Typic Hapludoll (Larson et al., 1972)	Residues added	2	11	334
	Cultivated, no residues	1	11	231
Ohio, USA: Ultic Hapludalf (Staley et al., 1988)	Conventional tillage	1	0	144
	No-tillage + manure	1	9	264
	No-tillage + manure	1	20	468
Sask., Canada: Gray Luvisols (Ellert, 1990)	Native forest	3	0	187
	Cultivated or pasture	4	85	196
Alberta, Canada: Gray Luvisols (Setsoafia, 1983)	Forage rotation	1	52	211
	Wheat/fallow	1	52	152
Sask., Canada: Black Chernozem (Bettany et al., 1980)	Native	1	0	497
	Cultivated	1	65	308
Australia: red-brown earths (McLachlan and De Marco, 1975)	Uncropped	10	0	131
	Cropped	10	?	94
	Unfertilized pasture	10	0	126
	Fertilized pasture	10	20+	220

New South Wales, Australia: Podzols (Freney et al., 1975)	Unfertilized pasture	1	0	106
	Fertilized pasture	1	45	248
Scotland: Gleysols (McLaren and Swift, 1977)	Permanent pasture	7	0	597
	Cultivated	7	100+	351
Oregon, USA: Typic Haploxeroll (Castellano and Dick, 1988)	Wheat/fallow	1	55	134
	Plus N fertilizer	1	55	114
	Plus pea vine	1	55	151
	Plus manure	1	55	161
Alberta, Canada: Typic Haploboroll (Chae and Krouse, 1986)	No manure	1	10	564
	60 Mg Manure $ha^{-1}yr^{-1}$	1	10	1101
	120 Mg Manure $ha^{-1}yr^{-1}$	1	10	1848
New Zealand: Udic Ustochrept (Haynes and Williams, 1992)†	Arable	1	11	174
	Wilderness	1	0	348
	Pasture (no P)	1	37	338
	Pasture (188 kg superphosphate $ha^{-1}yr^{-1}$)	1	37	447
	pasture (376 kg superphosphate $ha^{-1}yr^{-1}$)	1	37	452
British Columbia, Canada: Organic soil (Olson and Lowe, 1990)	Uncultivated	1	0	4700
	Cultivated	1	40	4480
Saskatchewan, Canada: Typic Cryoboralf (Gupta and Germida, 1988)	Check	1	5	260
	22 kg $S°ha^{-1}yr^{-1}$	1	5	284
	44 kg $S°ha^{-1}yr^{-1}$	1	5	381

Source: Adapted from Ellert, 1990.
†Values estimated from graph.

and sequence of crops can have an important impact on S dynamics in the soil.

C. Tillage Practices

In many agricultural systems, there has been a widespread adoption of reduced tillage practices. This transition may affect organic matter by minimizing the disruption of aggregates that protect organic substrates, maintaining plant litter near the soil surface, and altering soil moisture and temperature.

The impact of reduced tillage on S dynamics has not been fully described though several recent studies have sought to measure the impact of tillage on S fertility. Two studies in western Canada detected minimal effects of tillage treatment on soil sulfate after four or eight years (Grant and Lafond, 1994; Malhi et al., 1992). In a long-term study in Nebraska, similarly, total sulfate in the soil profile (60 cm) was not affected by tillage, but the no-till treatment retained higher sulfate and HI-reducible S concentrations near the surface (Tracy et al., 1990a). These authors hypothesized that tillage would hasten the onset of S deficiency. In a laboratory incubation of soils from the same site, S mineralization tended to be higher in no-till soil than in tilled soil, particularly in the surface soil layer (0–2.5 cm) (Tracy et al., 1990b). These findings are somewhat contrary to studies in Oregon (Rasmussen, 1996; Rasmussen and Douglas, 1992), which suggest that S deficiency was greater under no-till than conventional tillage.

Perhaps the most important effect of tillage on soil S status is simply the redistribution soil within the profile. As for C and N, S tends to accumulate near the surface in undisturbed systems, whereas tillage mixes the material into deeper layers (Staley et al., 1988). In soils where surface layers are deficient in available S, deep tillage may enhance S nutrition by making S in deeper layers more accessible (Cassel et al., 1996).

Tillage may also indirectly affect soil S status by affecting inputs and losses. Morra and Dick (1988) observed higher dry deposition of S under no-till (25 to 39 kg ha^{-1} yr^{-1}) than conventional tillage (14 to 26 kg ha^{-1} yr^{-1}). These differences were attributed to the effects of tillage on surface soil pH and moisture content. In the same study, S losses in runoff were shown to be higher from a conventional tillage treatment (26 kg ha^{-1}) than from a no-till treatment (11 kg ha^{-1}). These indirect effects may, conceivably be of greater magnitude than effects of tillage on organic S turnover.

D. Fertilizers

A wide range of S-containing fertilizers are commercially available (Hagstrom, 1986). These products include immediately available S sources (e.g., various sulfate salts) as well as sources, like elemental S, with appreciable residual benefits. As a result, S deficiencies can be readily alleviated through application of appropriate fertilizer forms.

Application of S-containing fertilizers not only increases concentrations of sulfate, but may also enhance the organic S content of soils, particularly in pastures on S-deficient soils (Freney, 1986). For example, Haynes and Williams (1992) reported that organic S in a soil (0–4 cm) under ryegrass-clover pasture was 16% higher in treatments receiving superphosphate than in a treatment receiving no fertilizer. The relative accumulation of S was greater than that observed for organic C. Benefits of S application on S status of cultivated soils have not been documented to the same extent, although there is some evidence of S accumulation in response to S fertilization (Castellano and Dick, 1990). Furthermore, laboratory studies have shown rapid incorporation of sulfate into organic forms, particularly ester sulfate (Ghani et al., 1993), suggesting the possibility of enhanced organic S from fertilizers.

The type of S fertilizer applied may influence the S dynamics within the soil. Addition of sulfate fertilizers provides a pulse of soluble S which is readily available for uptake by plants but is also highly susceptible to leaching (Curtin and Syers, 1990). Application of elemental S, in comparison, provides a more gradual release of sulfate into the soil, thereby reducing the risk of leaching losses (Korentajer et al., 1984; Santoso et al., 1995a) but increasing the lag time between time of application and time of plant S uptake.

In some instances, S application may also have detrimental influence on the long-term S-supplying capacity of agricultural soils. Studies in Saskatchewan showed that fertilization with elemental S increased total S, sulfate S, and HI-S in soil (Gupta and Germida, 1988; Gupta et al., 1988). At the same time, however, number of bacteria and fungal biomass were significantly reduced, apparently in response to soil acidification. These findings suggest that, at least in some cases, the secondary effects of S application may affect the nutrient turnover in soils.

Application of other nutrients may also affect the S cycle in agricultural soils. For example, Castellano and Dick (1988) reported a depletion of soil S in response to application of N. Most of this depletion occurred from the "residual S" fraction (total S − C-bonded S–HI-S). Ghani et al. (1992) similarly observed that application of N enhanced slightly the rate of S mineralization. As well, P application may reduce the capacity of soils to sorb sulfate (Haron and Hanson, 1988).

Liming of acid soils may result in the release of sorbed sulfate (Marsh et al., 1987; Santoso et al., 1995b). As a result, this practice may hasten the appearance of S deficiency under conditions of high leaching intensity (Korentajer et al., 1983). Addition of P fertilizer to limed soil may further accelerate S leaching (Lefroy et al., 1995).

E. Organic Amendments

Amendment of soils with animal manure increases the organic S concentration of the soil. In one study, annual application of manure at a rate of 180 mg ha^{-1} (wet weight basis) for 10 years increased total and HI-S of surface soil by about threefold over that of the unamended check (Chae and Krouse, 1986). This organic S, however, appears to turn over relatively slowly. Tabatabai and Chae (1991) found that manure-amended soils showed gradual linear S mineralization over time. In some cases, S mineralization in manure-amended soil was slower than that in unamended soils, indicating the immobilization of soil sulfate. Other studies have also concluded that animal manures and slurries are not a very good source of plant-available S in the short term (Eriksen et al., 1995c; Lloyd, 1994).

A number of recent studies have evaluated the impact of municipal waste applications on S dynamics in agricultural soils. Tabatabai and Chae (1991) observed a rapid initial S mineralization in sludge-amended soil, followed by a slow linear release. They observed large differences in S release from various sludge samples, partly related to the C/S ratios of the material. In another study (Perucci, 1990), application of municipal waste compost to soil was found to stimulate microbial biomass S as well as aryl sulfatase activity. The enhancement of sulfatase activity in amended soil may reflect the direct input of enzymes in the amendment, the stimulation of microbial activity, or the physical protection of enzymes by the organic material (Martens et al., 1992).

Amendment of soils with organic substrate may enhance release of volatile S (Banwart and Bremner, 1976; Minami and Fukushi, 1981). Lewis and Papavizas (1970) observed evolution of methane thiol, dimethyl sulfide, and dimethyl sulfide from the decomposition of cruciferous crop residues but not from corn residues. Staubes et al. (1989) observed enhanced emission of COS, DMS, and CS_2 from soil amended with manure.

F. Irrigation

Irrigation water is an important source of S in localized areas. For example, in southern Alberta the irrigation water may supply between 8 and 50 kg S ha^{-1} per 10 cm of water applied (Beaton and Soper, 1986; Bole, 1978). In some cases, therefore, the S inputs from irrigation water are severalfold the crop S requirements. A review of S concentrations in irrigation water in the United States similarly indicated that the water often had enough S to meet crop requirements (Olson and Rehm, 1986).

G. Example of Long-Term Study

A long-term fertility study in Sweden provides a good illustration of the impact of cropping practices on the net S balance (Kirchmann et al., 1996). Depending on cropping system and fertility amendment, the average annual change in total soil S ranged from -6 to $+13$ kg S ha^{-1} (Table 2). In cases where S applications exceed crop requirements, high leaching losses were observed. Thus, in this ecosystem at least, the selection of cropping treatments determines whether the soil S will eventually be depleted, or the agroecosystem results in net loss of S to adjacent environments.

V. SUMMARY AND CONCLUSIONS

Despite intensive research effort over the past decade, the chemical nature of the labile organic S pool, and the mechanism of its release, remains somewhat obscure. Although earlier approaches such as fractionation into bond-classes showed promise, recent studies have demonstrated inadequate resolution to adequately account for the complexity of organic S turnover.

One area that may deserve more attention is the release of S from plant litter and microbial debris. Because many biological tissues can store excess S, and because much of this S may be stored as inorganic and organic sulfate, short-term release from organic residues may be proportionally higher for S than for other nutrients like nitrogen. As a result, turnover times may be faster, and opportunities for management through cropping practices may be greater than that for other nutrients.

Most studies of S dynamics in agroecosystems have been performed under controlled conditions in the laboratory. While these studies have yielded much information about mechanisms and processes, the reliability of extrapolations to the field has not been fully established. Further investigations are required to extend present findings both in space and in time.

Table 2 Influence of Cropping Practices and Soil Amendment on the Sulfur Balance in a Long-Term Experiment in Central Sweden

Treatment	Atmospheric deposition	Seed input	Manure, fertilizer	Crop removal	Change in soil S	Leaching loss[a]
			kg S ha^{-1} yr^{-1}			
No N, fallow	10	0	27	0	−6.0	43.0
No N	10	0.1	27	8	−4.2	33.3
$Ca(NO_3)_2$	10	0.1	27	15	−2.0	24.1
$(NH_4)_2SO_4$	10	0.1	118	17	−2.2	113.3
Green manure	10	0.1	38	14	0.4	33.7
Animal manure	10	0.1	45	12	5.5	37.6
Sewage sludge	10	0.1	94	21	12.9	70.2
Peat	10	0.1	32	8	−1.9	36.0

[a]Determined by difference.
Source: Kirchmann et al., 1996.

Although the cycling of other nutrients, like C and N, has been extensively studied in long-term studies, often spanning many decades, this approach has not been adequately tapped for verification of long-term S trends.

A unique characteristic of the S cycle in agroecosystems is the impact of atmospheric inputs. Rates of deposition vary widely and, in some agroecosystems, atmospheric S inputs are sufficient to compensate for S lost in crop removal, leaching, and volatile S emission. For example, inputs of atmospheric S in rural regions may be in the range of 5 to 15 kg S ha^{-1} yr^{-1}. Thus a wheat-cropping system in semi-arid regions, with an annual S removal in grain of less than about 10 kg ha^{-1} and limited leaching, can continue almost indefinitely without depleting soil S reserves. In regions with higher S emissions, atmospheric S can support appreciably higher S removals, though diminishing emissions may eventually result in depletion of soil S reserves.

Appreciable progress has been made in recent decades in the development of fertilization practices to alleviate S deficiency. As a result, while greater emphasis on reducing S emissions may induce a greater extent of S deficiency, this depletion can be readily corrected by relatively low rates of fertilizer S.

REFERENCES

Anderson, D. W., S. Saggar, J. R. Bettany, and J. W. B. Stewart. (1981). Particle size fractions and their use in studies of soil organic matter: I. The nature and distribution of forms of carbon, nitrogen, and sulfur. Soil Sci. Soc. Am. J. 45: 767–772.

Andraski, T. W. and L. G. Bundy. (1990). Sulfur, nitrogen, and pH levels in Wisconsin precipitation. J. Environ. Qual. 19: 60–64.

Aneja, V. P. (1990). Natural sulfur emissions into the atmosphere. J. Air & Waste Manage. Assoc. 40: 469–476.

Aneja, V. P., and W. J. Cooper. (1989). Biogenic sulfur emissions. In *Biogenic Sulfur in the Environment. ACS Symposium Series No. 393* (ed. E. S. Saltzman and W. J. Cooper), pp. 2–13. American Chemical Society, Washington, DC.

Bailey, L. D. (1985). The sulphur status of eastern Canadian prairie soils: the relationship of sulphur, nitrogen and organic carbon. Can. J. Soil Sci. 65: 179–186.

Banerjee, M. R., S. J. Chapman, and K. Killham. (1993). Factors influencing the determination of microbial biomass sulphur in soil. Commun. Soil Sci. Plant Anal. 24: 939–950.

Banwart, W. L., and J. M. Bremner. (1976). Volatilization of sulfur from unamended and sulfate-treated soils. Soil Biol. Biochem. 8: 19–22.

Beaton, J. D., and R. J. Soper. (1986). Plant response to sulfur in western Canada. In *Sulfur in Agriculture. Agronomy Monograph No. 27.* (ed. M. A. Tabatabai), pp. 375–403. ASA-CSSA-SSSA, Madison, WI.

Bettany, J. R., H. H. Janzen, and J. W. B. Stewart. (1983). Sulphur deficiency in the prairie provinces of Canada. In *Proceedings International Sulphur '82 Conference*, vol. 2 (ed. A. I. More), pp. 787–799. British Sulphur Corporation Ltd., London.

Bettany, J. R., S. Saggar and J. W. B. Stewart. (1980). Comparison of the amounts and forms of sulfur in soil organic matter fractions after 65 years of cultivation. Soil Sci. Soc. Am. J. 44: 70–75.

Bettany, J. R. and J. W. B. Stewart. (1983). Sulphur cycling in soils. In *International Sulphur '82 Conference*, vol. 2 (ed. A. I. More), pp. 767–785. British Sulphur Corporation Ltd., London.

Biederbeck, V. O. (1978). Soil organic sulfur and fertility. In *Soil Organic Matter* (ed. M. Schnitzer and S. V. Khan), pp. 273–310. Elsevier Scientific Publishing Co., Amsterdam.

Bohn, H. L., N. J. Barrow, S. S. S. Rajan, and R. L. Parfitt. (1986). Reactions of inorganic sulfur in soils. In *Sulfur in Agriculture. Agronomy Monograph No. 27* (ed. M. A. Tabatabai), pp. 233–249. ASA-CSSA-SSSA, Madison, WI.

Bolan, N. S., D. R. Scotter, J. K. Syers, and R. W. Tillman. (1986). The effect of adsorption on sulfate leaching. Soil Sci. Soc. Am. J. 50: 1419–1424.

Bole, J. B. (1978). Fertilizer sulphur unnecessary on irrigated soils. Agriculture Canada Research Station, Lethbridge, Alberta.

Bole, J. B., and U. J. Pittman. (1984). Availability of subsoil sulphates to barley and rapeseed. Can. J. Soil Sci. 64: 301–312.

Brimblecombe, P., C. Hammer, H. Rodhe, A. Ryaboshapko, and C. F. Boutron. (1989). Human influence on the sulphur cycle. In *Evolution of the Global Biogeochemical Sulphur Cycle. Scientific Committee on the Problems of the Environment (SCOPE) No. 39* (ed. P. Brimblecombe and A. Y. Lein), pp. 77–121. John Wiley & Sons Ltd., New York.

Bristow, A. W., and E. A. Garwood. (1984). Deposition of sulphur from the atmosphere and the sulphur balance in four soils under grass. J. Agric. Sci., Cambridge 103: 463–468.

Cassel, D. K., E. J. Kamprath, and F. W. Simmons. (1996). Nitrogen-sulfur relationships in corn as affected by landscape attributes and tillage. Agron. J. 88: 133–140.

Castellano, S. D. and R. P. Dick. (1988). Distribution of sulfur fractions in soils as influenced by management of organic residues. Soil Sci. Soc. Am. J. 52: 1403–1407.

Castellano, S. D. and R. P. Dick. (1990). Cropping and sulfur fertilization influence on sulfur transformations in soil. Soil Sci. Soc. Am. J. 54: 114–121.

Castellano, S. D., and R. P. Dick. (1991a). Modeling sulfur transformations in soils. Soil Science 152: 448–454.

Castellano, S. D., and R. P. Dick. (1991b). Modified calibration procedure for the measurement of microbial sulfur in soil. Soil Sci. Soc. Am. J. 55: 283–285.

Chae, Y. M., and H. R. Krouse. (1986). Alteration of sulfur-34 natural abundance in soil by application of feedlot manure. Soil Sci. Soc. Am. J. 50: 1425–1430.

Chapman, S. J. (1987a). Microbial sulphur in some Scottish soils. Soil Biol. Biochem. 19: 301–305.

Chapman, S. J. (1987b). Partioning of ryegrass residue sulphur between the soil microbial biomass, other soil sulphur pools and ryegrass (*Lolium perenne*, L). Biol. Fertil. Soils 5: 253–257.

Cihacek, L. J., and J. M. Bremer. (1988). Capacity of soils for sorption of sulfur dioxide. Commun. Soil Sci. Plant Anal. 19: 1945–1964.

Cihacek, L. J. and J. M. Bremner. (1990). Capacity of soils for sorption of hydrogen sulfide. Commun. Soil Sci. Plant Anal. 21: 351–363.

Cihacek, L. J., and J. M. Bremner. (1992). Characterization of sulfur retained by soils exposed to sulphur dioxide. Commun. Soil Sci. Plant Anal. 23: 805–816.

Cihacek, L. J., and J. M. Bremner. (1993). Characterization of the sulfur retained by soils exposed to hydrogen sulfide. Commun. Soil Sci. Plant Anal. 24: 85–92.

Clapperton, M. J., and D. M. Reid. (1994). Effects of sulphur dioxide (SO_2) on growth and flowering of SO_2-tolerant and non-tolerant genotypes of *Phleum pratense*. Environ. Pollut. 86: 251–258.

Clarkson, D. T., and J. B. Hanson. (1980). The mineral nutrition of higher plants. Annual Review of Plant Physiology 31: 239–298.

Comfort, S. D., R. P. Dick, and J. Baham. (1991). Air-drying and pretreatment effects on soil sulfate sorption. Soil Sci. Soc. Am. J. 55: 968–973.

Coughenour, M. B., W. J. Parton, W. K. Lauenroth, J. L. Dodd, and R. G. Woodmansee. (1980). Simulation of a grassland sulfur-cycle. Ecol. Modell. 9: 179–213.

Courchesne, F. (1992). Relationships between soil chemical properties and sulfate sorption kinetics in Podzolic soils from Quebec. Can. J. Soil Sci. 72: 467–480.

Cowling, D. W., and L. H. P. Jones. (1970). A deficiency in soil sulfur supplies for perennial ryegrass in England. Soil Science 110: 346–354.

Cram, W. J. (1990). Uptake and transport of sulfate. In *Sulfur Nutrition and Sulfur Assimilation in Higher Plants* (ed. H. Renneberg, C. Brunold, L. J. De Kok and I. Stulen), pp. 3–11. SPB Academic Publishing, The Hague.

Curtin, D., and J. K. Syers. (1990). Extractability and adsorption of sulphate in soils. J. Soil Sci. 41: 305–312.

De Kok, L. J. (1990). Sulfur metabolism in plants exposed to atmospheric sulfur. In *Sulfur Nutrition and Sulfur Assimilation in Higher Plants* (ed. H. Rennenberg, C. Brunold, L. J. De Kok, and I. Stulen), pp. 111–130. SPB Academic Publishing, The Hague.

Duke, S. H., and H. M. Reisenauer. (1986). Roles and requirements of sulfur in plant nutrition. In *Sulfur in Agriculture. Agronomy Monograph No. 27* (ed. M. A. Tabatabai), pp. 123–168. ASA-CSSA-SSSA, Madison, WI.

Ellert, B. H. (1990). *Kinetics of nitrogen and sulfur cycling in Gray Luvisol soils. Ph.D. dissertation, University of Saskatchewan, Saskatoon, Canada. Catalogue #NN6821.* University Microfilms International, Ann Arbor, MI, USA.

Ellert, B. H., and J. R. Bettany. (1988). Comparison of kinetic models for describing net sulfur and nitrogen mineralization. Soil Sci. Soc. Am. J. 52: 1692–1702.

Ellert, B. H., and J. R. Bettany. (1992). Temperature dependence of net nitrogen and sulfur mineralization. Soil Sci. Soc. Am. J. 56: 1133–1141.

Elliott, E. T., and C. V. Cole. (1989). A perspective on agroecosystem science. Ecology 70: 1597–1602.

Eriksen, J., R. D. B. Lefroy., and G. J. Blair. (1995a). Physical protection of soil organic S studied by extraction and fractionation of soil organic matter. Soil Biol. Biochem. 27: 1011–1016.

Eriksen, J., R. D. B. Lefroy, and G. J. Blair. (1995b). Physical protection of soil organic S studied using acetylacetone extraction at various intensities of ultrasonic dispersion. Soil Biol. Biochem. 27: 1005–1010.

Eriksen, J., J. V. Mortensen, V. K. Kjellerup, and O. Kristjansen. (1995c). Forms and plant-availability of sulfur in cattle and pig slurry. Z. Pflanzenernahr. Bodenk. 158: 113–116.

Eriksen, J., J. V. Mortensen, J. D. Nielsen, and N. E. Nielsen. (1995d). Sulphur mineralisation in five Danish soils as measured by plant uptake in a pot experiment. Agriculture, Ecosystems and Environment 56: 43–51.

Ernst, W. H. O. (1990). Ecological aspects of sulfur metabolism. In *Sulfur Nutrition and Sulfur Assimilation in Higher Plants* (ed. H. Rennenberg, C. Brunold, L. J. De Kok, and I. Stulen), pp. 131–144. SPB Academic Publishing, The Hague.

Faller, N. (1970). Effects of atmospheric SO_2 on plants. Sulphur Institute Journal 6: 5–7.

Freney, J. R. (1986). Forms and reactions of organic sulfur compounds in soils. In *Sulfur in Agriculture. Agronomy Monographs No. 27* (ed. M. A. Tabatabai), pp. 207–232. ASA-CSSA-SSSA, Madison, WI.

Freney, J. R., G. E. Melville, and C. H. Williams. (1975). Soil organic matter fractions as sources of plant-available sulphur. Soil Biol. Biochem. 7: 217–221.

Freney, J. R., and C. H. Williams. (1983). The sulphur cycle in soil. In *The Global Biogeochemical Sulphur Cycle. Scientific Committee on the Problems of the Environment (SCOPE) No. 19* (ed. M. V. Ivanov and J. R. Freney), pp. 129–201. John Wiley & Sons, New York.

Ganeshamurthy, A. N., and N. E. Nielsen. (1990). Arylsulphatase and the biochemical mineralization of soil organic sulphur. Soil Biol. Biochem. 22: 1163–1165.

Germida, J. J., and H. H. Janzen. (1993). Factors affecting the oxidation of elemental sulfur in soils. Fert. Res. 35: 101–114.

Ghani, A., R. G. McLaren, and R. S. Swift. (1990). Seasonal fluctuations of sulphate and soil microbial biomass-S in the surface of a Wakanui soil. N. Z. J. Agric. Res. 33: 467–472.

Ghani, A., R. G. McLaren, and R. S. Swift. (1991). Sulphur mineralisation in some New Zealand soils. Biol. Fertil. Soils 11: 68–74.

Ghani, A., R. G. McLaren and R. S. Swift. (1992). Sulphur mineralisation and transformations in soils as influenced by additions of carbon, nitrogen and sulphur. Soil Biol. Biochem. 24: 331–341.

Ghani, A., R. G. McLaren, and R. S. Swift. (1993). The incorporation and transformations of ^{35}S in soil: Effects of soil conditioning and glucose or sulphate additions. Soil Biol. Biochem. 25: 327–335.

Grant, C. A., and G. P. Lafond. (1994). The effects of tillage systems and crop rotations on soil chemical properties of a Black Chernozemic soil. Can. J. Soil Sci. 74: 301–306.

Grundon, N. J. (1980a). Uptake of atmospheric sulfur by plants growing in an urban environment. In *Sulfur in Australia* (ed. J. R. Freney and A. J. Nicolson), pp. 251–255. Australian Academy of Science, Canberra.

Grundon, N. J. (1980b). Volatile losses of sulfur from intact plants. In *Sulfur in Australia* (ed. J. R. Freney and A. J. Nicolson), pp. 20–23. Australian Academy of Science, Canberra.

Grundon, N. J., and C. J. Asher. (1986). Volatile losses of sulfur by intact alfalfa plants. J. Plant Nutr. 9: 1519–1532.

Gupta, V. V. S. R., R. E. Farrell, and J. J. Germida. (1993). Activity of arylsulfatase in Saskatchewan soils. Can. J. Soil Sci. 73: 341–347.

Gupta, V. V. S. R. and J. J. Germida. (1988). Populations of predatory protozoa in field soils after 5 years of elemental S fertilizer application. Soil Biol. Biochem. 20: 787–791.

Gupta, V. V. S. R. and J. J. Germida. (1989a). Influence of bacterial-amoebal interactions on sulfur transformations in soil. Soil Biol. Biochem. 21: 921–930.

Gupta, V. V. S. R., and J. J. Germida. (1989b). Microbial biomass and extractable sulfate sulfur levels in native and cultivated soils as influenced by air-drying and rewetting. Can. J. Soil Sci. 69: 889–894.

Gupta, V. V. S. R., J. R. Lawrence, and J. J. Germida. (1988). Impact of elemental sulfur fertilization on agricultural soils. I. Effects on microbial biomass and enzyme activities. Can. J. Soil Sci. 68: 463–473.

Hagstrom, G. R. (1986). Fertilizer sources of sulfur and their use. In *Sulfur in Agriculture. Agronomy Monograph No. 27* (ed. M. A. Tabatabai), pp. 567–581. ASA-CSSA-SSSA, Madison, WI.

Haines, B., M. Black, and C. Bayer. (1989). Sulfur emissions from roots of the rain forest tree *Stryphnodendron excelsum.* In *Biogenic Sulfur in the Environment. ACS Symposium Series 393* (ed. E. S. Saltzman and W. J. Cooper), pp. 58–69. American Chemical Society, Washington, DC.

Haron, K. B., and R. G. Hanson. (1988). Sulfur supplying capacity of some Missouri soils. Soil Sci. Soc. Am. J. 52: 1657–1660.

Hartel, P. G., and B. L. Haines. (1992). Effects of potential plant CS_2 emissions on bacterial growth in the rhizosphere. Soil Biol. Biochem. 24: 219–224.

Haynes, R. J., and P. H. Williams. (1992). Accumulation of soil organic matter and the forms, mineralization potential and plant-availability of accumulated organic sulphur: Effects of pasture improvement and intensive cultivation. Soil Biol. Biochem. 24: 209–217.

He, Z. L., A. G. O'Donnell, J. Wu and J. K. Syers. (1994). Oxidation and transformation of elemental sulphur in soils. J. Sci. Food Agric. 65: 59–65.

Huang, L. and F. Murray. (1993). Effects of sulfur dioxide fumigation on growth and sulphur accumulation in wheat (*Triticum aestivum* cv. Wilgoyne (Ciano/Gallo)) under salinity stress. Agriculture, Ecosystems and Environment 43: 285–300.

Hue, N. V., and J. T. Cope, Jr. (1987). Use of soil-profile sulfate data for predicting crop response to sulfur. Soil Society of America Journal 51: 658–664.

Ivanov, M. V. (1983). Major fluxes of the global biogeochemical cycle of sulphur. In *The Global Biogeochemical Sulphur Cycle. Scientific Committee on the Problems of the Environment (SCOPE) No. 19* (ed. M. V. Ivanov and J. R. Freney), pp. 449–463. John Wiley & Sons, New York.

Kanda, K. and H. Tsuruta, (1995). Emissions of sulfur gases from various types of terrestrial higher plants. Soil Sci. Plant Nutr. 41: 321–328.

Kanda, K., H. Tsuruta and K. Minami. (1995). Emissions of biogenic sulfur gases from maize and wheat fields. Soil Sci. Plant Nutr. 41: 1–8.

Keer, J. I., R. G. McLaren, and R. S. Swift. (1990). Acetylacetone extraction of soil organic sulphur and fractionation using gel chromatography. Soil Biol. Biochem. 22: 97–104.

Kelly, D. P., and S. C. Baker. (1990). The organosulphur cycle: aerobic and anaerobic processes leading to turnover of C_1-sulphur compounds. Federation of European Microbiological Societies (FEMS) Microbiology Review 87: 241–246.

Kelly, D. P. and N. A. Smith. (1989). Organic sulfur compounds in the environment: Biogeochemistry, microbiology, and ecological aspects. Adv. Microb. Ecol. 11: 345–385.

Kirchmann, H., F. Pichlmayer, and M. H. Gerzabek. (1996). Sulfur balances and sulfur-34 abundance in a long-term fertilizer experiment. Soil Sci. Soc. Am. J. 59: 174–178.

Konopka, A. E., R. H. Miller, and L. E. Sommers. (1986). Microbiology of the sulfur cycle. In *Sulfur in Agriculture. Agronomy Monograph No. 27* (ed. M. A. Tabatabai), pp. 23–55. ASA-CSSA-SSSA, Madison, WI.

Korentajer, L., B. H. Byrnes, and D. T. Hellums. (1983). The effect of liming and leaching on the sulfur-supplying capacity of soils. Soil Sci. Soc. Am. J. 47: 525–530.

Korentajer, L., B. H. Byrnes, and D. T. Hellums. (1984). Leaching losses and plant recovery from various sulfur fertilizers. Soil Sci. Soc. Am. J. 48: 671–676.

Larson, W. E., C. E. Clapp, W. H. Piere, and Y. B. Morachan. (1972). Effects of increasing amounts of organic residues on continuous corn: II. Organic carbon, nitrogen, phosphorus, and sulfur. Agron. J. 64: 204–208.

Lefroy, R. D. B., W. Chaitep, and G. J. Blair. (1994). Release of sulfur from rice residues under flooded and non-flooded soil conditions. Aust. J. Agric. Res. 45: 657–667.

Lefroy, R. D. B., D. Santoso, and G. J. Blair. (1995). Fate of applied phosphate and sulfate in weathered acid soils under leaching conditions. Aust. J. Soil Res. 33: 135–151.

Lewis, J. A., and G. C. Papavizas. (1970). Evolution of volatile sulfur-containing compounds from decomposition of crucifers in soil. Soil Biol. Biochem. 2: 239–246.

Lloyd, A. (1994). Effectiveness of cattle slurry as a sulphur source for grass cut for silage. Grass Forage Sci. 49: 203–208.

Lou, G., and P. R. Warman. (1992). Labile ester sulphate in organic matter extracted from podzolic soils. Biol. Fertil. Soils 14: 267–271.

Lou, G. Q. J., and P. R. Warman. (1994). Characterization of ester sulphate in a gypsum-amended podzol using an immobilized sulphatase reactor. Biol. Fertil. Soils 17: 276–280.

Lynch, J. M., and J. M. Whipps. (1990). Substrate flow in the rhizosphere. Plant Soil 129: 1–10.

MacDonald, N. W., D. R. Zak, and K. S. Pregitzer. (1995). Temperature effects on kinetics of microbial respiration and net nitrogen and sulfur mineralization. Soil Sci. Soc. Am. J. 59: 233–240.

Mahler, R. L., G. A. Murray, and J. B. Swensen. (1993). Relationships between soil sulfate-sulfur and seed yields of winter rapeseed. Agron. J. 85: 128–133.

Malhi, S. S., D. W. McAndrew, and M. R. Carter. (1992). Effect of tillage and N fertilization of a Solonetzic soil on barley production and some soil properties. Soil & Tillage Res. 22: 95–107.

Marsh, K. B., R. W. Tillman, and J. K. Syers. (1987). Charge relationships of sulfate sorption by soils. Soil Sci. Soc. Am. J. 51: 318–323.

Martens, D. A., J. B. Johanson, and W. T. Frankenberger, Jr. (1992). Production and persistence of soil enzymes with repeated addition of organic residues. Soil Science 153: 53–61.

Mayland, H. F. and C. W. Robbins. (1994). Sulfate uptake by salinity-tolerant plant species. Commun. Soil Sci. Plant Anal. 25: 2523–2541.

Maynard, D. G., J. W. B. Stewart, and J. R. Bettany. (1983). Sulfur and nitrogen mineralization in soils compared using two incubation techniques. Soil Biol. Biochem. 15: 251–256.

Maynard, D. G., J. W. B. Stewart, and J. R. Bettany. (1984). Sulfur cycling in grassland and parkland soils. Biogeochemistry 1: 97–111.

Maynard, D. G., J. W. B. Stewart, and J. R. Bettany. (1985). The effects of plants on soil sulfur transformations. Soil Biol. Biochem. 17: 127–134.

McGill, W. B., and C. V. Cole. (1981). Comparative aspects of cycling of organic C, N, S and P through soil organic matter. Geoderma 26: 267–286.

McGrath, S. P., and F. J. Zhao. (1995). A risk assessment of sulphur deficiency in cereals using soil and atmospheric deposition data. Soil Use Manage. 11: 110–114.

McGrath, S. P., and F. J. Zhao. (1996). Sulphur uptake, yield responses and the interactions between nitrogen and sulphur in winter oilseed rape (*Brassica napus*). J. Agric. Sci., Cambridge 126: 53–62.

McLachlan, K. D., and D. G. De Marco. (1975). Changes in soil sulphur fractions with fertilizer additions and cropping treatments. Aust. J. Soil Res. 13: 169–176.

McLaren, R. G., A. Ghani, and R. S. Swift. (1988). Studies on sulphur mineralization from soils using an open incubation system. In *Towards the More Efficient Use of Soil and Fertilizer Sulphur.* (ed. R. E. White and L. D. Currie), pp. 66–75. Fertilizer and Lime Research Center, Massey University, Palmerston North, New Zealand.

McLaren, R. G., J. I. Keer, and R. S. Swift. (1985). Sulphur transformations in soils using sulphur-35 labelling. Soil Biol. Biochem. 17: 73–79.

McLaren, R. G., and R. S. Swift. (1977). Changes in soil organic sulphur fractions due to the long term cultivation of soils. J. Soil Sci. 28: 445–453.

McLeod, A. R. (1992). Factors influencing the yield response of cereals to sulphur dioxide in studies using an open-air fumigation system. Agriculture, Ecosystems and Environment 38: 11–20.

Milchunas, D. G., and W. K. Lauenroth. (1984). Sulfur deposition, cycling, and accumulation. In *The Effects of SO_2 on a Grassland. A Case Study in the Northern Great Plains of the United States. Ecological Studies* (ed. W. K. Lauenroth and E. M. Preston), pp. 61–95. Springer-Verlag, New York.

Minami, K. and S. Fukushi. (1981). Volatilization of carbonyl sulfide from paddy soils treated with sulfur-containing substances. Soil Sci. Plant Nutr. 27: 339–345.

Morra, M. J., and W. A. Dick. (1985). Production of thiocysteine (sulfide) in cysteine amended soils. Soil Sci. Soc. Am. J. 49: 882–886.

Morra, M. J., and W. A. Dick. (1988). Atmospheric sulfur deposition to agricultural land in northeastern Ohio. J. Environ. Qual. 17: 323–329.

Murray, F., R. Monk, K. Clarke, and M. Qifu. (1994). Growth responses of N and S deficient white clover and burr medic to SO_2, NO and NO_2. Agriculture, Ecosystems and Environment 50: 113–121.

Nguyen, M. L., and K. M. Goh. (1992). Nutrient cycling and losses based on a mass-balance model in grazed pastures receiving long-term superphosphate applications in New Zealand. 2. Sulphur. J. Agric. Sci., Cambridge 119: 107–122.

Noggle, J. C., J. F. Meagher, and U. S. Jones. (1986). Sulfur in the atmosphere and its effects on plant growth. In *Sulfur in Agriculture. Agronomy Monograph No. 27* (ed. M. A. Tabatabai), pp. 251–278. ASA-CSSA-SSSA, Madison, WI.

Nyborg, M., E. D. Solberg, S. S. Malhi, S. Takyi, P. Yeung, and M. Chaudhry. (1991). Deposition of anthropogenic sulphur dioxide on soils and resulting soil acidification. Dev. Plant Soil Sci. 45: 147–156.

O'Donnell, A. G., J. Wu, and J. K. Syers. (1994). Sulphate-S amendments in soil and their effects on the transformation of soil sulphur. Soil Biol. Biochem. 26: 1507–1514.

Olson, B. M., and L. E. Lowe. (1990). Effects of intensive vegetable production on chemical properties and nutrient mineralization of a British Columbia Humisol. Can. J. Soil Sci. 70: 445–460.

Olson, R. A., and G. W. Rehm. (1986). Sulfur in precipitation and irrigation waters and its effects on soils and plants. In *Sulfur in Agriculture. Agronomy Monograph No. 27* (ed. M. A. Tabatabai), pp. 279–293. ASA-CSSA-SSSA, Madison, WI.

Pasricha, N. S., and R. L. Fox. (1993). Plant nutrient sulfur in the tropics and subtropics. Adv Agron. 50: 209–269.

Perrott, K. W., and S. U. Sarathchandra. (1990). Seasonal variations in soil S flush and possible contributions from plant roots in the measurement of soil microbial sulfur, phosphorus, potassium and nitrogen. Aust. J. Soil Res. 28: 747–753.

Perucci, P. (1990). Effect of the addition of municipal solid-waste compost on microbial biomass and enzyme activities in soil. Biol. Fertil. Soils 10: 221–226.

Pirela, H. J., and M. A. Tabatabai. (1988). Sulfur mineralization rates and potentials of soils. Biol. Fertil. Soils 6: 26–32.

Poortinga, A. M., and L. J. De Kok. (1995). Utilization of atmospheric H_2S by plant foliar tissue: its interaction with sulfate assimilation. Z. Pflanzenernahr. Bodenk. 158: 59–62.

Rasmuson, A., T. Gimmi, and H. Fluhler. (1990). Modeling reactive gas uptake, transport, and transformation in aggregated soils. Soil Sci. Soc. Am. J. 54: 1206–1213.

Rasmussen, P. E. (1996). Effect of nitrogen, sulfur, and phosphorus sufficiency on nutrient content in white winter wheat. Commun. Soil Sci. Plant Anal. 27: 585–596.

Rasmussen, P. E., and C. L. Douglas, Jr. (1992). The influence of tillage and cropping-intensity on cereal response to nitrogen, sulfur, and phosphorus. Fert. Res. 31: 15–19.

Rennenberg, H. (1984). The fate of excess sulfur in higher plants. Annual Review of Plant Physiology 35: 121–153.

Rennenberg, H. (1989). Synthesis and emission of hydrogen sulfide by higher plants. In *Biogenic Sulfur in the Environment. ACS Symposium Series 393* (ed. E. S. Saltzman, and W. J. Cooper), pp. 44–57. American Chemical Society, Washington, DC.

Rennenberg, H., B. Huber, P. Schroder, K. Stahl, W. Haunold, H.-W. Georgii, S. Slovik, and H. Pfanz. (1990). Emission of volatile sulfur compounds from spruce trees. Plant Physiol. 92: 560–564.

Rhodes, L. H., and J. W. Gerdemann. (1978a). Hyphal translocation and uptake of sulfur by vesicular-arbuscular mycorrhizae of onion. Soil Biol. Biochem. 10: 355–360.

Rhodes, L. H., and J. W. Gerdemann. (1978b). Influence of phosphorus nutrition on sulfur uptake by vesicular-arbuscular mycorrhizae of onion. Soil Biol. Biochem. 10: 361–364.

Roberts, T. L., J. R. Bettany, and J. W. B. Stewart. (1989). A hierarchical approach to the study of organic C, N, P, and S in western Canadian soils. Can. J. Soil Sci. 69: 739–749.

Roppel, C. J. (1991). Plant available nitrogen and sulfur from canola residue. M.Sc. thesis, University of Saskatchewan.

Rovira, A. D. (1969). Plant root exudates. Bot. Rev. 35: 35–57.

Ryaboshapko, A. G. (1983). The atmospheric sulphur cycle. In *The Global Biogeochemical Sulphur Cycle. Scientific Committee on the Problems of the Environment (SCOPE) No. 19* (ed. M. V. Ivanov and J. R. Freney), pp. 203–296. John Wiley & Sons, New York.

Saggar, S., J. R. Bettany, and J. W. B. Stewart. (1984). Incorporation of sulphur, nitrogen and carbon into soil organic matter following incubation. J. Indian Soc. Soil Sci. 32: 20–25.

Saggar, S. K., J. R. Bettany, and J. W. B. Stewart. (1981). Measurement of microbial sulphur in soil. Soil Biol. Biochem. 13: 493–498.

Santoso, D., R. D. B. Lefroy and G. J. Blair. (1995a). A comparison of sulfur extractants for weathered acid soils. Aust. J. Soil Res. 33: 125–133.

Santoso, D., R. D. B. Lefroy, and G. J. Blair. (1995b). Sulfur and phosphorus dynamics in an acid soil/crop system. Aust. J. Soil Res. 33: 113–124.

Sawyer, J. E., and S. A. Ebelhar. (1995). Sulfur nutrition of winter wheat varieties with till and no-till planting on highly weathered Alfisol soils. Commun. Soil Sci. Plant Anal. 26: 1257–1271.

Schmidt, A., and K. Jager. (1992). Open questions about sulfur metabolism in plants. Annu. Rev. Plant Physiol. Plant Mol. Biol. 43: 325–349.

Schoenau, J. J., and J. R. Bettany. (1987). Organic matter leaching as a component of carbon, nitrogen, phosphorus, and sulfur cycles in a forest, grassland, and gleyed soil. Soil Sci. Soc. Am. J. 51: 646–651.

Schoenau, J. J., and J. R. Bettany. (1989). 34_S natural abundance variations in prairie and boreal forest soils. J. Soil Sci. 40: 397–413.

Schoenau, J. J., and J. J. Germida. (1992). Sulphur cycling in upland agricultural systems. In *Sulphur Cycling on the Continents: Wetlands, Terrestrial Ecosystems and Associated Water Bodies. Scientific Committee on the Problems of the Environment (SCOPE) No. 48* (ed. R. W. Howarth, J. W. B. Stewart, and M. V. Ivanov), pp. 261–277. John Wiley & Sons, New York.

Sears, P. D., V. C. Goodall, and R. H. Jackman. (1965). Pasture growth and soil fertility. IX. Repeated cropping of a soil previously under high-quality permanent pasture. N. Z. J. Agric. Res. 8: 497–510.

Setsoafia, K. M. (1983). Comparison of nitrogen and sulphur mineralization in soils. M.Sc. thesis, University of Alberta.

Smith, J. L., and E. A. Paul. (1990). The significance of soil microbial biomass estimations. In *Soil Biochemistry*, vol. 6. *Soils, Plants, and the Environment* (ed. J.-M. Bollag and G. Stotzky), pp. 357–396. Marcel Dekker, Inc., New York.

Sparling, G. P., and P. L. Searle. (1993). Dimethyl sulphoxide reduction as a sensitive indicator of microbial activity in soil: The relationship with microbial biomass and mineralization on nitrogen and sulphur. Soil Biol. Biochem. 25: 251–256.

Staley, T. E., W. M. Edwards, C. L. Scott, and L. B. Owens. (1988). Soil microbial biomass and organic component alterations in a no-tillage chronosequence. Soil Sci. Soc. Am. J. 52: 998–1005.

Staubes, R., H.-W. Georgii, and G. Ockelmann. (1989). Flux of COS, DMS and CS_2 from various soils in Germany. Tellus 41B: 305–313.

Stewart, J. W. B., D. W. Anderson, E. T. Elliott, and C. V. Cole. (1990). The use of models of soil pedogenic processes in understanding changing land use and climatic change. In *Soils on a Warmer Earth. Developments in Soil Science* (ed. H. W. Scharpenseel, M. Schomaker, and A. Ayoub), pp. 121–131. Elsevier, Amsterdam.

Stewart, J. W. B., and C. V. Cole. (1989). Influences of elemental interactions and pedogenic processes in organic matter dynamics. Plant Soil 115: 199–209.

Tabatabai, M. A. (1984). Importance of sulphur in crop production. Biogeochemistry 1: 45–62.

Tabatabai, M. A., and Y. M. Chae. (1991). Mineralization of sulfur in soils amended with organic wastes. J. Environ. Qual. 20: 684–690.

Tan, Z., R. G. McLaren, and K. C. Cameron. (1994). Comparison of sulfur mineralisation in open and closed incubation systems and from field moist and conditioned soils. Commun. Soil Sci. Plant Anal. 25: 3191–3198.

Terman, G. L. (1978). Atmospheric sulphur—the agronomic aspects. The Sulphur Institute, Washington, DC.

Thompson, J. F., I. K. Smith, and J. T. Madison. (1986). Sulfur metabolism in plants. In *Sulfur in Agriculture. Agronomy Monograph No. 27* (ed. M. A. Tabatabai), pp. 57–121. ASA-CSSA-SSSA, Madison,WI.

Till, A. R., G. J. Blair, and R. C. Dalal. (1982). Isotopic studies of the recycling of carbon, nitrogen, sulfur and phosphorus from plant material. In *Cycling of Carbon, Nitrogen, Sulfur and Phosphorus in Terrestrial and Aquatic Ecosystems.* (ed. J. R. Freney and I. E. Galbally), pp. 51–59. Springer-Verlag, Berlin.

Tisdale, S. L., R. B. Reneau, Jr., and J. S. Platou. (1986). Atlas of sulfur deficiencies. In *Sulfur in Agriculture. Agronomy Monograph No. 27* (ed. M. A. Tabatabai), pp. 295–322. ASA-CSSA-SSSA, Madison, WI.

Tracy, P. W., D. G. Westfall, E. T. Elliott, and G. A. Peterson. (1990a). Tillage influence on soil sulfur characteristics in winter wheat-summer fallow systems. Soil Sci. Soc. Am. J. 54: 1630–1634.

Tracy, P. W., D. G. Westfall, E. T. Elliott, G. A. Peterson, and C. V. Cole. (1990b). Carbon, nitrogen, phosphorus, and sulfur mineralization in plow and no-till cultivation. Soil Sci. Soc. Am. J. 54: 457–461.

Trudinger, P. A. (1986). Chemistry of the sulfur cycle. In *Sulfur in Agriculture. Agronomy Monograph No. 27* (ed. M. A. Tabatabai), pp. 1–22. ASA-CSSA-SSSA, Madison, WI.

Trust, B. A., and B. Fry. (1992). Stable sulphur isotopes in plants: a review. Plant, Cell and Environment 15: 1105–1110.

Wainwright, M. (1984). Sulfur oxidation in soils. Adv Agron. 37: 349–396.

Wesely, M. L., D. R. Cook, and R. L. Hart. (1985). Measurements and parameterization of particulate sulfur dry deposition over grass. J. Geophys. Res. 90: 2131–2143.

Williams, C. H., and J. Lipsett. (1961). Fertility changes in soils cultivated for wheat in southern New South Wales. Aust. J. Agric. Res. 12: 612–629.

Wilson, L. G., R. A. Bressan, and P. Filner. (1978). Light-dependent emission of hydrogen sulfide from plants. Plant Physiol. 61: 184–189.

Winner, W. E., C. L. Smith, G. W. Koch, H. A. Mooney, J. D. Bewley and H. R. Krouse. (1981). Rates of emission of H_2S from plants and patterns of stable isotope fractionation. Nature 289: 672–673.

Wu, J., A. G. O'Donnell, Z. L. He, and J. K. Syers. (1994). Fumigation-extraction method for the measurement of soil microbial biomass-S. Soil Biol. Biochem. 26: 117–125.

Wu, J., A. G. O'Donnell, and J. K. Syers. (1993). Microbial growth and sulphur immobilization following the incorporation of plant residues into soil. Soil Biol. Biochem. 25: 1567–1573.

Young, J. R., E. C. Ellis, and G. M. Hidy. (1988). Deposition of air-borne acidifiers in the western environment. J. Environ. Qual. 17: 1–26.

Zhao, F. and S. P. McGrath. (1994). Comparison of sulphur uptake by oilseed rape and the soil sulphur status of two adjacent fields with different soil series. Soil Use Manage. 10: 47–50.

Zhong, T., and M. J. Morra. (1994). Thiosulfate production in cysteine-amended soil. Soil Sci. Soc. Am. J. 58: 829–835.

3

Dynamics of Sulfur Transformations in Grazed Pastures

Surinder Saggar
Landcare Research NZ Ltd., Palmerston North, New Zealand

Michael J. Hedley and
Sathien Phimsarn*
Fertilizer and Lime Research Centre, Massey University, Palmerston North, New Zealand

I. INTRODUCTION

Losses of key nutrients need to be quantified and replaced to sustain agricultural production systems. New management practices are required that reduce nonproduct nutrient loss. There is a particular requirement to do this for mobile nutrients like sulfur (S) and nitrogen (N). To design efficient practices for S fertilization of pasture systems is difficult because of the complex nature of nutrient redistribution in grazed pasture systems. A pasture ecosystem comprises soils that anchor and supply water and nutrients to plants; animals that graze pasture plants; residues of plants and animals and microflora; and micro- and macrofauna that decompose these residues. These functional entities are linked by food webs, in their broadest sense, which operate with the hydrological cycle controlling the flows of energy and mineral nutrients like S [1]. Within the pasture ecosystem these components are responsible for recycling nutrients from animal and plant res-

*Current address: Ministry of Agriculture and Cooperatives, Chatucchak, Bangkok, Thailand

idues to plant available forms again. The pasture ecosystem in turn receives and loses energy and nutrients to the atmosphere and hydrological systems.

In grazed systems, a subcycle exists that rapidly moves S from soil to plant and back to soil, either directly in plant residues, or indirectly via the excreta of grazing animals (Fig. 1). From the larger global sulfur cycle S enters the pasture subcycle mostly as sulfate (SO_4^{2-}) through rainfall [2], dry deposition [3], or fertilizers. Inputs through rainfall and dry deposition vary with distance from coast and proximity to industrialized areas [2,3]. In environments remote from industrialized areas, as in Australia, New Zealand, and South America, such inputs may be less than 1–2 kg ha^{-1} yr^{-1} rising to 8–10 kg ha^{-1} yr^{-1} near coastlines. In industrialized areas, inputs derived from burning or processing fossil fuels may be as high as 20–30 kg ha^{-1} yr^{-1} [3]. Sulfate entering the soil is immobilized into organic S by plant and microbial processes [4]. Soil organic S is mostly insoluble in water and complexes with soil minerals. In this form it represents the largest pool in the pasture S cycle and a major reserve of plant-available S. Decomposer organisms in the soil, mostly microflora, remineralize organic S to SO_4^{2-}, the plant-available form. Factors that limit plant and microbial growth—most notably, temperature, the availability of water, and nutrient deficiencies—control the rate of S cycling. In this regard, deficiencies of SO_4^{2-} in the plant root zone represent a key limitation to the rate of S cycling in pastures. The major factors affecting the amount of SO_4^{2-} in the root zone are frequency and intensity of fertilizer application, plant uptake rates, adsorption to and desorption from soil minerals, mineralization and immobilization of organic S, and rainfall and drainage. To determine whether the S nutrition of a pasture soil is sustainable or to improve the efficiency of fertilizer use, a thorough understanding of factors influencing the transformation and fate of S are required. The objectives of this chapter are to review and develop an understanding of the nature of S in, and factors affecting its circulation through, and losses from, soil-plant-animal systems, with particular emphasis on temperate legume-based pasture systems. This area has also been reviewed by Nguyen and Goh (5), to provide a sound basis for estimating the fertilizer S requirements of grazed pastures.

II. FORMS OF SOIL S

Our understanding of the forms and amounts of soil S is limited by current analytical and separation techniques. The specific S compounds associated with organic and inorganic soil components cannot be identified without chemical treatment and separation or extraction from soil. Currently soil S

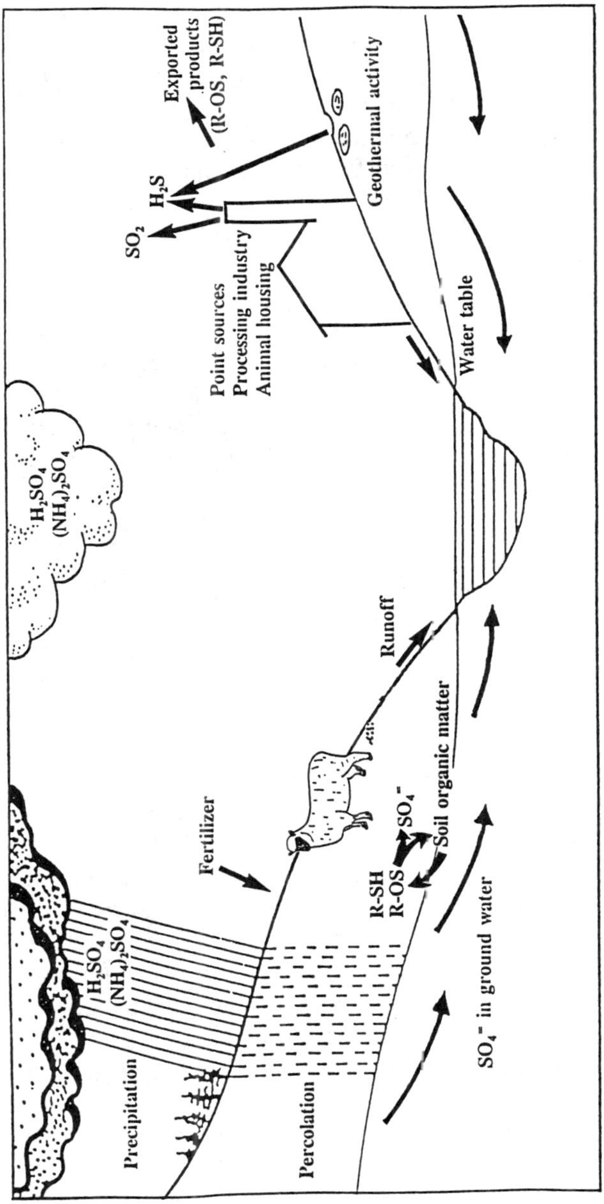

Figure 1 Pathways of S loss associated with pastoral farming.

forms can only be described by the dominant fractions identified by extraction. These descriptions will change as new techniques provide more information on the natural chemical structures of soil organic S.

In the late 1970s, realization of the importance of S in crop nutrition and the lack of understanding of soil S as a source of S for plants, spawned several studies of the nature and forms of S in soils [4,6–12]. Soil S includes both organic and inorganic compounds, the proportions and forms of which vary according to soil type [7,8], depth in profile [13], climate, and cultural conditions [14,15].

Inorganic soil S comprises chiefly soluble, adsorbed, insoluble, and co-precipitated sulfates and sulfides (Fig. 2). Elemental S and other compounds of lower oxidation state than SO_4^{2-} are less prevalent in the soil environment, being present mostly as transitory reaction intermediates [16]. In aquic [17] temperate soil environments, the dominant inorganic form is SO_4^{2-} unless oxygen deficit (perhaps due to a long period of waterlogging) causes the soil redox potential to drop. Only when redox potentials drop below -100 mV is SO_4^{2-} likely to be used as a terminal electron acceptor for microbial respiration [18].

Most agricultural or mineral soils have S contents ranging from 50 to 1000 mg kg^{-1} in the surface 15 cm [3,19]. Sulfate-S (SO_4^{2-}) is the form most available to plants, but it represents a relatively small proportion of total S, less than 25% of total S in most agricultural soils [11] and more commonly less than 5%. This proportion fluctuates seasonally, depending on the net balance between addition from rainfall, irrigation water, mineralization of organic matter, applied fertilizers, and losses through leaching and uptake by plants and microorganisms [20,21]. For example, Nguyen [22], Cornforth et al. [23], and Nguyen et al. [24,25] found that in permanent pasture soils in New Zealand, the amounts of extractable soil SO_4^{2-} present in spring were lower than in autumn, possibly due to increased winter leaching loss of SO_4^{2-}, and the slow rate of organic S mineralization during the colder winter period. Also in New Zealand soils, Ghani et al. [26] found that amounts of soil SO_4^{2-} can decrease in a matter of days, particularly after rainfall events causing drainage. Anion-exchange capacities of soils, particularly subsoils, strongly influence the amounts of SO_4^{2-} retained by soils [27,28] during drainage events.

The major proportion of the total S in most agricultural surface soils is in organic forms [12,19,29–31], mainly because, unlike SO_4^{2-}, soil organic S is mostly insoluble in water and not susceptible to leaching losses. The organic S mainly originates as plant and animal residues, which are subsequently decomposed and re-metabolized by soil microorganisms. The main forms of organic S deposited in soils include S-containing amino acids and sulfonates, in which S is directly bonded to carbon (C—S), and

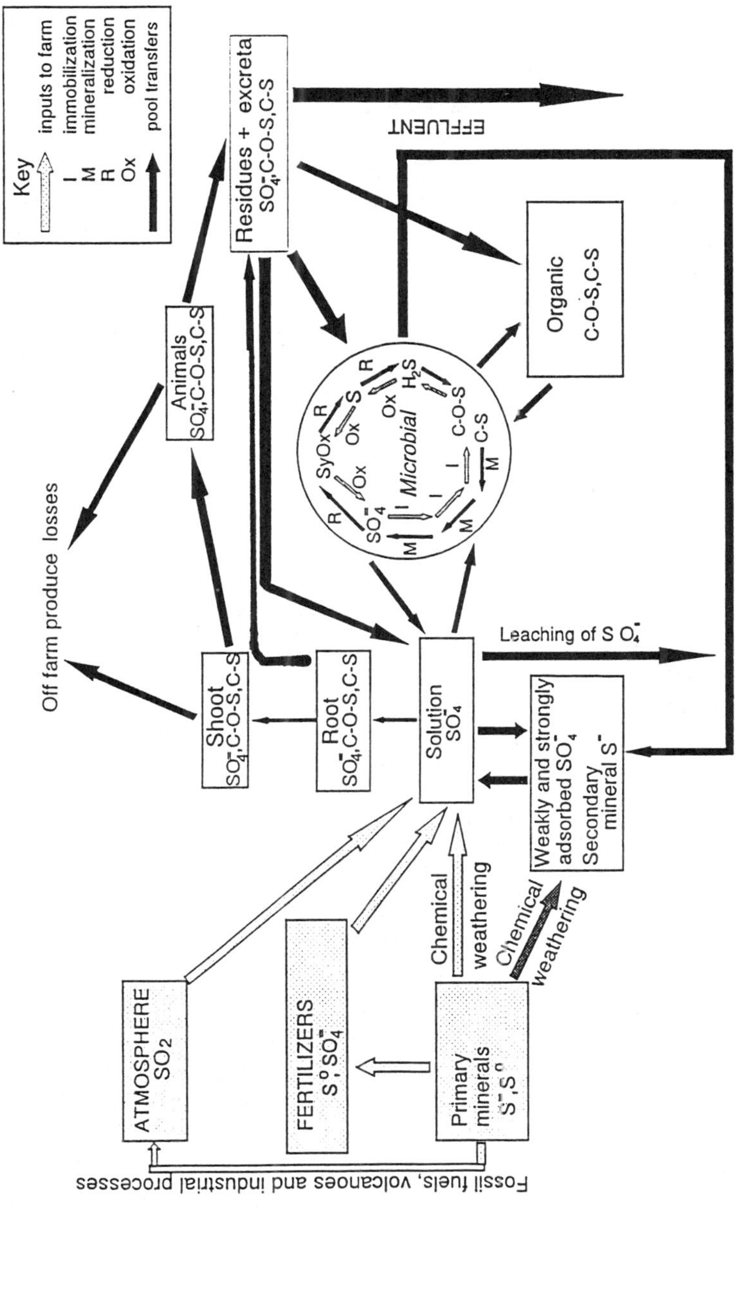

Figure 2 A conceptual flow diagram of the main forms and transformations of S in the soil-plant-animal system. (M. J. Hedley, 1990. Lecture notes. Soil Science Department, Massey University, New Zealand.)

the true organic esters of sulfuric acid (C—O—S), in which S is bonded to oxygen in the form of C—O—SO_{3-} linkages. Sulfamates may also be found in which S occurs in the form of N—O—SO_{3-} and N—SO_3^- groups. Mechanisms of ester-SO_4^{2-} formation, transformation, and its significance in the S cycle were reviewed by Fitzgerald [29,30].

Ester-SO_4^{2-} is generated predominantly by soil microflora [30,32] that metabolize organic residues in the presence of adequate S [33–35]. Soil carbon-bonded S may be derived directly from both leaf litter and root inputs, as well as microbial protein synthesis.

Although most soil S may be present in the organic form, on an annual basis only a small percentage of this fraction (as little as 2%) may enter the active S cycling pool, which supplies plant uptake [36–43].

The amount of soil organic S is closely associated with the amounts of soil C and N. This association is expected because organic matter provides the major nonleachable reserve of S and N in soils. Over a large number of top soils, the mean C:N:S ratio was 130:10:1.3 [19]; however, the amount of S in relation to C and N varies between soils. The difference may be attributed to differences in soil parent material, soil-forming factors (climate, vegetation, leaching intensity, drainage, and temperature), cultivation, and soil management practices [19].

Soil microbial processes are important in governing the nature of soil organic S because all organic residues entering the soil undergo at least partial microbial decomposition. Laboratory studies have shown that the relative proportions of carbon-bonded and ester-SO_4^{2-} formed by microorganisms depend, in part, on the availability of carbon and SO_4^{2-}, or the C:S ratio of the organic substrate supporting microbial growth [34,44,45]. Carbon-bonded S predominates with carbon-rich substrate, and ester-SO_4^{2-} with SO_4^{2-} rich substrate.

There is conflicting evidence on the agronomic value of differentiating between carbon-bonded S and ester-SO_4^{2-}. In mineralization studies using Canadian soils, Lowe [46,47] considered carbon-bonded S to be of little value as a source of mineralizable S; in contrast, Ghani et al. [44,45], using New Zealand soils, found that most of the SO_4^{2-} generated from soil incubation studies originated from carbon-bonded S. Freney et al. [48] also showed that over 9 months, most of the available S (60%) removed by plants in a pot experiment came from carbon-bonded S, although there were changes in all organic fractions. In New Zealand hill country soils regularly fertilized with SO_4^{2-}, hydriodic acid (HI) reducible S provided a larger portion of mineralizable S than carbon-bonded S forms [40], but this was not the case on less fertile sites. Thus neither HI-reducible nor carbon-bonded S can be considered to be globally the dominant form of labile organic S in soils. The forms of organic S mineralized depend on the previous soil management.

Although assays of SO_4^{2-}, C—S, and C—O—S are technically simple, characterization of S in soil based on bonding relationships with carbon does not always provide biologically meaningful fractions [15,49–51]. This is because chemical fractionation of soil ignores biological and physical constraints that place similar chemical forms in biologically distinct sections of the soil S subcycle, or in locations physically inaccessible to soil organisms. For example, S from living microbial cells and roots is grouped chemically with S from partially decomposed and completely transformed plant and animal residues. Some residues will be inaccessible to plant roots in tightly bound aggregates of soil particles and the organic S in these residues is also protected from microorganisms and degradative enzymes [52]. More research is needed before chemical characterization of soil S represents a useful method of assessing the availability of soil organic S to plants.

Recognition of the importance of soil microbial populations in cycling and transforming S in the plant root zone has led to efforts to measure microbial pool size and activity as approaches to solving the riddle of S availability in soils [34,53,54]. Saggar et al. [33] developed a method for measuring microbial biomass S in soil, which represented about 2.8% of the total S in Canadian prairie soils. Microbial biomass S values representing 1.2% and 2.2% of total S in forest soils of the United States [55], 0.9% and 2.6% in Scottish soils [39], and 0.4% and 1.8% in New Zealand soils [26] have subsequently been reported. McLaren et al. [56] estimated that the amount of soil microbial biomass S measured by Saggar's technique [33] was the same as that involved in the "active" organic cycling pool derived by calculation from the dilution of $^{35}S—SO_4^{2-}$ in a 75-day soil incubation study. Furthermore, Chapman [39,43] suggested that microbial biomass S can form a significant proportion of the organic S pool that is involved in S cycling and is potentially available to plants. Ghani et al. [26] showed that the amounts of microbial biomass S in range of New Zealand soils were fluctuating within seasons according to soil moisture and temperature changes. Pasture soils had larger amounts of soil microbial biomass S (1.5 times) than cultivated and fallow soils. In general, soil SO_4^{2-} concentrations in soils were related to amounts of microbial biomass S concentration. The significance of microbial biomass S in soil is seen in its position as a boundary between inorganic and organic S, the catalyst of the interconversion between the two. The size of the microbial pool, which ranges from 3 to 24 mg S kg^{-1} soil [54], or its activity, however, seems unrelated to S availability in pasture soils.

Currently there seems no single method of estimating the availability of S in pasture soils to plants by characterizing soil S forms chemically. Such methods are more applicable to cultivated well-mixed soil horizons than permanent pasture soils where an undisturbed soil profile places physical constraints on nutrient availability that cannot be interpreted by soil-testing methods.

III. CHEMICAL AND BIOLOGICAL TRANSFORMATIONS OF S IN SOIL

A pool of mobile SO_4^{2-} ions (Fig. 2) in soil solution is a central requirement for the cycling of S in the soil-plant system. This pool is depleted and replenished by a number of abiotic and biotic processes. Abiotic processes include physicochemical reactions with soil mineral surfaces, such as sorption-desorption and precipitation-dissolution. These rapid processes mostly control the SO_4^{2-} concentration in soil solution. Biotic processes, which include plant and microbial assimilation of S into insoluble organic forms and decomposition of plant litter, roots and animal excreta by soil microflora and fauna, remain key processes in controlling the amounts of S cycling through the SO_4^{2-} pool.

A. Abiotic Processes

Depletion of solution SO_4^{2-} concentration by plant uptake (Section IV.A) or leaching (Section VI) promotes desorption of SO_4^{2-} adsorbed on soil mineral surfaces. In this context, adsorbed SO_4^{2-} in the plant root zone provides an immediate, plant-available reserve of S in soil [57].

1. Adsorption of Sulfate

Sulfate or possibly HSO_4^- ions are adsorbed on surfaces of various soil colloids. The capacity of soils to adsorb SO_4^{2-} varies widely and is dependent on several soil properties, the most important being the soil clay content [58] and nature of clay minerals [59]. These properties determine the nature and extent of active oxide surfaces [60]. Quantities of exchangeable cations [61] and soil pH [62] influence the degree of surface positive charge which is partly responsible for anion sorption. In calcareous soils, SO_4^{2-} is adsorbed by $CaCO_3$ and presumably microcrystals of gypsum are formed. In noncalcareous soils, FeOH and AlOH groups provide the main sites for adsorption. Heavily weathered iron- and aluminium-rich soils usually have greater adsorption capacities than less weathered soils [63]. Sulfate adsorption capacity frequently increases with soil depth and this plays an important role in reducing SO_4^{2-} leaching [28,37,38,64–66] and minimizes the luxury uptake of SO_4^{2-} by plants [67]. Increased SO_4^{2-} sorption capacity with depth can reflect greater concentrations of amorphous Fe and Al oxides. These are formed either by weathering caused by fluctuating water levels (i.e., frequent oxidation and reduction) or clay movement to depth. Alternatively, lower soil layers may have lower concentrations of other anions competing for SO_4^{2-} sorption sites (see later section).

2. Adsorption Mechanisms

Our current understanding of anion sorption mechanisms has been reviewed by Barrow [68] and Bohn et al. [69]. In general, the mechanisms of anion adsorption in soils can be divided into two types: nonspecific, where the anion is held electrostatically as a counter-ion in the diffuse double layer next to a positively charged colloid surface [70], and specific or chemi-sorption or ligand exchange (all define the same reaction), in which a bidentate complex is formed where SO_4^{2-} bridges across two Al or Fe atoms to give a six-membered ring following the displacement of two ligands (mainly OH^- or H_2O) [71–73]. Both mechanisms occur simultaneously in a given soil horizon.

Sulfate adsorption is extremely rapid [74] and the initial charge of the adsorption site could be either positive or neutral [69]. However, unlike phosphate, SO_4^{2-} adsorption does not appear to disrupt hydrous oxide polymers, break Fe—OH or Al—OH bonds, or displace structural silicates [72,73]. A recent study by Curtin and Syers [75] supports an alternative adsorption hypothesis postulated by Uehara and Gillman [76] in which SO_4^{2-} actually enters the Stern or compact layer, thereby inducing further positive charge on the surface through OH^- release, rather than chemisorption, in which the adsorbed ion effectively becomes part of the surface.

The amount of SO_4^{2-} adsorbed on the soil surface is dependent not only on the number and type of adsorption sites, as described above, but also on the concentration of SO_4^{2-}in soil solution. The equilibrium between the quantity of adsorbed SO_4^{2-} and soil solution concentration can usually be described adequately by mathematical functions called adsorption isotherms: Freundlich, Langmuir, Tempkin [70,77]. These adsorption isotherms are described by Johnson and Mitchell (Ch. 7).

3. Desorption of Sulfate

Most adsorbed SO_4^{2-} can be readily desorbed as soil solution concentrations are depleted. The amount of readily desorbed SO_4^{2-} tends to decrease with time, particularly at higher temperatures [78] because of either conversion to a more tightly held form [78] or as microbially immobilized forms [79]. Aylmore et al. [80] showed that SO_4^{2-} sorption was essentially reversible on kaolinites but almost irreversible on Fe and Al oxides. Curtin and Syers [81], while investigating the release of adsorbed SO_4^{2-}, were able to quantitatively remove SO_4^{2-} by extraction with an indifferent electrolyte (1M NaCl). This raises questions about the susceptibility of this SO_4^{2-} to leaching, particularly from urine patches in the pasture cycle where chloride concentrations are high [82].

4. Plant Availability of Adsorbed Sulfate

Adsorbed SO_4^{2-}, through desorption, is available to plants and microorganisms and in many soils provides the major source of plant-available S [57,83–85]. In other more strongly weathered tropical soils, S deficiency has been observed in crops, despite the soil containing several thousand kg ha^{-1} of SO_4^{2-}—S within the root zone [28]. The low plant availability of adsorbed SO_4^{2-} is associated with high Fe and Al oxide content and/or low soil pH. Soils with a low Fe and Al oxide content adsorb and release SO_4^{2-} into soil solution more rapidly than do soils with high Fe and Al oxide content [86]. Management practices such as liming and phosphate additions are known to release some of the adsorbed SO_4^{2-} [61,77,87]. The mechanism of SO_4^{2-} release is through ligand exchange and decreased positive charge on the soil surface. This released SO_4^{2-} may be used by plants and micro-organisms, or may be leached, depending upon the rainfall regime [88].

The agronomic value of subsoil SO_4^{2-} to pasture and deep-rooting crops such as lucerne has been recognized by various workers [89–91], but again its availability to plants depends partly upon the soils' Fe and Al oxide content and depth of root growth. In general it would appear that although subsoil SO_4^{2-} can provide S to plants, the size of its contribution depends upon soil chemical and physical factors [8,91] as well as plant rooting depth. Although it is difficult to judge the benefit of subsoil SO_4^{2-} [69], an increased ability of soils to retain SO_4^{2-} may reduce fertilizer S requirements of pastures [88,92].

B. Biotic Processes

Immobilization and mineralization, general terms for enzymatically catalyzed synthesis and hydrolysis of organic S, are the main transformations occurring in soils and regulate the internal cycling of S from one soil S form (pool) to another [19]. A more complete discussion of the mechanisms and metabolic pathways of various S transformations is available elsewhere [29,30,93–96]. A summary of these processes is given below.

1. Mineralization

Mineralization of organic S is an important process for generating plant-available SO_4^{2-} within all soils, particularly permanent pasture soils (Fig. 3), especially when compared quantitatively with the atmospheric and weathering inputs of S [97]. Sorn-srivichai [98] showed that if S transformations in a pasture soil (Typic Fragiaqualfs) were in steady state, the rate of mineralization would be about 19 kg S ha^{-1} $year^{-1}$. This was equivalent to about 2–5% of the soil organic S being net mineralized per year and was the single largest input of plant-available S. Using $^{35}SO_4^{2-}$-labelled

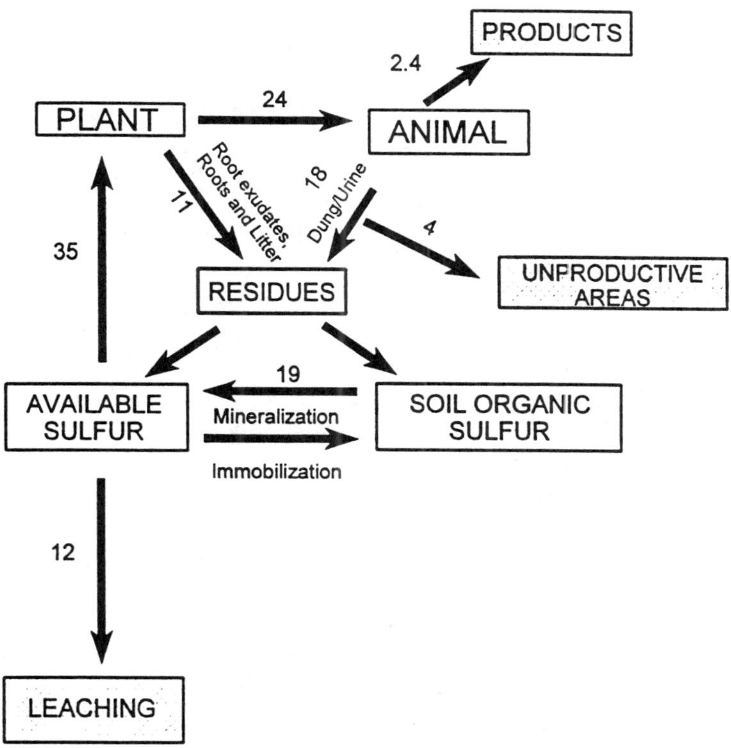

Figure 3 Sulfur cycle for grazed pastures. (Adapted from Ref. 97; transfer rates between pools are expressed as kg S ha^{-1} $year^{-1}$.)

fertilizer Sakadevan et al. [41] showed that approximately 80% of the S taken up by pastures annually fertilized with SSP (12% S) at 12.5 to 37.5 kg S ha^{-1} was derived from the mineralization of soil organic S.

Although many details of S mineralization are still poorly understood [96], it is well established that the rate of S mineralization is controlled by factors influencing the growth of microorganisms and their output of extracellular hydrolytic enzymes, such as moisture [99], temperature [100], pH [101,102], food (energy) supply [103–105] as influenced by soil depth [106], the presence and absence of plants (plant roots) [83,107], organic matter addition [34], and the amount of SO_4^{2-} present [107]. Similar factors have been shown to influence the level of sulfatase enzymes in soil [108–111].

Conflicting evidence exists on the nature of the soil S mineralized. Freney et al. [112], McLaren and Swift [113], McLaren et al. [114], and Ghani et al. [44,45] have suggested that carbon-bonded S is readily transformed to HI-reducible S, which is then mineralized. Bettany et al. [115], however, were unable to demonstrate mineralized S was derived from the HI-reducible S fraction. It has been observed that the protection of organic S against decomposition was dependent on both the nature of organic compounds and on the association with clay in soil aggregates. Recently, Erikson et al. [116] have indicated that HI reducible S fraction in the physically unprotected fraction is important for supply of S to plants.

McGill and Cole [117] and David et al. [106] considered that mineralization can be divided into two processes, biological and biochemical mineralization, definitions that are directed to the mineralization of carbon-bonded and ester-SO_4^{2-}, respectively. This concept proposes that carbon-bonded S mineralization is the result of an energy demand for carbon by microorganisms, which subsequently release inorganic S during oxidation of organic carbon; ester-SO_4^{2-} mineralization is regulated by both energy demand of microbes (to a lesser degree than in C-bonded S mineralization) and microbial nutritional need for SO_4^{2-} [117,118] (i.e., simple hydrolysis of SO_4^{2-} esters may satisfy the microbial demand for S). Although this concept of separating biological and biochemical mineralization may be useful in constructing computer simulations of S mineralization in soil, it is more likely that a number of microbially produced extracellular enzymes act together to hydrolyze complex soil organic matter into simple amino acids, sugars and nutrient ions that can then be utilized. The act of cleaving a SO_4^{2-} ester may be as much a part of producing a less ionic, more transportable, energy source, as it is for releasing SO_4^{2-}. In addition, it is doubtful whether extracellular enzyme activity in soil is the factor limiting mineralization reactions because enzyme activity levels are mostly in excess of those expected from available substrate concentrations [119].

Most of the information available on S mineralization has been obtained from short- to medium-term laboratory incubations designed primarily as rapid methods for predicting the amounts of S available to plants. How relevant these results are to mineralization rates of S in field soils is unclear. Maynard et al. [120], Syers et al. [3], and Pirela and Tabatabai [121] found that the apparent rates of S mineralization are very much influenced by the incubation technique used. In addition, the results obtained by incubation procedures may be biased by soil pretreatment designed to remove adsorbed SO_4^{2-} to encourage S mineralization [44,45].

Under field conditions very little is known about S mineralization rates, except for net rates that are represented in the sum of plant uptake and SO_4^{2-} leaching losses, and changes in amounts of extractable soil SO_4^{2-}. It appears that no reliable method has yet been developed for assessing the "potential" mineralization of S in field soils [51]. This is a difficult task because the rate of this microbially mediated process will fluctuate with soil physical conditions as well as being spatially dependent upon opportunistic microbial populations growing on recently added (or recently disturbed) organic substrate.

Blair et al. [122] have developed a soil test, a hot KCl extraction, which purports to measure a fraction of organic S that is relatively labile and utilized by plants. The test requires further evaluation before it is determined whether it successfully assesses the potential amount of plant-available S in pasture soils. Watkinson and Kear [123] also found that measuring organic S in the phosphate extract of either 0.01M Ca $(H_2PO_4)_2$ or 0.01M K H_2PO_4 leads to the improved index of the susceptibility of pasture growth to S deficiency. Giddens et al. [124] showed that S tests that purport to directly or indirectly measure the readily mineralizable fraction of organic S in soils correlate significantly and positively with pasture yield response.

2. Immobilization

Microbial involvement in nutrient immobilization has been the subject of increasing interest as a mechanism for increasing the nutrient retention capacity of soils. In the short term (one year), the amount and rate of fertilizer S incorporation into the soil organic pool appears to be of practical importance in relation to the size of fertilizer inputs [19,125]. The microbially mediated incorporation of SO_4^{2-}—S into organic matter has been well documented [93,96,126]. In the long term, the rate of net immobilization can be quantitatively estimated from increases in soil organic S [127–132].

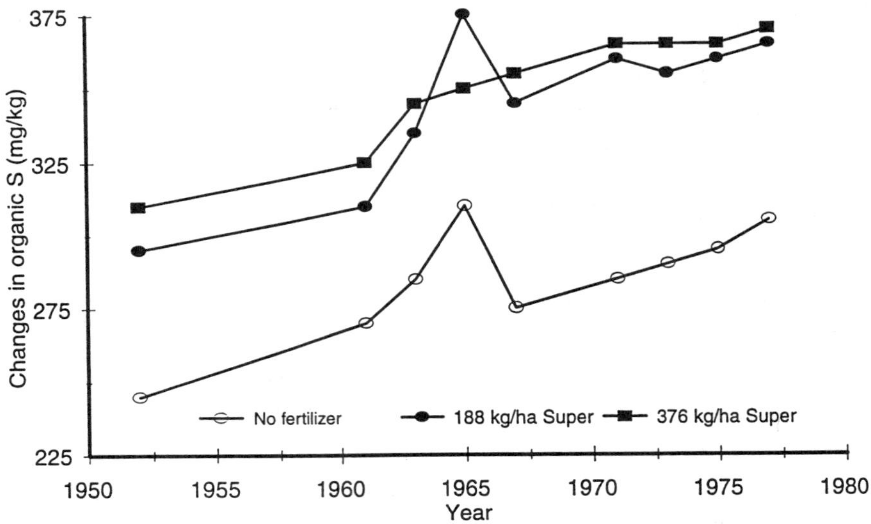

Figure 4 Changes in soil S status in pastures receiving long-term applications of superphosphate. (Adapted from Ref. 146.)

Immobilization processes have been investigated in soil incubations under laboratory and glasshouse conditions in the presence and absence of plants, and under field conditions through the use of ^{35}S-labeled fertilizer SO_4^{2-} [34,44,125]. This has enabled investigators to simultaneously monitor the opposing processes, mineralization and immobilization, and provide a more meaningful interpretation of the fate of applied S. The initial immobilization rates of SO_4^{2-}—S have been correlated with the amounts converted to microbial biomass S and C:S ratio of the amendments [133–135].

There are two major pathways for immobilization of SO_4^{2-}, either by direct immobilization of solution SO_4^{2-} via microbial uptake or via plant uptake, where the SO_4^{2-} is reduced to organic S compounds. The primary mechanism of incorporation of SO_4^{2-}—S into soil organic matter via microbial biomass [135] and the extent of immobilization is controlled by both the amount of available S and the availability of C substrate [34,134]. Microbial and plant residues (or subsequent animal residues) are eventually incorporated into soil organic matter. The ^{35}S tracer evidence showed only limited direct incorporation of S into humic acid [34,136,137]. Considerable incorporation, however, occurred after plant residues were allowed to decompose in the soil. The relative importance of plants is to be expected because they are the primary source of fixed C entering the soil, whereas soil microorganisms have a major catabolic role.

The amount of SO_4^{2-}—S immobilized varies considerably and can range from 0 to 100% of the available SO_4^{2-} pool depending on a number of factors. These include soil and litter (or excreta) characteristics, the concentration of SO_4^{2-} added, the presence and absence of plant roots (or another readily utilizable carbon source), and environmental conditions. Laboratory incubation studies suggest that the concentrations of SO_4^{2-} in soil solution and the amount and quality of C source are important in determining the extent of S mineralization, or the nature of the S immobilized; i.e., C—S or C—O—S. In field soils, however, the availability of SO_4^{2-} has a more direct effect on the rate of carbon input. For example, in the case of grazed pastures, the pasture plant is the primary source of carbon entering the soil—directly, as plant litter and roots plus their exudates, or indirectly, as dung from the grazing animal. The growth of the plant, and therefore the rate of carbon fixation and organic S synthesis, are limited in part by SO_4^{2-} availability in soils. The plant can be considered the primary agent for nutrient immobilization, whereas microorganisms and soil animals are mainly responsible for mineralization. The fact that laboratory incubations have focused attention on the size of the microbial populations, and carbon and SO_4^{2-} supply, as the most important factors influencing the extent of immobilization, may be an artifact of studying the incomplete organic cycle. More S must be immobilized during periods of plant growth than after its decay. Not only is there more carbon in living plants than their decomposition products and associated decomposers but the C:S ratio in plants is usually wider (>200) than in the decomposers (e.g., microorganisms (<50) [33,39,44,105]. In other words, the rate of immobilization is governed by the potential of plants to fix C. The C:S ratio or S content of the residues will, however, influence the speed at which mineralization occurs and the final form of residual soil S (decomposer products complexed with soil colloids).

Long-term immobilization processes are responsible for the success of New Zealand's grazed pasture system [127,130] and current pastoral productivity. Fertilizer elements such as P and S applied to pasture stimulate clover growth; N is fixed and builds up soil organic matter levels. Large amounts of organic S (and more importantly N) have been accumulated in New Zealand pasture soil fertilized with single superphosphate relative to virgin or unfertilized soil [6,127,138].

The factors affecting long-term rates of organic S accumulation in soils, such as climate, soil characteristics, plant species and addition of fertilizers, are evident from periodic analysis of soil S content in the pastoral soil studies of Walker et al. [130] and Jackman [131,132] (Fig. 4). Once an equilibrium state is reached between the formation (immobilization) and decomposition (mineralization) of organic S, the turnover of organic S contributes large amounts of available S for pasture growth [92,98]. It can be considered that in well-developed and well-fertilized pasture soils, immobilization is an important process for storing S. In contrast, this process may be considered as either a temporary or permanent loss of available S in less-developed pasture soils. Walker [127] and Walker et al. [130] pointed out that additions of S and P during the development stage of grass-clover pasture increased soil organic forms of these nutrients dramatically. Twenty-five years after pasture establishment and fertilization with single superphosphate, organic S accumulated at the rate of 7–18 kg S ha^{-1} $year^{-1}$ [130–132]. Similar increases in soil organic S were measured recently by Lambert et al. [139] and Saggar et al. [140,141], who recorded rates of soil organic S increase of 2.6–5.6 kg S ha^{-1} $year^{-1}$ at 125 and 375 kg SSP ha^{-1} $year^{-1}$ for low-sorption, and 3.0 to 34.6 kg S ha^{-1} $year^{-1}$ at 125 to 1250 kg SSP ha^{-1} $year^{-1}$ for high-sorption North Island (New Zealand) hill country soils (Table 1). Hingston, [142], Barrow [143], and Lewis et al. [144] have measured rates of 4.4 to 10.0 kg S ha^{-1} $year^{-1}$ for Australian soils fertilized with 125 to 300 kg ha^{-1} $year^{-1}$ of single superphosphate. The increase in organic S in pasture soils may result mainly from accumulation of plant residues, dung, and urine from grazing animals. Saggar et al.[140,141] have shown that in hill country soils the accumulation of soil S is not spatially uniform because more soil S may accumulate on areas receiving proportionally more animal excreta than in other areas less well suited to animal camping. For example, on steep

Table 1 Accumulation of Soil S in Two Soils Differing in Sorption Capacity

Soil	S applied as SSP[a] (kg ha^{-1} $year^{-1}$)	Rate of accumulation (kg S ha^{-1} $year^{-1}$)
Low sorption (0–150 mm)	125	2.6
	375	5.6
High sorption (0–150 mm)	125	3.0
	250	10.0
	375	15.1
	625	21.3
	1250	34.6

[a]During a period of 12 and 8 years in low- and high-sorption soils, respectively.
Source: Calculated from Refs. 140 and 141.

slopes (Fig. 5) exposed to unfavorable climatic conditions, soil organic S may be continuously depleted, irrespective of S fertilizer application rate.

McLachlan and De Marco [145] showed that under a pasture regime, most of the applied S accumulated in organic forms with greater accumulation of C-S. There was an increase in the C-S fraction with increasing level of S applied. About two-thirds of the applied S was incorporated into C-S. Accumulation of HI-reducible S was only weakly related to the level of S applied. Furthermore, the authors showed that C-S was remineralized to available S through HI-reducible S [114] when the soils were brought under cultivation for cropping. In New Zealand, Nguyen et al. [146] and Nguyen and Goh [138] also found that after 35 years of superphosphate application (21 kg S ha^{-1} $year^{-1}$) to irrigated pasture there were large

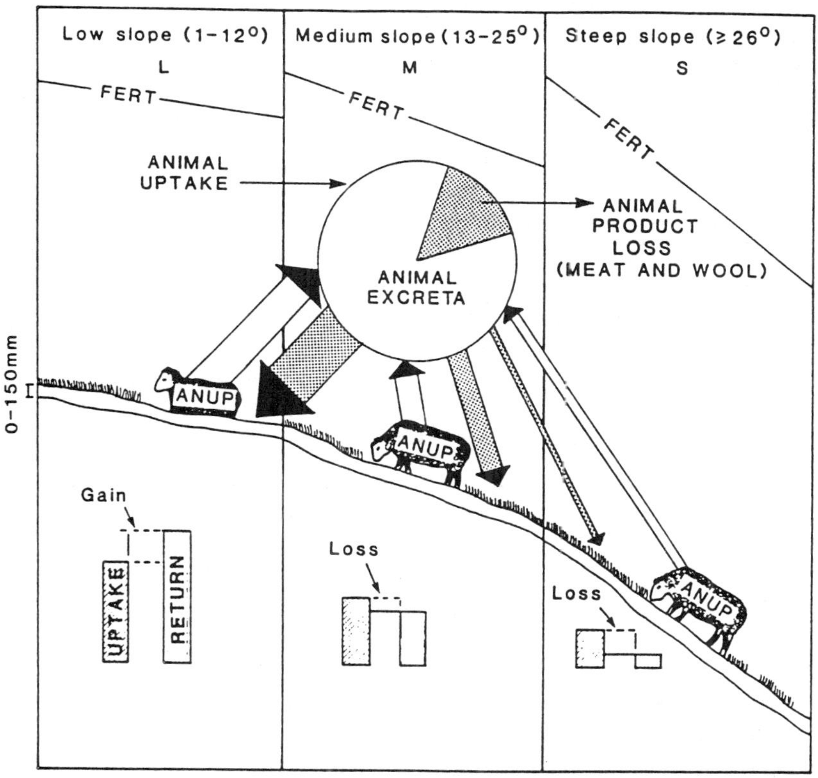

Figure 5 Schematic illustration of nutrient transfers in a grazed hill-country pasture. (From Ref. 140.) The diagonal lines with FERT represent the increase in land area caused by increasing slope. Arrow width represent approximate S flows.

accumulations of organic S in soil, 50–60% as C-S and 40–50% as C—O—S. However, no further increase in the organic fractions were caused by increasing annual inputs from 21 to 42 kg S ha^{-1}. This was presumably because plant biomass production was limited by secondary factors that limited the fixed carbon available for S immobilization. In many New Zealand legume-based pastures, N availability limits overall pasture production even when S and P fertilizers are adequately supplied [147]. The low efficiency of N use results partly from low rates of N fixation and partly from N transfer and leaching losses accelerated by the grazing animal [148]. The same inefficiencies that influence the fate of fertilizer S are discussed in Section V.

IV. PLANT PROCESSES

The rate of pasture plant growth is the main factor controlling the rate of S cycling in pasture systems. In this respect, factors limiting pasture growth, such as temperature, availability of water, and nutrients (particularly nitrogen, phosphorus, and potassium) have strong influences on the rate of S cycling. Climatic factors being equal, the major factor limiting S cycling in legume-based pastures is the rate at which nitrogen is fixed and becomes available for grass growth. Phosphorus (P) and S play key roles in stimulating legume growth, and single superphosphate (SSP, 9–10% P and 11–12% S) use has been the backbone of effective pastoral farming in Australia and New Zealand.

When the concentration of SO_4^{2-} in soil solution is increased, irrespective of whether S is the key nutrient limiting pasture yield, plant S uptake tends to increase [24,25,38,140,141,149–151]. Rennenberg [151] argued that the plant root cells are not equipped to prevent an uptake of excess SO_4^{2-}. Therefore, root cells do not appear to cope with excess SO_4^{2-} in the soil by avoidance of its uptake.

A. Plant Uptake

Soil SO_4^{2-}, the major source of plant-available S, moves to plant roots mostly by mass flow [152,153] and enters the plant in the root hair region. The most important factors governing the plant uptake of S are soil moisture content (plant-available water), SO_4^{2-} concentration in soil solution, soil SO_4^{2-} buffer capacity, plant root density distribution, and the actual evapotranspiration from the sward. Computer simulation models developed to apply these simple concepts provide reasonable predictions of pasture

S uptake in a range of soil fertility conditions in New Zealand [35]. For most soils where pasture production ranges from 17 to 5 t DM ha^{-1} $year^{-1}$, the S uptake may range from 60 kg S ha^{-1} $year^{-1}$ to below 8 kg S ha^{-1} $year^{-1}$, depending on soil fertility, climate, and sward composition. The spatial variability of S uptake can be short range, depending on the influence of slope, aspect, and soil moisture content as well as the influence of stock grazing and camping behavior on soil fertility. For example, New Zealand hill pastures on Typic Fragiaqualfs have mean annual pasture production ranging from 17.2 t ha^{-1} on campsites to 12.4, 9.5, 7.3 t ha^{-1} on easy, moderate, and steep slopes, respectively. Pastures on each of these slope units varied in S concentration from 0.40% on campsites to 0.22% on steep slopes (where the soil extractable levels were 40–50 and 20–40 mg S kg^{-1}, respectively). This resulted in pasture S uptake values ranging from 69 kg S ha^{-1} $year^{-1}$ on campsites to 16 kg S ha^{-1} $year^{-1}$ on steep slopes [140,141].

Obviously a number of factors, including plant species, plant age, rooting depth, and soil moisture content in the profile and potential evapotranspiration from the sward, interact to produce the differences in plant S uptake observed by various authors [see 5]. Currently insufficient measurements have been made to be able to rank the importance of each factor on plant S uptake, but it can be simply calculated from measurements of excreta return (Section V.B.3) that the grazing animal has the greatest influence on spatial variability of soil S in developed pastures [140,141] and on the variability observed in pasture plant S uptake. Herbage samples taken for elemental analysis and diagnostic purposes should consider the local environment and in particular the pattern of grazing animal excreta return.

B. Organic S Synthesis

The majority of plant S (90–97%) occurs as proteins and there is no evidence that SO_4^{2-} esters are metabolically essential. Nevertheless, ester compounds such as choline SO_4^{2-}, phenol SO_4^{2-}, polysaccharide SO_4^{2-}, lipid SO_4^{2-}, and glucosinolates, however, do occur in plants in small concentrations [154].

A more complete discussion of the biosynthesis and roles of S-containing compounds in plants is provided by Thompson et al. [154,155]. It is evident from these reviews that although organic S compounds can be synthesized in roots, little SO_4^{2-} is transformed in the roots. Thus most of the S is translocated to the shoots as inorganic SO_4^{2-} where SO_4^{2-} reduction and incorporation of S into organic forms occurs in the chloroplast.

This reduced S, principally in the form of the cysteine-containing tripeptide glutathione, is then redistributed throughout the plant. Pasture plant S primarily in the carbon-bonded form has two fates: (1) ingestion by the grazing animal and (2) uneaten plant material senesces (either through age or saprophytic attack) and is returned to the soil as litter, which is decomposed by soil organisms.

V. ANIMAL PROCESSES

Grazing animals exert both direct and indirect effects on pasture production and nutrient cycling [148,156–159]. In a grazed pasture only a proportion of the herbage is ingested; the rest has the potential to become decomposing plant litter. The majority of the nutrients ingested in by the animals are excreted in dung and urine. Therefore, grazing animals may affect pasture production through grazing, treading, scorching by excreta return, and rejection of herbage [158,159]. Plant photosynthetic area is reduced by grazing, particularly when pasture cover drops below 1200 kg dry matter ha^{-1} and photophosphorylation (ATP synthesis) is reduced, leading to less fixed C being translocated to the roots. Root elongation and root respiration may therefore be slower [160,161], and oxidative phosphorylation reduced, decreasing the supply of ATP to drive active nutrient uptake and other metabolic processes such as SO_4^{2-} reduction. Current information on pasture regrowth rates indicates that grazing animals (at sustainable stocking rates) has greater impact on S flows through pasture ingestion and excretal return than through moderating pasture S uptake by controlling the photosynthetic potential of the sward [Dr. A. D. Mackay, personal communication]. Therefore, in the section that follows, emphasis is placed on the influence of the animal on the fate of plant S rather than on the synthesis of plant S.

A. Return and Decomposition of Plant Litter

The amount of litter returned to the soil depends on pasture production and on grazing animals and grazing management [162]. In well-managed New Zealand pastoral farming, about 80–90% of the pasture production is ingested by the grazing animal [5,88,148,163]. The pasture production and the percentage of pasture ingested usually declines as land slope increases or soil fertility becomes poorer. The amount of litter remaining on hill country pasture was estimated to be about 2.5–3.0, 1.2–1.9, and 1.1–1.6 t ha^{-1} $year^{-1}$ on campsites (the resting grounds for the grazing animals), moderate slopes, and steep slopes [164,165]. The amount of root mass added annually in grazed pasture has been estimated to be about 4–5 t ha^{-1} [166]. Additionally, dry matter resulting from root exudation may amount to 0.2–1.0 t ha^{-1} $year^{-1}$ [167].

In general, a pasture growing 15,000 kg of dry matter ha^{-1} $year^{-1}$, with a S concentration in mixed herbage of 0.3% and 70% of the pasture being utilized by grazing animals, will have an uneaten litter return of 13.5 kg S ha^{-1}. With root dry matter turnover of 4000 kg ha^{-1} $year^{-1}$ at 0.2% S in roots, then a total of 21.5 kg S is returned in plant remains.

Little is known about S transformations during the breakdown of litter into soil organic matter, the latter depending on the associated microflora and fauna of the decomposer system. The degree of physical contact between litter and decomposer macro- and microorganisms can also affect litter decomposition [168]. Initially, trampling by stock and breakage by frost may be important in allowing contact of litter with soil. Subsequently, soil insects, earthworms, and soil microorganisms are responsible for mixing and decomposition of the litter fragments in soil.

In the short term (1 year), much litter S remains immobilized in soil organic matter. Studies involving the incubation of ^{35}S-labeled plant residues in soil [137] showed that up to 30% of the ^{35}S label was incorporated into soil humic acids. Later work [43,115,169–171] showed that under controlled conditions up to 40% of ^{35}S-labeled ryegrass/clover litter can be re-utilized by clover and ryegrass plants over 12 months of growth, the rest being incorporated into soil organic S. Litter decomposition followed two steps: a rapid initial process lasting 1–5 weeks, when the soluble material was removed and the more labile components were mineralized, followed by a slow process, lasting 10–24 months, in which tissues more resistant to decomposition were mineralized. Boswell [149] found that more than 80% of S released from plant litter was rapidly reimmobilized into soil organic S.

B. Ingestion and Excreta Return by the Grazing Animal

Sulfur cycling via grazing animals is considered to avoid the slow processes of decomposition of plant residues and effectively increases the turnover rate of S in the plant-soil cycle [1,172]. This increase in the rate of S cycling occurs because:

1. The passage of plant material through animal short circuits the decomposition cycle. Plant S is returned to soil more rapidly through animals than through litter derived from plant senescence [156,172,173].

2. In addition, a large proportion (50–80%) of S in the excreta, particularly urine, is immediately plant-available as SO_4^{2-} [172].

1. Metabolism of S

Pasture S is metabolized in the grazing ruminant by a variety of biochemical pathways (Fig. 6). In the rumen, both the dietary inorganic and some organic S are reduced to sulphide-S, which is then utilized by the rumen microflora to synthesize S amino acids for microbial protein [174–176]. Sulfur as microbial protein then enters the small intestine, is broken down by proteolytic enzymes, and is absorbed as S amino acids to be utilized by the animal tissue. The unabsorbed S is excreted in the feces as organic S. The bulk of the products of S metabolism within the animal (free or esterified SO_4^{2-}, taurine, thiosulfate, sulfocyanates, etc.) are excreted in the urine.

For both sheep and cattle, the fecal excretion of S is usually constant per unit of dry matter eaten [177] with approximately 0.1 g of S being excreted in the dung per 100 g of feed eaten. Urine S content, however, is variable, balancing the relatively constant dung S content (Table 2). The proportion of excreta S in urine varies from 6 to 90% [177–181] depending upon pasture S content. The urine of animals grazing S-sufficient pasture contains 50–60% of the excreted S.

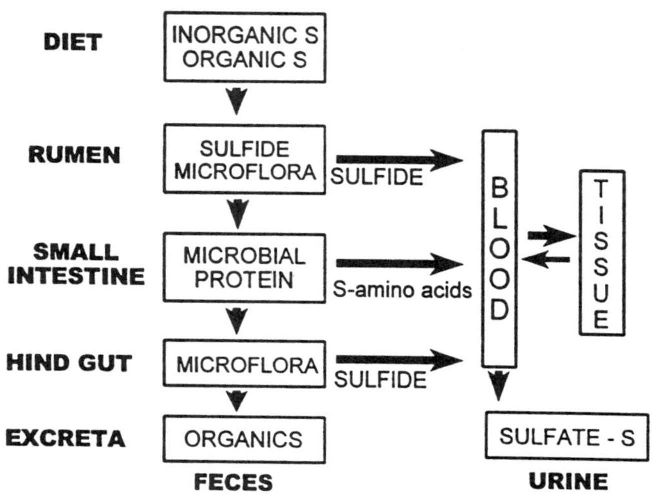

Figure 6 Schematic illustration of S transformations in the ruminant. (Adapted from Ref. 234.)

Table 2 Sulfur Excretion by Sheep on Varied Dietary S

		S excretion (g day^{-1})		
		Urine		Feces
Treatment	S intake (g day^{-1})	Total S	Sulfate-S	Total S
Control	0.82	0.48	0.44	0.42
Ammonium sulfate	1.86	1.28	1.12	0.36
Sodium sulfate	2.75	1.42	1.22	0.38

Source: Calculated from Ref. 180.

The nature of S in the dung may reflect rumen microflora activity rather than the form of plant S ingested. Most of the S in dung remains in organic combination [177,181] although 9–14% inorganic S is present [182,183]. Most of this organic S in dung is resistant to rapid mineralization [173,177,184]. Bird [185] reported that of the labeled ^{35}S (SO_4^{2-}) infused into the rumen of sheep, 87–94% of dung S was carbon bonded S and 4.1–5.4% was ester-SO_4^{2-}, whereas inorganic SO_4^{2-} represented only 0.5–4% of dung S.

Sulfur is generally present in urine in three forms: SO_4^{2-}—S (80–85%), ester-SO_4^{2-} (5–8%), and carbon-bonded (amino acids) (10–14%) [186]. The urine from sheep fed spring grass/clover pasture had little carbon-bonded S [187]. The amount of S in urine varied from 0.27 to 7.58 g day^{-1} depending on the S status of the diet, and SO_4^{2-} supplementation always increased excretion of urinary S [188]. This S is either immediately available to plants or in a form that is readily hydrolyzed to SO_4^{2-} [173,177,187–190]. Increases in the rate of S cycling caused by grazing and excreta deposition are probably attributable mostly to the S deposited in urine.

2. Excretion and Distribution

Grazing animals deposit excreta unevenly across the pasture [140,141, 149,157,163,172,191–195] due to their mobility. The distribution of excreta in time and space is generally affected by stocking rate, camping and grazing pattern, type of animal, and the amount and frequency of excretion [1]. This means that S ingested over a large grazing area is voided on a small area, thus concentrating the S at that site. Favored camping sites become nutrient rich, and nutrient availability can exceed pasture growth potential; in contrast, unfavored areas such as slopes become nutrient poor

This redistribution leads to loss of effective nutrient and has been termed the nutrient transfer loss [88]. Losses of S by leaching can be higher from favored camping areas [140,141,149,196–198]. The cycling of S through animals is spatially less effective than through plant root and shoot decay because animals cause some nutrient to be transferred to areas where the nutrients are unproductive (an area already rich in nutrients or raceways and yards). The camping effect is most pronounced in sheep-grazed hill country and may result in one-fourth of the dung being deposited in campsite areas that occupy only less than 5% of the land area [199]. This may cause up to 50% of the total estimated loss of S. On flat ring-fenced pastures, transfer losses are smaller, accounting for only 30% of the total loss.

3. Excreta Decomposition

There has been little published information on the rate of release of S from dung with the exception of pioneering work by Barrow [173], then Boswell [149] and recent work by Sakadevan et al. [182]. Barrow [173] showed that the amount of S mineralized was closely related to the S content of the dung and the proportion of S mineralized was less than with plant material. Dung S is more resistant to decomposition than S in plant material [149,173]. Barrow [173] also reported that there was likely to be net immobilization below a dung S content of 0.22%; similar results were likely to occur with plant litter with S content below about 0.12%.

Dung mineralization rates are more rapid in crushed samples than pad samples [193,200]. In the presence of arthropods such as earthworms [201] and dung beetles [202], mineralization rates are enhanced through comminution [203]. Dung decomposition rates are slow in dry conditions compared to cold moist conditions because water is required as the transport medium for microbial decomposers and their extracellular digestive enzymes and the products of decomposition. In New Zealand, Rowarth [193] reported that dung samples decomposed (physical disappearance) within 20 days in cool moist winter periods, whereas longer periods (up to 60 days) were observed during summer.

Boswell [149] followed the fate of ^{35}S-labeled sheep urine and dung and decomposition of unlabeled dung under field conditions. The mean rate of release of S from dung (0–37 days) was about 4.5 mg S g^{-1} S day^{-1}, which was much lower than from plant litter (16.2 mg S g^{-1} S day^{-1}). Sulfur released from dung was immobilized as soil organic S. This agreed with the results reported by Barrow [173]. Movement of S from dung and urine to soil organic S was as rapid as from plant litter. In particular, about 68% of the urine S was incorporated into soil organic S within 6 days. This S can be subsequently mineralized and recovered by plant [204]. In general, more than 80% of the S released was converted into organic forms. Only 5% and 16% of dung and urine S were recovered by plants in 100 days of total pasture growth, when total pasture uptake was 15.8 kg ha^{-1}. These results agree with those of Sakadevan et al. [182], who showed that under adequate soil S status only 1.6% and 20% of dung and urine were recovered by a year's pasture growth.

Both dung and urine have also been shown to affect other soil properties where they were returned, particularly at campsites. They have been shown to increase soil organic matter, cation exchange capacity, inorganic N, Olsen P and exchangeable potassium [205], calcium and magnesium [184,206], and soil pH [184,194]. As the soil pH increases, greater desorption of SO_4^{2-} may occur [59,60,77,184] and mineralization of soil organic S may increase [101,104,105,202,207]. This favors accelerated leaching losses of S from excreta-affected soil [88,140,141,149] and leads many researchers to consider that nutrient cycling through grazing animals is an inefficient process resulting in considerable loss of effective nutrients from productive areas of fields [140,141,198,208].

C. Sulfur Removal in Animal Products

The amounts of nutrients removed by grazing animals in wool, meat, and milk are small in comparison with the total amount of nutrient cycling in the pasture ecosystem [1] (Figs. 7a and 7b). Only 10–15% of ingested S [172,188] is used by animals in the production of body tissues, milk, and wool. The remaining 85–90% is returned to the soil in animal excreta. Sinclair [92] and Saggar et al. [140] also estimated that the amount of S or P removed in sheep products accounted for up to 15% of the S and P taken up by animals. Their calculation assumed that one stock unit (SU) is equivalent to a 50–60 kg ewe producing one lamb per year. The losses are calculated from 35 kg lamb at 0.15% S weaned year^{-1}; 5 kg greasy wool year^{-1} with 3.5–4% S; and a replacement rate of 50–60 kg ewe live weight at 0.13–0.15% S every 4 years. This accounted for about 2.7 kg S SU^{-1} year^{-1}. Similar calculations for dairy cattle provide comparable figures [209].

The general fate of pasture S ingested by grazing animals can be represented by litter, dung, urine, and amounts retained in animal tissue. Of the total S in pasture plants, about 10–30% is incorporated in litter, 25–30% in dung, 30–40% in urine, and about 10–15% held in animal products. Of the total S in litter, dung, and urine returned to the soil, it appears that approximately 5–10%, less than 5% and 20–30%, respectively, will be re-utilized in the short term by pasture plants. The remainder (about 60–75% of total S returned to the soil) is mostly immobilized into

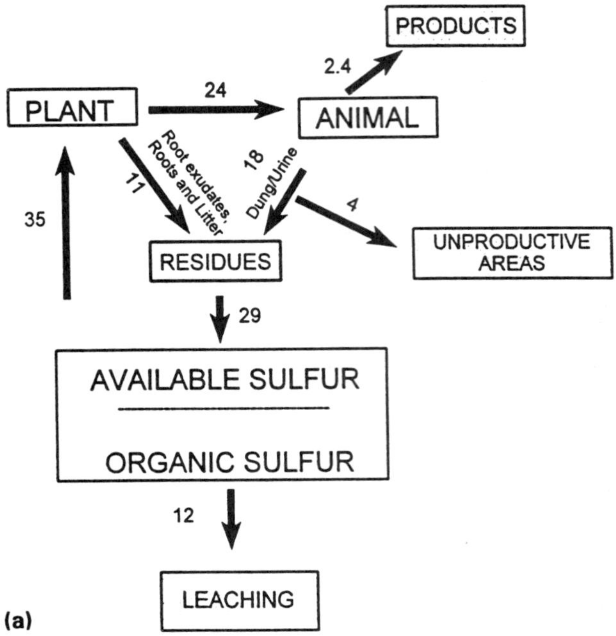

Figure 7 Model of the S cycle for an established pasture grazed by (a) dairy cattle and (b) sheep. (From Refs. 35 and 97.) (Transfer rates between pools are expressed as kg S ha^{-1} $year^{-1}$).

soil organic matter. Urine deposition is expected to stimulate the greatest increase in plant S uptake or leaching in the short term. Sulfur incorporated in carbon-rich plant litter and particularly, dung is expected to be only slowly recovered by the plant. Such products, in the short term, appear to be responsible for promoting immobilization of soil SO_4^{2-}.

VI. LEACHING OF SULFATE

Sulfate leaching is the largest single cause of loss of S from most New Zealand pasture soils [88,140,141].

Tabatabai [210] considered that both physical and chemical soil properties affect the movement of the SO_4^{2-} ion within the soil; physical properties determine the rate and pattern of water movement and diffusion of ions, whereas chemical properties determine the exchange and reactions of the ion with the soil constituents, SO_4^{2-} buffer capacity, and susceptibility

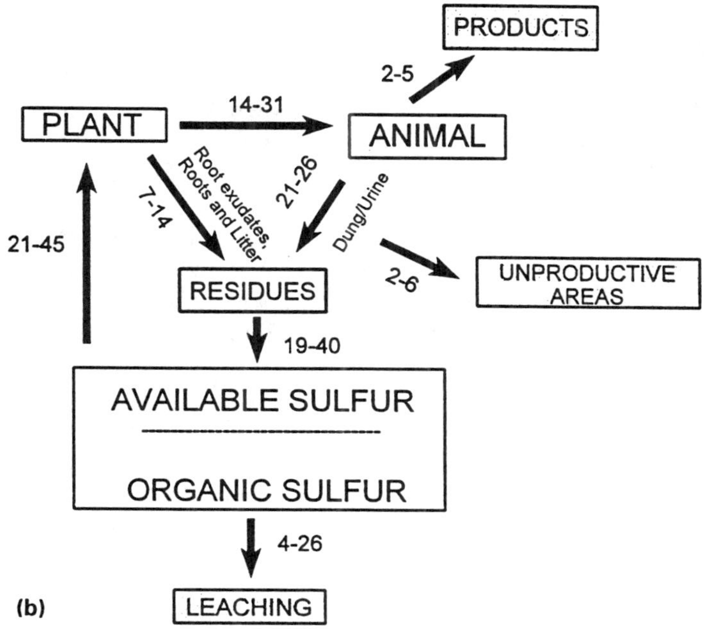

(b)

to leaching loss. The combination of these properties determines the final distribution of SO_4^{2-} within the soil profile. Sulfate is not strongly adsorbed by many well-developed pasture top soils because they have high organic matter content, which tends to carry net negative charge. These top soils are well fertilized with phosphate, which outcompetes SO_4^{2-} for anion sorption sites [211].

Laboratory, glasshouse, and lysimeter studies have been employed most frequently in investigations of S movement in soil, rather than direct field studies. Rare exceptions for pasture soils are the field studies of Smith et al. [212], Heng et al. [213], and Sakadevan et al. [182,183]. Laboratory and glasshouse studies have been primarily useful in determining the relative influence of soil properties on the extent of S leaching and the relative susceptibility of various S fertilizers to leaching.

Losses of SO_4^{2-} in laboratory and glasshouse studies are unlikely to reflect actual losses in the field. Sieving, bulking and repacking alter natural soil physical properties such as soil structure, porosity, and stratification, particularly in permanent pasture soil. The amount, velocity, and pattern of soil-water movement, the soil solution SO_4^{2-} concentration, and hence the extent of S leaching is altered [62]. Repacked pots may cause more or less solute/soil interaction than that occurring in field soils. "Edge-effects" associated with the soil container interface may promote "preferential" water movement [214] and increase S leaching loss. Often, increased amounts of SO_4^{2-} are detected after air-drying [101,173]. These are probably associated with fine root and microbial cell lysis [215]. Also, air-drying soils enhances leaching of S [216].

Undisturbed or intact soil cores have also been used in leaching studies [182,217–220]. In contrast to repacked soil columns or lysimeters, it has been shown that preferential flow through natural soil macro-pores such as earthworm channels or continuous cracks can markedly increase the rate of leaching of surface-applied fertilizer S and urine S and K [82,219]. Leaching trends for S in undisturbed soil cores were similar to P, in which there were high initial losses and about 40–70% of the S, applied as gypsum, was lost [218]. These results indicated that in conditions of macropore flow physical soil parameters may be more important than soil anion retention characteristics [82].

Direct measurement of field leaching losses have been undertaken by several researchers [142,182,212,213,221–224]. Smith et al. [212], in tile-drained (45-cm depth) plots on a pastoral soil (Typic Fragiaqualfs) in the North Island, New Zealand, found that on the unfertilized and nonirrigated plot 7.5 kg SO_4^{2-} S ha^{-1} $year^{-1}$ was lost through winter drainage. An additional 6.7 kg SO_4^{2-} S ha^{-1} $year^{-1}$ (total 14.2 kg S ha^{-1} $year^{-1}$ was lost during irrigation in the summer. About 10% of the S from the applied fertilizer (single superphosphate at 45 kg S ha^{-1}) was lost through the drainage system. Similar results were also obtained on the same soils by Heng et al. [213], who found that winter drainage losses of S could be reduced by applying S^0 rather than superphosphate in the autumn (Table 3).

Table 3 Losses of S in Drainage from Sulfur Superphosphate (SSP) and Elemental Sulfur (S°) Fertilized Farmlets

Treatment	1988	1989
SSP	17.0	9.4
S°	3.4	3.5

Source: Ref. 213.

Other authors have calculated field leaching losses indirectly through the recovery of isotopically labeled fertilizer [33,41] or by input-output mass balance calculations [140,141]. In general, it can be concluded that:

1. The amount and timing of rainfall that results in drainage is very important. Heng et al. [213] showed that 330 mm of drainage resulted in 17 kg ha^{-1} of S leached whereas 120 mm of drainage resulted only in 8.1 kg S ha^{-1} loss. In New Zealand, autumn or early winter events commonly contain high SO_4^{2-} concentrations relative to late winter or early spring events [213].
2. For soil with similar rainfall and SO_4^{2-} retention capacities, the penetration of leached S will be greater in a soil with a lower water-holding capacity [65,66] and will increase with increase in volume of a urination event [i.e., compare dairy (~2 liter urine per event) with sheep (~150 mL urine per event) grazed pasture in Fig. 8].
3. Greater leaching of SO_4^{2-} is likely to occur in low SO_4^{2-} retention soils than higher SO_4^{2-} retention soils [27,64,65,221,225] but this will be dependent on the rate of S application or excreta deposition [140,141,198]. Urine deposition on soils prone to cracking

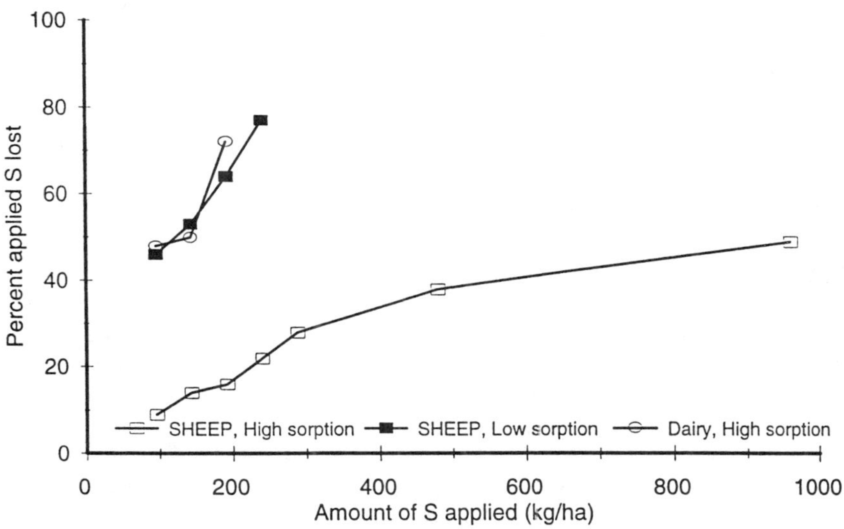

Figure 8 Sulfur losses from superphosphate-fertilized pastures expressed as percent of accumulated amounts of S applied. (From Ref. 198.)

or which have high macro-porosity is likely to lead to large S losses from the pasture root zone via macro-pore flow.

4. Fertilizer S applications in drier, warmer, periods when plants grow rapidly result in less leaching loss than autumn applications, when cooler, wetter periods cause slow plant growth [64,65].

VII. MODELING RESEARCH ON S IN PASTORAL SOILS

Agriculture in Australia and New Zealand is dominated by legume-based pastoral farming and most of the recent developments in pastoral S research have originated in these two countries.

Low soil S status limits the long-term productivity of legume-based pastures, through limiting legume growth and biological N fixation. There is an ongoing requirement for S fertilizers to sustain these pastures, particularly in aquic [17] environments where S leaching losses are large and costly in terms of lost fertilizer inputs.

The focus of recent research has been to improve the efficiency of S fertilizer use in grazed pasture through providing quantitative information on the pasture S cycle. The information is required for the design of fertilizer forms and management practices that will reduce S loss and increase S recycling.

Efficient use of S occurs when the rate at which S is cycled in the soil-plant-animal system is maximized and when the number of cycles completed are large compared to the rate of nonproduct losses from the cycle. Although, conceptually, the mechanisms of S cycling and loss from pastures are well understood, few studies have been conducted that quantify the rates of these processes sufficiently for scientists to be able to predict the long-term S fertilizer requirements of pastoral systems.

At present, in New Zealand, the S fertilizer requirements of pastures can be estimated for given soil groups by using a computerized model that calculates S mass balances for various pasture systems (Ministry of Agriculture and Fisheries–MAF, Soil Fertility Service–SAF model) [88]. The model assumes the major transfers of S occurring in, and losses from, a grazed pasture are in an approximate quasi-steady state. The model recommends that sufficient S is applied as fertilizer to replace product loss, animal transfer loss, and leaching loss. Soil SO_4^{2-} tests calibrated to pasture growth are used to adjust the fertilizer inputs to produce near-maximum pasture growth. The model estimates that S fertilizer requirements for pasture production for the North Island, New Zealand, range from 13 to 28 kg ha^{-1} year^{-1} [226]. This S model is regarded as the best basis for S fertilizer recommendations in New Zealand at present [227] but is open to improvement as new information is gained. Boswell [228] argued that although the model is correct in its broad assumptions on inputs and outputs, it is not completely successful in describing the detail of these inputs and losses in all situations. It is not a dynamic model, although it recognises that older improved pastures, where organic S has accumulated, have lower fertilizer requirements. This is accounted for by using a Pasture-Development Index. The model has a S leaching index (S.I.) which accounts for the effect of the total amount of rainfall on runoff and leaching losses but does not consider the intensity or timing relative to S fertilizer inputs. This may result in over- or underestimation of the loss factor. To improve the accuracy of the S model, more information is required on the fate of fertilizer S in grazed pastures, particularly the extent to which fertilizer S is leached or conserved as organic S in established pastures and how much of this is subsequently made available to plants in relation to the pasture development (i.e., the status of soil reserves).

To be useful to natural resource scientists and farmers, this information must be constructed into computer simulation models that can predict the flows and fate of S, and other nutrients, in pasture systems as influenced by climate, soils, and management. A fertilizer recommendation consisting of amount, form, timing and frequency of application should be an output of the model. Another output could be a stock management policy to minimize nutrient leaching losses. Figure 9 shows the conceptual components that should be in this model; namely plant uptake, S leaching, mineralization and immobilization, animal intake and animal transfer. How far has research progressed in obtaining sufficient information to construct such a model?

A. Modeling Plant S Uptake and S Leaching Loss

McCaskill and Blair [229–231] have developed a pseudo-mechanistic computer simulation model for predicting perennial pasture growth and S uptake. The model calculates pasture growth rate as a function of genetic potential achievable under optimum growth conditions, with reduction factors for temperature, moisture stress, light interception, and nutrient (N, P, S) restrictions. Sulfur uptake is simulated by a method based on diffusion theory and plant uptake kinetics. The model is mostly conceptual and still requires adjustments to simulate increased S uptake during spring. Although the accuracy of the model has not been verified, it is the first attempt at a comprehensive model. The mineralization and breakdown of S esters by plant and microbial esterases in response to nutrient stress, S movement in the soil profile, S uptake from depth and pasture growth, all need improved simulation [229] for the model to be more accurate.

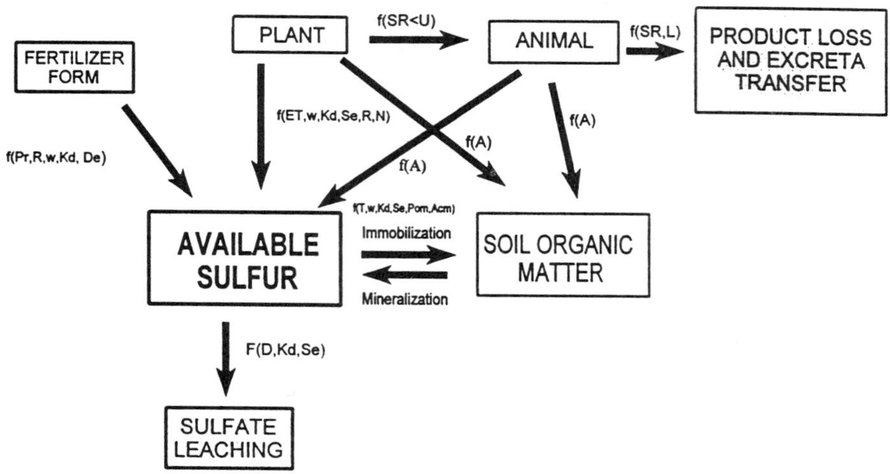

Figure 9 A simple conceptual dynamic model showing the S cycle. f = function of a variable listed; R = rainfall; T = temperature; Et = transpiration; w = soil water volumetric content; D = drainage; Se = extractable sulfate; Kd = soil sulfate adsorption coefficient; De = diffusion coefficient of sulfate; Pr = particle size plus oxidation or solubility constant; N = growth-limiting nutrient; SR = stocking rate; L = landslope; U = pasture utilization; A = animal excretion pattern including SR and L; Aom = animal organic matter, quality; Pom = plant organic matter, quality.

Using an alternative approach to McCaskill and Blair [229–231], simple models of pasture S uptake have been developed [35,232,233] using the relationships between actual daily evapotranspiration (AET), average root density in the soil profile, soil solution SO_4^{2-} levels predicted from extractable soil SO_4^{2-} levels, and soil SO_4^{2-} buffer capacities. The models assume that all soil-limited evapotranspiration originates from the sward. This is a safe assumption for dense swards covering the soil. This upward flow of soil water (AET) draws with it S from the soil solution in proportion to the soil solution SO_4^{2-} concentration, root density, and the water content in a series of finite soil horizons. The soil solution S concentration is predicted from knowing initial amounts of extractable soil SO_4^{2-} and the SO_4^{2-} sorption buffer capacity in each soil horizon. Soil solution concentrations and drainage volumes from each soil horizon are used to predict daily S leaching loss. Daily immobilization and mineralization of S in a particular soil horizon are simulated, albeit unsuccessfully, by both processes being controlled by AET, root density, and soil solution SO_4^{2-} concentration in that horizon. Increases in SO_4^{2-} concentration accelerate immobilization and retard mineralization.

Using a more complicated water balance model and a similar method of estimating soil solution SO_4^{2-} concentrations, Heng [232] produced a convection-dispersion model, to predict S leaching from a permanent pasture. In addition, she found that a transfer function model could predict the SO_4^{2-} concentration in drainage quite successfully, assuming a log-normal probability density function of solute travel pathway lengths and considering adsorption of SO_4^{2-} [232]. Both models were able to demonstrate the effect of different types of S fertilizers on SO_4^{2-} leaching losses. The models were, however, unable to predict the SO_4^{2-} leaching during periods of increased mineralization or after superphosphate application.

Long-term simulations by models described in these two sections are currently inaccurate simulations of actual plant S uptake or leaching loss, mainly because of the lack of effective submodels describing the immobilization of fertilizer S and the mineralization of soil organic S. These processes represent the largest fluxes of S in the pasture system, which are highly dependent upon seasonal and climatic variation.

B. Modeling Animal Transfers and the Resultant S Loss

In grazed pasture ecosystems, the deposition of animal excreta represents a major pathway for the aboveground returns of nutrient to the soil. There is high variability associated with uneven return of dung and urine. The cause of most of the uneven return of dung and urine are shelter, slope, aspect, watering facilities, and grazing management [5,148]. By making measurements on pasture production, plant nutrient content, percentage ingestion, and the return of sheep excreta [234] in each landscape unit within a paddock, the influence of the aforementioned features on nutrient redistribution can be simulated [140,141].

In areas studied by Gillingham [164] and Saggar et al. [140,141], the slope and aspect of the land dominated the pattern of excreta return (Figures 10a and 10b). Simple regression analysis was used to construct a model that described the influences of slope and aspect on the returns of plant S and P to the soil. Leaching losses of S were estimated from the difference in accumulation of topsoil (150 mm depth) P and S in each landscape unit. All P applied over 15 years, except that lost in animal products, could be accounted for in the top 150 mm of soil. On average, across all landscape units only 23–54% of the applied fertilizer S could be accounted for. Use of the model at this and other sites [140,141,209] has highlighted the large losses of S from single superphosphate fertilized pasture on soils of low S retention and the relatively smaller losses on soils of high S retention (Figure 8). Losses of S by SO_4^{2-} leaching on both soils, however, make up a considerable proportion of the fertilizer S, suggesting

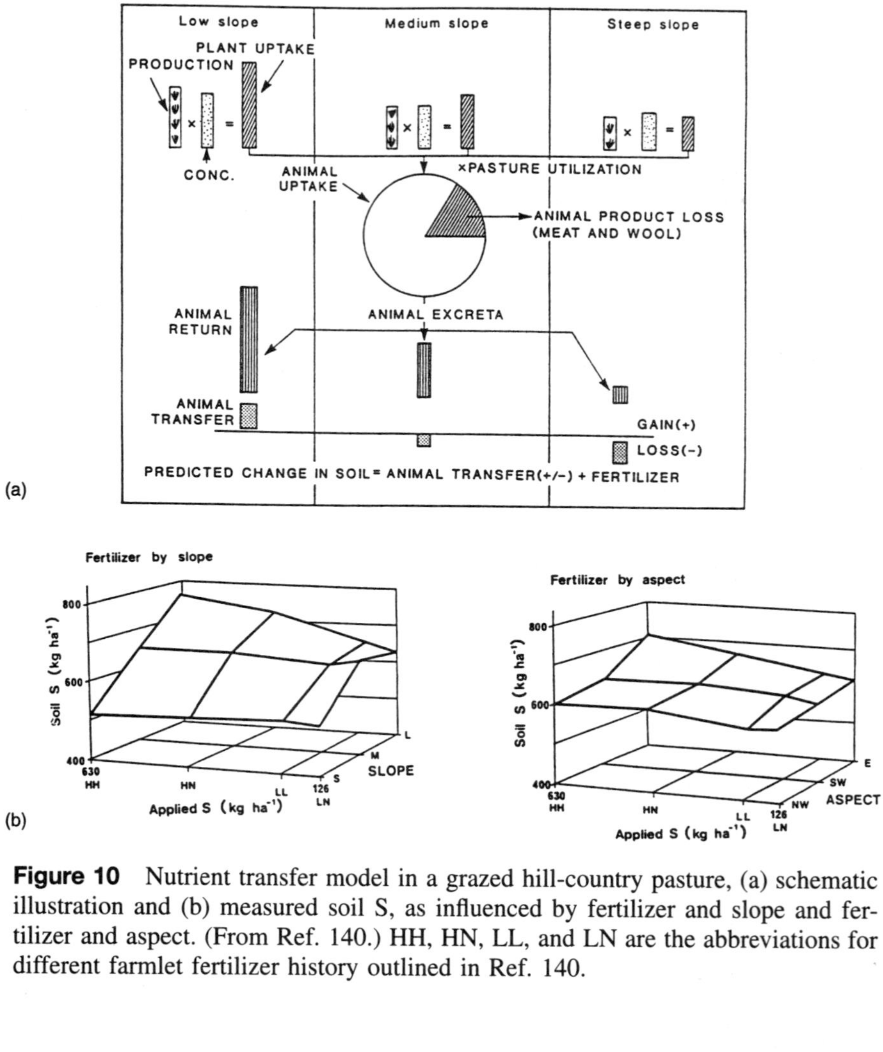

Figure 10 Nutrient transfer model in a grazed hill-country pasture, (a) schematic illustration and (b) measured soil S, as influenced by fertilizer and slope and fertilizer and aspect. (From Ref. 140.) HH, HN, LL, and LN are the abbreviations for different farmlet fertilizer history outlined in Ref. 140.

that S applied as single superphosphate is not being utilized efficiently in these pastoral systems, especially where high rates of fertilizer are applied.

Although S redistribution in the long term can be predicted from the behavior of grazing animals, the shorter-term fate of excretal S, through dung decomposition or immobilization of SO_4^{2-} in urine patches, presents problems in simulation akin to those found in modeling soil organic matter decomposition. Unless successful simulation of organic S turnover in pasture soils can be achieved, dynamic models of S cycling in pasture systems will remain inaccurate and ineffective as tools in evaluating new forms of fertilizer management.

It is important that mineralization rates of soil organic matter and its spatial and temporal variation in the soil are known to accurately predict both plant uptake and SO_4^{2-} leaching. Similar conclusions were drawn by Heng [232] in the development and comparison of three separate models for the leaching of SO_4^{2-} in pasture soils.

C. Future Research Requirements

Further research is needed before the short-term fluctuations in net S mineralization rates can be accurately modeled. This requires effective description of the immobilization of fertilizer S and the mineralization of soil organic S.

VIII. SUMMARY AND CONCLUSIONS

Sulfate is a predominant form of S that is highly mobile in terrestrial ecosystems. It is essential for the nutrition of higher plants, particularly in protein synthesis. The primary nitrogen fixation step in legume-based pastures is dependent on adequate S nutrition. In the absence of S fertilization or atmospheric S inputs derived from industrial pollution, the removal of S in animal products and loss by leaching from pastures can lead to S deficiencies in the pasture, showing that soil S reserves are not sustainable. The current lack of sustainability of soil S in aquic environments is to some extent a function of the grazing system—i.e., as the productivity increases, either the rate of S cycling increases or the amount of S cycled increases; either way, the amount of S leaking from the cycle as SO_4^{2-} is likely to increase. Leaching losses of S in such systems are high and strategies are required to reduce such losses if the overall productivity is to be maintained. Appropriate strategies must be based on a thorough quantitative understanding of the S cycle in pastures.

The major S constituents in the pastoral ecosystem include inorganic SO_4^{2-} C—bonded S, and ester SO_4^{2-}, with organic fractions composing the major soil S pools. The persistence of plant-available S in soils is therefore dependent on rainfall, drainage, SO_4^{2-} retention characteristics of soils, and the extent to which added S is converted to organic S reserves and remineralized to plant-available SO_4^{2-}. Incorporation of fertilizer S into organic S may conserve S against leaching losses. Leaching or surface runoff under field conditions lead to inefficient use of SO_4^{2-} based fertilizers (e.g., superphosphate). The use of slow-release forms of S such as elemental S (S°) is considered to reduce these initial leaching losses in some climatic situations; in New Zealand, little information is available on the comparative fate of SO_4^{2-}- and S°-based fertilizers in soils.

Although considerable laboratory and glasshouse research has studied the influence of soil properties on S uptake, immobilization and retention and the mineralization of organic S, few of the findings can be extrapolated with confidence to effectively explain the cycling of S under field soil conditions.

The role of animals in cycling and nutrient loss is relatively well understood and can be effectively modeled. Leaching can also be modeled. Simulation models highlight the inadequacy of our current understanding of the provision of plant-available S through the decomposition (mineralization) of organic soil and excretal S. This prevents the effective simulation of the pastoral S cycle in dynamic fertilizer requirement models.

ACKNOWLEDGMENTS

The authors are grateful to Drs. K. R. Tate and R. Lee, Landcare Research, and P. H. Williams New Zealand Institute for Crop and Food Research, for comments on various aspects of the text and thank A. West, N. Collins, and M. R. Bretherton, Soil Science Department, Massey University, and Elaine Hagenaars, Landcare Research, for assistance with figures.

REFERENCES

1. S. R. Wilkinson and R. W. Lowrey, Cycling of mineral nutrients in pasture ecosystem, *Chemistry and Biochemistry of Herbage* (G. W. Buttler and R. W. Bailey, eds.), *2*:247 (1973).
2. S. F. Ledgard and M. P. Upsdell, Sulphur inputs from rainfall throughout New Zealand, *N. Z. J. Agric. Res.*, *34*:105 (1991).
3. J. K. Syers, R. J. Skinner, and D. Curtin, "Soil and fertilizer sulphur in UK Agriculture," Proceedings The Fert. Soc. London, 264: 43 pp. (1987).
4. J. R. Freney, Form and reactions of organic sulfur compounds in soils, *Sulphur in Agriculture* (M. A. Tabatabai, ed.), Agron Monogr. 27, Am. Soc. Agron., Madison, Wisc., pp. 207–232 (1986).
5. M. L. Nyugen and K. M. Goh, Sulphur cycling and its implications on sulphur fertiliser requirements of grazed grassland ecosystems, *Agric. Ecosys. Environ.*, *49*:173 (1994)
6. A. J. Metson, "Sulphur status of New Zealand soils," Proc. Tech. Conf. NZFMA, New Zealand, *12*: pp. 73–84 (1969).

7. A. J. Metson, Sulphur in New Zealand soils 1. A review of sulphur in soils with particular reference to adsorbed sulphate-sulphur, *N. Z. J. Agric. Res.*, *22*:95 (1979).
8. A. J. Metson, Sulphur in New Zealand Soils 3. Sulphur distribution in soils derived from parent materials of predominantly sedimentary origin, N. Z. Soil Bureau Scientific Report No. 41: 42 pp. (1979).
9. A. J. Metson, Sulphur in New Zealand Soils 4. Sulphur distribution in soils derived from parent materials of predominantly volcanic origin, N. Z. Soil Bureau Scientific Report No. 42: 37 pp. (1979).
10. C. H. Williams, The chemical nature of sulphur compounds in soils, *Sulphur in Australasian Agriculture* (K. D. McLachlan, ed.), Sydney Univ. Press, Sydney, pp. 21–30 (1975)
11. R. L. Halstead and P. J. Rennie, The effects of sulfur on soils in Canada, *Sulfur and Its Organic Derivatives in Canadian Environment*, Nat. Res. Council, Bull. NRCC, 15015, Ottawa (1977).
12. V. O. Biederbeck, Soil organic sulfur and fertility. *Soil Organic Matter* (M. Schnitzer and S. U. Khan, eds.), Elsevier Scientific Publishing Co., New York, pp. 273–310 (1978).
13. C. H. Williams, The chemical nature of sulphur in some New South Wales soils, *Handbook on Sulphur in Australian Agriculture* (K. D. McLachlan, ed.), CSIRO, Melbourne (1974)
14. J. R. Bettany, J. W. B. Stewart, and S. Saggar, The nature and forms of sulfur in organic matter fractions of soils selected along an environmental gradient, *Soil Sci. Soc. Am. J.*, *43*:981 (1979).
15. J. R. Bettany, S. Saggar, and J. W. B. Stewart, Comparison of the amounts and forms of sulfur in soil organic matter fractions after 65 years of cultivation, *Soil Sci. Soc. Am. J.*, *44*:70 (1980).
16. P. A. Trudinger, The biological sulphur cycle, *Biogeochemical Cycling of Mineral Forming Elements* (P. A. Trudinger and D. J. Swaine, eds.), Elsevier Scientific Publishing Co., New York, pp. 293–313 (1979).
17. Soil Survey Staff, *Keys to Soil Taxonomy*, SMSS Technical Monograph no. 6, Blacksburg, Virginia, 422 pp. (1990).
18. K. R. Reddy, T. C. Fiejtel, and W. H. Patrick Jr. Effect of soil redox conditions on microbial oxidation of organic matter, *Role of Organic Matter in Modern Agriculture, Developments in Plant and Soil Sciences* (Y. Chen and Y. Avnimelech, eds.), Martinus Nijhoff Publishers, pp. 117–156 (1986)
19. J. R. Freney and C. H. Williams, The Sulphur Cycle in Soil. *Global Biogeochemical Sulphur Cycle* (M. V. Ivanov and J. R. Freney, eds.), John Wiley & Sons, pp. 129–201. (1983).
20. C. H. Williams, Seasonal fluctuations in mineral sulphur under subterranean clover pasture in Southern New South Wales, *Aust. J. Soil Res.*, *6*: 131 (1968).
21. A. O. Stuanes, Yearly and seasonal variation in extractable soil sulphate at two locations in Norway. *Acta Agric. Scand.*, *35*:33 (1985).

22. M. L. Nguyen, An Evaluation of S Topdressing Strategies in Eastland Pasture. M. Agr. Diss., Massey Univ., Palmerston North, New Zealand (1982).
23. I. S. Cornforth, G. S. Smith, and D. M. Wheeler, "Soil testing and plant analysis for sulphur," Proc. Sulphur in New Zealand Agriculture (P. E. H. Gregg and J. K. Syers, eds.), Massey University, Palmerston North, New Zealand, pp. 49–60 (1983).
24. M. L. Nguyen, R. W. Tillman, and P. E. H. Gregg, Evaluation of sulphur status of East Coast pasture of North Island, New Zealand 1. Soil and herbage survey, *N. Z. J. Agric. Res.*, *32*:263 (1989).
25. M. L. Nguyen, R. W. Tillman, and P. H. E. Gregg, Evaluation of sulphur status of East Coast pasture of the North Island, New Zealand 2. Field trials, *N. Z. J. Agric. Res.*, *32*:273 (1989).
26. A. Ghani, R. G. McLaren, and R. S. Swift, Seasonal fluctuations of sulphate and soil microbial biomass-S in the surface of a Wakanui soil, *N. Z. J. Agric. Res.*, *33*:467 (1990)
27. D. E. Hogg, Studies on leaching losses of sulphur from pumice soils, yellow brown loams and sands, *N. Z. Grassland Assoc.*, *27*:123 (1965).
28. R. L. Fox, M. Ashgar, and W. Cable, "Sulfate accretions in soils of tropics," Proc. Sulfur in South East Asian and South Pacific Agriculture (G. J. Blair, ed.), pp. 39–53. (1983).
29. J. W. Fitzgerald, Sulfate ester formation and hydrolysis: a potentially important yet often ignored aspect of the sulfur cycle of aerobic soils, *Bact. Rev.*, *40*:698 (1976).
30. J. W. Fitzgerald, Naturally occurring organo sulfur compounds in soils, *Sulfur in the Environment, part II Ecological Impacts* (J. O. Nriagu, ed.), Wiley, New York, pp. 391–443. (1978).
31. F. J. Stevenson, *Cycle of Soil*, John Wiley & Sons, New York (1986).
32. M. B. David, M. J. Mitchell, and S. C. Schnindler, "Dynamics of organic and inorganic sulfur constituents in hardwood forest soils and their relationships to nutrient cycling," Proc. 6th Forest Soils Conf. (E. I. Stone, ed.), Knoxville, Tennessee, pp. 221–245 (1984).
33. S. Saggar, J. R. Bettany, and J. W. B. Stewart, Measurement of microbial sulfur in soil, *Soil Biol. Biochem.*, *13*:493 (1981).
34. S. Saggar, J. R. Bettany, and J. W. B. Stewart, Sulfur transformation in relation to carbon and nitrogen in incubated soils, *Soil Biol. Biochem.*, *13*: 499 (1981).
35. K. Sakadevan, Factors Influencing the Transformation and Fate of Sulphur and Nitrogen in Grazed Hill Country Pasture. Ph.D. Diss., Massey Univ., Palmerston North, New Zealand (1991).
36. A. R. Till, and P. F. May, Nutrient cycling in grazed pastures 4. The fate of sulphur-35 following its application to small area in grazed pasture. *Aust. J. Agric. Res.*, *22*:391 (1971).
37. K. M. Goh and P. E. H. Gregg, Field study on the fate of radioactive sulphur fertilizer applied to pastures, *Fert. Res.*, *3*:337 (1982).

38. K. M. Goh and P. E. H. Gregg, Field studies with radioactive sulphur-labelled gypsum fertilizer III. Sulphur pools and their significance, *N. Z. J. Agric. Res.*, *25*:135 (1982).
39. S. J. Chapman, Microbial sulphur in some Scottish soils, *Soil Biol. Biochem.*, *19*:301 (1987).
40. K. Sakadevan, M. J. Hedley, and A. D. Mackay, Sulphur cycling in New Zealand hill country pastures. I. Laboratory sulphur, nitrogen and carbon mineralization studies, *J. Soil Sci.*, *44*:73 (1993).
41. K. Sakadevan, M. J. Hedley, and A. D. Mackay, Sulphur cycling in New Zealand hill country pastures. I. The fate of fertilzer sulphur, *J. Soil Sci.*, *44*: 615 (1993).
42. K. Sakadevan, M. J. Hedley, and A. D. Mackay, Mineralisation and fate of soil sulphur in hill pastures, *N. Z. J Agric. Res.*, *36*:271 (1993).
43. S. J. Chapman, Partitioning of ryegrass residue sulphur between the soil microbial biomass, other soil sulphur pools and ryegrass (Lolium perenne L), *Biol. Fert. Soils*, *5*:253 (1987).
44. A. Ghani, R. G. McLaren, and R. S. Swift, Sulphur mineralisation and transformation in soils as influenced by additions of carbon, nitrogen and sulphur, *Soil Biol. Biochem.*, *24*:331 (1992).
45. A. Ghani, R. G. McLaren, and R. S. Swift, Sulphur mineralization in some New Zealand soils, *Biol. Fert. Soils*, *11*:58 (1991).
46. L. E. Lowe, An approach to the study of the sulfur status of soils and its application to selected Quebec soils, *Can. J. Soil Sci.*, *44*:176 (1964).
47. L. E. Lowe, Sulfur fractions of selected Alberta soil profiles of the Chernozem and Podzolic order, *Can. J. Soil Sci.*, *45*:297 (1965).
48. J. R. Freney, G. E. Melville, and C. H. Williams, Soil organic matter fractions as sources of plant-available sulphur, *Soil Biol. Biochem.*, *4*:227 (1975).
49. J. R. Bettany, J. W. B. Stewart, and E. H. Halstead, Sulfur fractions and carbon, nitrogen and sulfur relationships in grassland, forest and associated transitional soils, *Soil Sci. Soc. Am. Proc.*, *37*:915 (1973).
50. K. M. Goh, "Sulphur transformation in field soils—a review summary," Proc. Towards the More Efficient Use of Soil and Fertilizer Sulphur (R. E. White and L. D. Currie, eds.), Massey University, Palmerston North, New Zealand, pp. 78–94 (1988).
51. R. S. Swift, R. G. McLaren, and J. I. Keer, "Some method for the extraction and fractionation of soil organic sulphur," Proc. Towards the More Efficient Use of Soil and Fertilizer Sulphur (R. E. White and L. D. Currie, eds.), Massey University, Palmerston North, New Zealand, pp. 55–65 (1988).
52. J. Erikson, R. D. Lefroy, and G. J. Blair, Physical protection of soil organic S studied using acetylacetone extraction at various intensities of ultrasonic dispersion, *Soil Biol. Biochem.*, *27*:1005 (1995).
53. J. Wu, A. G. O'Donnell, Z. L. He, and J. K. Syres, Fumigation-extraction method for the measurement of microbial biomass-S. *Soil Biol. Biochem.*, *26*:117 (1994).

54. M. R. Benerjee and S. J. Chapman, The significance of microbial biomass sulphur in soil, *Biol. Fert. Soils*, *22*:116 (1996).
55. J. E. Strick and J. P. Nakas, Calibration of microbial sulfur technique for use in forest soils, *Soil Biol. Biochem.*, *16*:289 (1984).
56. R. G. McLaren, J. I. Keer, and R. S. Swift, Sulphur transformations in soils using sulphur-35 labelling, *Soil Biol. Biochem.*, *17*:73 (1985).
57. R. L. Fox, R. A. Olsen, and H. F. Rhoades, Evaluation of the sulfur status of soils by plant and soil tests, *Soil Sci. Soc. Am. Proc.*, *28*:243 (1964).
58. I. E. Ensminger, Some factor affecting the adsorption of sulfate by Alabama soils, *Soil Sci. Soc. Am. Proc.*, *19*:259 (1954).
59. E. J. Kamprath, W. L. Nelson, and J. W. Fitts, The effect of pH, sulfate and phosphate concentrations on the adsorption of sulfate by soils, *Soil Sci. Soc. Am. Proc.*, *21*:463 (1965).
60. M. E. Harward, T. T. Chao, and S. C. Fang, The sulfur status and sulfur supplying power of Oregon soils, *Agron. J.*, *54*:101 (1962).
61. T. T. Chao, M. E. Harward, and S. C. Fang, Cationic effects on sulfate adsorption by soils. *Soil Sci. Soc. Am. Proc.*, *27*:35 (1963).
62. M. E. Harward and H. M. Reisenauer, Reactions and movement of inorganic soil sulfur, *Soil Sci.*, *101*:326 (1966).
63. L. C. Blakemore, T. W. Collie, and A. J. Metson, Sulphur, N. Z. Soil Bureau Bull. No. 26(2), pp. 103–109 (1968).
64. P. E. H. Gregg and K. M. Goh, Field studies with radioactive sulphur-labelled gypsum fertilizer I. Soil factors affecting the movement of fertilizer sulphur, *N. Z. J. Agric. Res.*, *21*:593 (1978).
65. P. E. H. Gregg, and K. M. Goh, Field studies with radioactive sulphur-labelled gypsum fertilizer II. Climatic and management factors affecting the movement of fertilizer sulphur in soils, *N. Z. J. Agric. Res.*, *22*:425 (1979).
66. P. E. H. Gregg and K. M. Goh, Field studies with radioactive sulphur-labelled gypsum fertilizer IV. Recovery by pasture and soil component, *N. Z. J. Agric. Res.*, *25*:345 (1982).
67. N. J. Barrow, Reaction of fertilizer sulphate in soils, *Sulphur in Australasian Agriculture* (K. D. McLachlan, ed.), Sydney Univ. Press., pp. 50–54 (1975).
68. N. J. Barrow, Reaction of anions and cations with variable-charge soils, *Adv. Agron.*, *38*:183 (1985).
69. H. L. Bohn, N. J. Barrow, S. S. S. Rajan, and R. L. Parfitt, Reaction of inorganic sulfur in soils, *Sulphur in Agriculture* (M. A. Tabatabai, ed.), Agron. Monogr. 27, Amer. Soc. Agron., Madison, Wisc., pp. 233–250 (1986).
70. K. B. Marsh, R. W. Tillman, and J. K. Syres, Charge relationships of sulfate sorption by soils, *Soil Sci. Soc. Am. J.*, *51*:318 (1987).
71. R. L. Parfitt and R. S. C. Smart, The mechanism of sulfate adsorption on iron oxides, *Soil Sci. Soc. Am. J.*, *42*:48 (1978).
72. S. S. S. Rajan, Sulfate adsorbed on hydrous alumina, ligand displaced, and changes in surface charge, *Soil Sci. Soc. Am. J.*, *42*:39 (1978).

73. S. S. S. Rajan, Adsorption and desorption of sulfate and charge relationships in allophanic clays, *Soil Sci. Soc. Am. J.*, *43*:65 (1979).
74. N. S. Bolan, D. R. Scotter, J. K. Syers, and R. W. Tillman, The effect of adsorption on sulphate leaching, *Soil Sci. Soc. Am. J.*, *50*:1419 (1987).
75. D. Curtin and J. K. Syers, Mechanism of sulphate adsorption by two tropical soils, *J. Soil Sci.*, *41*:295 (1990).
76. G. Uehara and G. Gillman, The Mineralogy, Chemistry and Physics of Tropical Soils with Variable Charge Clays, Westview Tropical Agri. Sr. No. 4. Westview Press, Boulder, Co. (1981).
77. N. S. Bolan, D. R. Scotter, J. K. Syers, and R. W. Tillman, Effect of liming and phosphate application on sulphate leaching, *J. Soil Sci.*, *39*:493 (1988).
78. N. J. Barrow, and T. C. Shaw, The slow release reactions between soils and anions: 7: Effect of time and temperature on contact between adsorbing soil and sulphate, *Soil Sci.*, *124*:347 (1977).
79. F. E. Sanders and P. B. H. Tinker, Adsorption of sulfate by sandy soils (calcic cambisol), *Geoderma*, *13*:317 (1975).
80. L. A. G. Aylmore, M. Karim, and J. P. Quirk, Adsorption of sulfate ions by soil constituents, *Soil Sci.*, *103*:10 (1967).
81. D. Curtin and J. K. Syres, Extractability and adsorption of sulphate in soils, *J. Soil Sci.*, *41*:305 (1990).
82. P. H. Williams, M. J. Hedley, and P. E. H. Gregg, The effect of preferential flow of dairy cow urine and simulate rainfall on movement of potassium through undisturbed top soil cores, *Aust. J. Soil Res.*, *28*:857 (1990).
83. J. R. Freney and K. Spencer, Soil sulphate changes in the presence and absence of growing plants, *Aust. J. Agric. Res.*, *11*:339 (1960).
84. N. J. Barrow, Studies on extraction and availability to plants of adsorbed plus soluble sulfate, *Soil Sci.*, *104*:242 (1967).
85. S. M. Hasan, R. L. Fox, and C. C. Boyd, Solubility and availability of sorbed sulfate in Hawaiian soils, *Soil Sci. Soc. Am. Proc.*, *34*:897 (1970).
86. N. J. Barrow, Effects of adsorption of sulfate by soils on the amount of sulfate present and its availability to plants, *Soil Sci.*, *108*:193 (1969).
87. A. J. Metson and L. C. Blakemore, Sulphate retention by New Zealand soils in relation to the competitive effect of phosphate, *N. Z. J. Agric. Res.*, *21*: 243 (1978).
88. A. G. Sinclair and W. H. M. Saunders, Sulphur, Fertilizer and Lime Recommendations for Pastures and Crops in New Zealand (I. S. Cornforth and A. G. Sinclair, eds.), N. Z. Ministry of Agriculture and Fisheries, Wellington, pp. 15–17 (1984).
89. L. E. Ensminger, Sulfur in relation to soil fertility, Alabama Agric. Exp. Station, Bull. 312, 19 pp. (1958).
90. L. C. Blakemore, T. E. Ludecke, M. L. Leamy, and A. J. Metson, A study of the pattern of sulphur response shown by dry-land lucerne on some brown-grey earths, *N. Z. J. Agric. Res.*, *12*:333 (1969).

91. P. E. H. Gregg, K. M. Goh, and D. W. Bash, Isotope studies on the uptake of sulphur by plants 2. Uptake from various soil depths at several field sites, *N. Z. J. Agric. Res., 20*:299 (1977).
92. A. G. Sinclair, "Problems in modelling sulphur requirements in New Zealand," Proc. Sulphur in New Zealand Agriculture (P. E. H. Gregg and J. K. Syers, eds.), Massey Univ., Palmerston North, New Zealand, pp. 106–117 (1983).
93. J. R. Freney, Sulphur containing organics, *Soil Biochemistry* (A. D. McLaren and G. H. Peterson, eds.) Marcel Dekker Inc., New York (1967).
94. J. R. Postgate, Inorganic sulphur chemistry, *The Sulphur Cycle* (G. Nickless, ed.), Elsevier Scientific Publishing Co., New York, pp. 259–272 (1968).
95. A. B. Roy and P. A. Trudinger, *The Biochemistry of Inorganic Compounds of Sulphur*, Cambridge Univ. Press, Cambridge, 400 pp. (1970).
96. N. M. Scott, Sulphur in soils and plant, *Soil Organic Matter and Biological Activity Development in Plant and Soil Sciences* (D. Vaughan and R. E. Malcolm, eds.), Martinus Nijhoff/Dr W Junk Publishers, *16*:379 (1985).
97. R. W. Tillman, "Cycling of sulphur in soil-plant-animal systems: An overview," Proc. Sulphur in New Zealand Agriculture (P.E.H. Gregg and J. K. Syres, eds.), Massey University, Palmerston North, New Zealand, pp. 5–8 (1983).
98. P. Sorn-srivichai, Mineralization of Soil Organic Sulphur in the Tokomaru Silt Loam. M.Agr.Sci. Diss., Massey Univ., Palmerston North, New Zealand (1980).
99. I. A. Chaudhry and A. H. Cornfield, Effect of moisture content during incubation of soil treated with organic material on changes in sulphate and sulphide levels, *J. Sci. Food Agri., 18*:38 (1967).
100. I. A. Chaudhry and A. H. Cornfield, Effect of temperature on sulphate levels in aerobic and sulphide levels in anaerobic soil, *J. Sci. Food Agri., 18*:82 (1967).
101. C. H. Williams, Some factors affecting the mineralization of organic sulphur in soils, *Pl. Soil, 26*:205 (1967).
102. R. S. Swift, "Mineralization and immobilization of sulphur in soils," Proc. Sulphur in New Zealand Agriculture (P. E. H. Gregg and J. K. Syers, eds.), Massey University, Palmerston North, New Zealand, pp 18–28 (1983).
103. N. J. Barrow, The effect of varying the nitrogen, sulphur, and phosphorus content of organic matter on its decomposition, *Aust. J. Agric. Res., 11*:317 (1960).
104. N. J. Barrow, Stimulated decomposition of soil organic matter during the decomposition of added organic material, *Aust. J. Agric. Res., 11*:331 (1960).
105. N. J. Barrow, A comparison of the mineralization of nitrogen and of sulphur from decomposing organic materials, *Aust. J. Agric. Res., 11*:960 (1960).
106. M. B. David, S. C. Schindler, M. J. Mitchell, and J. E. Strick, Importance of organic and inorganic sulfur to mineralization processes in forest soil, *Soil Biol. Biochem., 15*:671 (1983).

107. D. G. Maynard, J. W. B. Stewart, and J. R. Bettany, The effects of plants on soil sulfur transformations, *Soil Biol. Biochem.*, *17*:127 (1985).
108. M. A. Tabatabai and J. M. Bremner, Factors affecting arylsulfatase activities, *Soil Sci. Soc. Am. Proc.*, *34*:427 (1970).
109. T. W. Speir, Studies on climo-sequence of soils in Tussock grasslands 2. urease, phosphatase, and sulphatase activities with other properties including plant available sulphur, *N. Z. J. Sci.*, *20*:159 (1977).
110. R. Lee and T. W. Speir, Sulphur uptake by ryegrass and its relationships to inorganic and organic sulphur levels and sulphatase activity in soil, *Pl. Soil*, *53*:407 (1977).
111. G. A. Speir and W. B. McGill, Effect of phosphorus addition and energy supply on acid phosphatase production and activity in soils, *Soil Biol. Biochem.*, *11*:3 (1979).
112. J. R. Freney, G. E. Melville, and C. H. Williams, Organic sulphur fractions labelled by addition of ^{35}S-sulphate to soil, *Soil Biol. Biochem.*, *3*:133 (1971).
113. R. G. McLaren and R. S. Swift, Changes in soil organic matter fractions due to the long term cultivation of soil, *J. Soil Sci.*, *28*:445 (1977).
114. R. G. McLaren, A. Ghani, and R. S. Swift, "Studies on sulphur mineralization from soils using open incubation," Proc. Towards the More Efficient Use of Soil and Fertilizer Sulphur (R. E. White and L. D. Currie, eds.), Massey University, Palmerston North, New Zealand, pp. 66–75 (1988).
115. J. R. Bettany, J. W. B. Stewart, and E. H. Halstead, Assessment of available soil sulphate in an [cf11]35S growth chamber experiment, *Can. J. Soil Sci.*, *54*:309 (1974).
116. J. Erikson, R. D. Lefroy, and G. J. Blair, Physical protection of soil organic S studied by extraction and fractionation of soil organic matter, *Soil Biol. Biochem.*, *27*:1011 (1995).
117. W. B. McGill and C. V. Cole, Comparative aspects of cycling of organic C, N, S and P through soil organic matter, *Geoderma*, *26*:267 (1981).
118. R. Lee, L. C. Blakemore, B. K. Daly, E. J. Gibson, T. W. Speir, and V. A. Orchard, Sulfur supply to ryegrass during a pot trial and correlations with soil biological activity: the influence of two difference methods of determining the adsorbed sulphate status of soils, *Communs. Soil Sci. Pl. Anal.*, *16*: 97 (1985).
119. R. G. Burns, *Soil Enzymes*, Academic Press, London, pp. 386 (1978).
120. D. G. Maynard, J. W. B. Stewart, and J. R. Bettany, Sulfur and nitrogen mineralization in soil compared using two incubation techniques, *Soil Biol. Biochem.*, *15*:251 (1983).
121. H. J. Pirela and M. A. Tabatabai, Sulfur mineralization rates and potential of soils, *Biol. Fert. Soils*, *6*:26 (1988).
122. G. J. Blair, N. Chinoim, R. D. B. Lefroy, G. C. Anderson, and G. J. Crocker, A soil sulphur test for pastures and crops, *Aust. J. Soil Res.*, *29*:619 (1991).
123. J. H. Watkinson and M. J. Kear, "A soil test for mineralisable organic sulphur," Proc. Fertilizer Requirements of Grazed Pasture and Field

Crops—macro- and micro-nutrients (L. D. Currie and P. Loganathan, eds.), Massey University, Palmerston North, New Zealand, pp. 144–149 (1995).

124. K. M. Giddens, S. Saggar, M. D. Craighead, and P. L. Searle, "Evaluation of three recently developed soil sulphur tests in realtion to pasture response," Proc. Fertilizer Requirements of Grazed Pasture and Field Crops—macro- and micro-nutrients (L. D. Currie and P. Loganathan, eds.), Massey University, Palmerston North, New Zealand, pp. 182–189 (1995).
125. P. E. H. Gregg, Field Investigations into the Fate of Fertilizer Sulphur Added to Pasture-Soil Systems. Ph.D. Diss., Univ. of Canterbury, New Zealand (1976).
126. T. C. Strickland and J. W. Fitzgerald, Incorporation of sulphate-sulphur into organic matter extracts of litter and soil: Involvement of ATP sulphurylase, *Soil Biol. Biochem.*, *17*:779 (1985).
127. T. W. Walker, Changes induced in the soil by pastoral farming, *Soils of New Zealand*, Part 1, Soil Bureau Bulletin 26(1), N. Z. DSIR, pp. 104–109 (1968).
128. T. W. Walker and A. F. R. Adams, Studies on soil organic matter I. influence of phosphorus content of parent materials on accumulations of carbon, nitrogen, sulphur and organic phosphorus in grassland soils, *Soil Sci.*, *85*:307 (1958).
129. T. W. Walker and A. F. R. Adams, Studies on soil organic matter 2. influence of increased leaching at various stages of weathering on level of carbon, nitrogen, sulphur, and organic and total phosphorus, *Soil Sci.*, *87*:3 (1959).
130. T. W. Walker, B. K. Thupa, and A. F. R. Adams, Studies on soil organic matter 3. Accumulation of carbon, nitrogen and sulphur organic and total phosphorus in improved grassland soil, *Soil Sci.*, *87*:135 (1959).
131. R. H. Jackman, Accumulation of organic matter in some New Zealand soils under permanent pasture, I Pattern of change of organic carbon, nitrogen, sulphur and phosphorus, *N. Z. J. Agric. Res.*, *7*:445 (1964).
132. R. H. Jackman, Accumulation of organic matter in some New Zealand soils under permanent pasture, II. Rate of mineralization of organic matter and the sulphur and phosphorus, *N. Z. J. Agric. Res.*, *7*:472 (1964).
133. J. Wu, A. G. O'Donnell, and J. K. Syres, Microbial growth and sulphur immobilization following the incorporation of plant residue into soil. *Soil Biol. Biochem.*, *25*:1567 (1993).
134. J. Wu, A. G. O'Donnell, and J. K. Syres, Influence of glucose, nitrogen and plant residues on the immobilization of sulphate-S in soil. *Soil Biol. Biochem.*, *27*:1363 (1995).
135. O'Donnell, J. Wu, and J. K. Syres, Sulphate-S amendments in soil and their effects on the transformation of soil sulphur. *Soil Biol. Biochem.*, *26*:1507 (1994).
136. J. R. Freney and F. J. Stevenson, Organic sulphur transformations in soils, *Soil Sci.*, *101*:307 (1966).

137. W. H. Scharpenseel and R. Krausse, Radiochromatographic investigations on the cycle of sulfate and the S-amino acid cysteine and methionine, *Z. Pflanzenernahr, Dung Bodenk, 101*:11 (1963).
138. M. L. Nguyen and K. M. Goh, Accumulation of soil sulphur fractions in grazed pasture receiving long-term superphosphate application, *N. Z. J. Agric. Res., 33*: 111 (1990).
139. M. G. Lambert, D. A. Clark, D. A. Costall, and P. J. Budding, "Phosphorus and sulphur accumulation after long-term superphosphate application to grazed hill country," Proc. Towards the More Efficient Use of Soil and Fertilizer Sulphur (R. E. White and L. D. Currie, eds.), Massey University, Palmerston North, New Zealand, pp. 95–100 (1988).
140. S. Saggar, A. D. Mackay, M. J. Hedley, M. G. Lambert, and D. A. Clark, A nutrient transfer model to explain the fate of phosphorus and sulphur in grazed hill-country pasture, *Agric. Ecosys. Environ., 30*:295 (1990).
141. S. Saggar, M. J. Hedley, A. G. Gillingham, J. S. Rowarth, S. Richardson, N. S. Bolan, and P. E. H. Gregg, Predicting the fate of fertilizer sulphur in grazed hill country pasture by modelling the transfer and accumulation of soil phosphorus, *N. Z. J. Agric. Res., 33*:129 (1990).
142. F. J. Hingston, The loss of applied phosphorus and sulphur from soils under pasture in W.A., *J. Aust. Inst. Agric. Sci., 25*:209 (1959).
143. N. J. Barrow, The accumulation of soil organic matter under pasture and its effect on soil properties, *Aust. J. Exp. Agric. Anim. Husb., 9*: 437 (1969).
144. D. C. Lewis, A. L. Clarke, and W. B. Hall, Accumulation of nutrients and changes in soil properties of sandy soil under fertilized pasture in South-eastern South Australia 2. Total sulphur and nitrogen, organic carbon and pH, *Aust. J. Agric. Res., 25*:203 (1987).
145. K. D. McLachlan and D. G. De Marco, Changes in soil sulphur fractions with fertilizer additions and cropping treatment, *Aust. J. Soil Res., 13*:169 (1975).
146. M. L. Nguyen, D. S. Rickard, and S. D. McBride, Pasture production and changes in phosphorus and sulphur status in irrigated pastures receiving long-term application of superphosphate, *N. Z. J. Agric. Res., 32*:245 (1989).
147. P. B. Lynch, The use of nitrogen fertilisers in New Zealand Agriculture: Summary and practical recommendations, *Nitrogen Fertilisers in New Zealand Agriculture* (P. B. Lynch, ed.), NZ Institute of Agricultural Science, Ray Richards Publishers, Auckland, pp. 185–206 (1982).
148. R. J. Haynes and P. H. Williams, Nutrient cycling and soil fertility in the grazed pasture ecosystem, *Adv. Agron., 49*:119 (1993).
149. C. C. Boswell, An Assessment of Phosphorus and Sulphur Interactions on Aspects of the Sulphur Cycle in Pasture. Ph.D. Diss., Univ. New England, Australia (1983).
150. C. C. Boswell and B. Swanney, "The effect of particle size, rate and frequency of application of elemental sulphur on the release of sulphate to lucerne in the upper Waitaki basin," Proc. Towards the More Efficient Use

of Soil and Fertilizer Sulphur (R. E. White and L. D. Currie, eds.), Massey University, Palmerston North, New Zealand, pp. 159–168 (1988).

151. H. Rennenberg, The fate of excess sulphur in higher plants, *Annu. Rev. Pl. Physiol.*, *35*:121 (1984).

152. P. H. Nye and P. B. H. Tinker, Solute movement in the soil-root system, *Studies in Ecology No. 4*, Scientific Publications, 342 pp. (1977).

153. S. A. Barber, *Soil Nutrient Bioavailability, A mechanistic approach*, John Wiley and Sons: 398 pp. (1984).

154. J. F. Thompson, I. K. Smith, and D. P. Moore, Sulfur in plant nutrition, *Sulfur in Nutrition* (O. H. Muth and J. E. Oldfield, eds.), AVI Publishing Co. Westport, CT, pp. 80–96 (1970).

155. J. F. Thomson, I. K. Smith, and J. T. Madison, Sulfur metabolism in plants, *Sulfur in Agriculture* (M. A. Tabatabai, ed.), Agron. Monogr. 27, Amer. Soc. Agron., Madison, Wisc., pp. 59–122 (1986).

156. N. J. Barrow, Some investigation of the effect of grazing on the nutrition of pasture, *J. Aust. Inst. Agric. Sci.*, *33*:254 (1967).

157. R. G. Petersen, H. L. Lucal, and W. W. Woodhouse, Distribution of excreta by freely grazing cattle and its effect on pasture fertility, *Agron. J.*, *48*:440 (1956).

158. M. L. Curll, The grass and clover content of pastures grazed by sheep, *Herb. Absts.*, *52*:403 (1982).

159. E. A. Lantinga, J. A. Keuning, J. Groenwold, and P. J. A. G. Deenen, Distribution of excreta nitrogen by grazing cattle and its effect on sward quality, herbage production and utilization, *Animal Manure on Grassland and Fodder Crops: fertiliser or waste*. (H G. van der Meer, R. J. Unwin, T. A. van Dijk and G. C. Ennik, eds.) Nijhoff, Dordrecht, The Netherlands, pp. 103–117 (1987).

160. J. L. Davidson and F. L. Milthorpe, Leaf growth in Dactylus glomerate following defoliation, *Ann. Bot.*, *30*:174 (1966).

161. J. L. Davidson and F. L. Milthorpe, The effect of defoliation on the carbon balance in Dactylus glomerate, *Ann. Bot.*, *30*:185 (1966).

162. K. L. King and K. J. Hutchison, The effects of sheep stocking intensity on the abundance and distribution of meso-fauna in pasture, *J. Appl. Eco.*, *13*: 41 (1976)

163. A. G. Gillingham and C. During, Pasture production and transfer of fertility within long-established hill pasture, *N. Z. J. Exp. Agric.*, *1*:227 (1973).

164. A. G. Gillingham, Phosphorus uptake and return in grazed, steep-hill country, 1. Pasture production dung and litter accumulation, *N. Z. J. Agric. Res.*, *23*: 313 (1980).

165. A. G. Gillingham, J. K. Syers, and P. E. H. Gregg, Phosphorus uptake and return in grazed, steep-hill country, 1. Above ground components of the phosphorus cycle, *N. Z. J. Agric. Res.*, *23*:323 (1980).

166. M. A. J. Goedewaagen and J. S. Schuurman, Root production by agricultural crops on arable land and grassland as source of organic matter in soil, *Trans. 4th Int. Cong. Soil Sci.*, *2*:28 (1950).

167. S. Shamoot, L. McDonald, and W. W. Batholomew, Rhizo-deposition of organic debris in soil, *Soil Sci. Soc. Am. Proc.*, *32*:817 (1968).
168. E. A. Holland and D. C. Coleman, Litter placement effects on microbial and organic matter dynamics in an agroecosystem, *Ecology*, *68*:425 (1987).
169. A. R. Till and G. J. Blair, The utilization by grass of sulphur and phosphorus from clover litter, *Aust. J. Agric. Res.*, *29*:235 (1978).
170. K. M. Goh and P. E. H. Gregg, Re-utilization by perennial ryegrass (Lolium perenne L.) of labelled fertilizer sulphur incorporated in field-growth tops and roots of pasture plants added to soils, *Fert. Res.*, *1*:73 (1980).
171. A. R. Till, G. J. Blair, and R. C. Dalal, Isotopic studies of the recycling of carbon, nitrogen, sulphur and phosphorus from plant material, *The Cycling of Carbon, Nitrogen, Sulphur, and Phosphorus in Terrestrial and Aquatic Ecosystems* (I. E. Galbally and J. R. Freney, eds.), Springer, Berlin, pp. 51–59 (1982).
172. A. R. Till, Sulphur cycle in grazed pasture, *Sulphur in Australasian Agriculture* (K. D. McLachlan, ed.), Sydney Univ. Press, pp. 68–76 (1975).
173. N. J. Barrow, Mineralization of nitrogen and sulphur from sheep faeces, *Aust. J. Agric. Res.*, *12*:644 (1961).
174. A. C. Bray and A. R. Till, Metabolism of sulphur in the gastro-intestinal tract, *Digestion and Metabolism in Ruminant* (I. W. McDonald and A. C. I. Warner, eds.), The Univ. New England Publishing Unit, Armidale, 243 pp. (1975).
175. J. M. Gawthorne and C. J. Nader, The effect of molybdenum on the conversion of sulphate to sulphide and microbial protein-sulphur in the rumen of sheep, *Br. J. Nutr.*, *35*:11 (1976).
176. N. P. McMeniuman, D. Ben-Ghedalia, and R. Elliot, Sulphur and cystine incorporation into rumen microbial protein, *Br. J. Nutr.*, *36*:571 (1976).
177. N. J. Barrow and L. J. Lambourne, Partitioning of excreted nitrogen, sulphur and phosphorus between the faeces and urine of sheep being fed pasture, *Aust. J. Agric. Res.*, *12*:461 (1962).
178. T. W. Walker, The sulphur cycle in grassland soils, *J. Br. Grassland Soc.*, *12*:10 (1957).
179. A. C. Bray and J. A. Hemsley, Sulphur metabolism in sheep. IV, The effect of a varied dietary sulphur content on some body fluid sulphate levels and on the utilization of urea supplemented roughage by sheep, *Aust. J. Agric. Res.*, *20*:759–773 (1969).
180. J. L. L'Estrange, P. Upton, and D. M. McAleese, Sulphur metabolism in the ruminant, *Sulphur in Agriculture*, The Agric. Inst., Dublin pp. 177–194 (1970).
181. P. T. Doyle and R. J. Moir, Sulfur and methionine metabolism in sheep, 1. first approximation of sulfur pools in and sulfur flows from the recticulo-rumen, *Aust. J. Biol. Sci.*, *32*:51 (1979).
182. K. Sakadevan, M. J. Hedley, and A. D. Mackay, Influence of sheep excreta on pasture uptake and leaching losses of sulfur, nitrogen and potassium from a grazed hill country pasture. *Aust. J. Soil Res.*, *31*:151 (1993).

183. K. Sakadevan, M. J. Hedley, and A. D. Mackay, An *in situ* mini lysimeter with a removable ion exchange resin trap for measuring nutrient losses by leaching from grazed pastures. *Aust. J. Soil Res.*, *32*:1389 (1994).
184. C. During and W. C. Weeda, Some effect of cattle dung on soil properties, pasture production and nutrient uptake, 1: Dung as a source of phosphorus, *N. Z. J. Agric. Res.*, *16*:423 (1973).
185. P. R. Bird, Sulphur metabolism and excretion studies in ruminants, 2: Organic and inorganic sulphur excretion by sheep after intraruminalor intraduodenal infusion of sodium sulfate, *Aust. J. Biol. Sci.*, *24*: 1329 (1971).
186. V. I. Georgievskii, The physiological role of macronutrients, *Mineral Nutrition of Animals* (V. I. Georgievskii, B. N. Annenkov and V. T. Samokhin, eds.) Butterworth and Co. (Publishers) Ltd., London, pp. 91–170 (1981).
187. P. H. Williams and R. J. Haynes, Cycling of P and S through soil-plant-animal system under intensively grazed grass-clover pastures, *Trans. 14th Int. Cong. Soil Sci.*, *IV*: 276 (1990).
188. P. M. Kennedy, Utilization and excretion of sulphur in cattle fed on tropical roughage, *Aust. J. Agric. Res.*, *25*:1015 (1974).
189. C. During, *Fertilizers and Soils in New Zealand Farming*. 3rd edition. Govt. printer, Wellington, New Zealand, pp. 355 (1984).
190. C. During and K. J. McNaught, The effect of cow urine in growth of pasture and uptake of nutrients, *N. Z. J. Agric. Res.*, *4*:591 (1961).
191. P. D. Sears, Soil fertility and pasture growth, *J. Br. Grassland Soc.*, *5*:267 (1950).
192. E. J. Hilder, The distribution of plant nutrients by sheep at pasture, *Proc. Aust. Soc. Anim. Prod.*, *5*:241 (1964).
193. J. S. Rowarth, Phosphorus Cycling in Grazed Hill-Country Pasture. Ph.D. Diss., Massey Univ., Palmerston North, New Zealand (1987).
194. P. H. Williams, The Fate of Potassium in Grazed Dairy Pastures. Ph.D. Diss., Massey Univ., Palmerston North, New Zealand (1988).
195. J. D. Morton and D. B. Baird, Spatial distribution of dung patches under sheep grazing, *N. Z. J. Agric. Res.*, *33*:285 (1990).
196. P. H. Williams and R. J. Haynes, Balance sheet of phosphorus, sulphur and potassium in a long-term grazed pasture supplied with superphosphate, *Fert. Res.*, *31*:51 (1992).
197. P. H. Williams and R. J. Haynes, Transformations and plant uptake of urine-sulphate in urine-affected areas of pasture soil, *Pl. Soil*, *145*:167.
198. M. J. Hedley, R. W. Tillman, and P. R. Ball, "The environment: nutrient losses from New Zealand agriculture," Proc. NZFMRA Conf., The Challenge for Fertilizer Research in 1990, New Zealand, pp. 62–91 (1990).
199. E. J. Hilder, "Distribution of excreta by sheep at pasture," Proc. 10th Int. Grassland Cong., Helsinki. pp. 977–981 (1966).
200. S. M. Bromfield and O. L. Jone, The effect of sheep on recycling of phosphorous in hayed-off pasture, *Aust. J. Agric. Res.*, *21*:699 (1970).
201. K. P. Barley, Earthworms and decay of plant litter and dung: A review, *Proc. Aust. Soc. Anim. Prod.*, *5*:236 (1964).

202. G. F. Bornemissza and C. H. Williams, An effect of dung beetle activity on plant yield, *Pedobiol.*, *10*:1 (1970).
203. A. MacFadyen, "The role of fauna in decomposition processes in grasslands," Scientific Proc. Royal Dublin Society, Sr. A., 6(11):197 (1978).
204. P. H. Williams and R. J. Haynes, Fate of 35S-labelled urine sulphate in urine affected areas of pasture under field conditions, *J. Agric. Sci.*, *121*:83 (1993).
205. P.H. Williams, M.J. Hedley, and P.E.H. Grergg, Uptake of potassium and nitrogen by pasture from urine-affected soil, *N. Z. J. Agric. Res.*, *32*:415 (1989).
206. B. N. MacDiarmid and B. R. Watkin, The cattle dung patch, 2. The effect of dung patch, on the chemical status of soil, and ammonia nitrogen losses from the patch, *J. Br. Grassland Soc.*, *27*:43 (1972).
207. J. G. White, Mineralization of nitrogen and sulfur in sulfur deficient soil, *N. Z. J. Agric. Res.*, *2*:255 (1959).
208. K. M. Goh and M. L. Nguyen, "Effect of grazing animals on the plant availability of Sulphur fertilizers in grazed pastures," Proc. N. Z. Grassland Assoc., 52:181 (1990)
209. A. Ghani, M. J. Hedley, A. H. C. Roberts, A. D. Mackay, S. Saggar, and A. G. Gillingham, "Use of a simplistic nutrient transfer model in explaining the fate of added P and S in a dairy pasture," Middle East Sulphur Symp., Cairo, Egypt, p. 27 (1990).
210. M. A. Tabatabai, Physicochemical fate of sulfate in soils, *Int. J. Air Poll. Contr. Hazardous Waste Management*, *37*:34 (1987).
211. N. S. Bolan, J. K. Syres, and R. W. Tillman, Ionic strength effects on surface charge and adsorption of phosphate and sulphate by soils, *J. Soil Sci.*, *37*: 379 (1986).
212. C. M. Smith, P. E. H. Gregg, and R. W. Tillman, Drainage losses of sulphur from a yellow-grey earth soil, *N. Z. J. Agric. Res.*, *26*:363 (1983).
213. L. K. Heng, R. E. White, N. S. Bolan, and D. R. Scotter, Leaching losses of major nutrients from a mole-drained soil under pasture, *N. Z. J. Agric. Res.*, *34*:325 (1991).
214. A. R. Till and T. P. McCabe, Sulphur leaching and lysimeter characterization, *Soil Sci.*, *121*:44 (1976).
215. G. P. Sparling, K. N. Whale, and A. J. Ramsay, Quantifying the contribution from soil microbial biomass to extractable P level from fresh and air-dried soils, *Aust. J. Soil Res.*, *23*:613 (1985).
216. K. I. Peverill, G. P. Briner, and L. A. Douglas, Changes in extractable sulphur and potassium levels in soil due to oven drying and storage, *Aust. J. Soil Res.*, *13*:67 (1975).
217. K. I. Peverill and L. A. Douglas, Use of undisturbed cores of surface soil for investigating leaching losses of sulphur and phosphorus, *Geoderma*, *18*: 193 (1976).
218. K. I. Peverill, L. A. Douglus, and N. B. Greenhill, Leaching losses of applied P and S from undisturbed cores of some Australian surface soils, *Geoderma*, *19*:91 (1977).

219. C. A. D. McLay, K. C. Cameron, and R. G. McLaren, "Sulphate leaching in undisturbed soil monoliths and repacked soil columns," Proc. Towards the More Efficient Use of Soil and Fertilizer Sulphur (R. E. White and L. D. Currie, eds.), Massey University, Palmerston North, New Zealand, pp. 129–136 (1988).
220. P. H. Williams, P. E. H. Gregg, and M. J. Hedley, Fate of potassium in dairy cow urine applied to intact cores, *N. Z. J. Agric. Res.*, *33*:151 (1990).
221. G. P. Gillman, Studies on some deep sandy soils in Cape York Peninsula, North Queensland, 3. Losses of applied phosphorus and sulphur, *Aust. J. Exp. Agric. Anim. Husb.*, *13*:418 (1973).
222. M. B. Jones, W. E. Martin, and W. A. Williams, Behaviour of sulfate sulfur and elemental sulfur in three California soils in lysimeters, *Soil Sci. Soc. Am. Proc.*, *32*:535 (1968).
223. R. D. Rhue and E. J. Kamprath, Leaching losses of sulfur during winter months when applied as gypsum, elemental S or prilled S, *Agron. J.*, *65*:603 (1973).
224. C. During and M. Cooper, Sulphate nutrition and movement in a soil with high sulphate sorption characteristics, *N. Z. J. Exp. Agric.*, *2*:45 (1974).
225. D. E. Hogg and M. R. J. Toxopeus, Studies on the retention of sulphur by yellow-brown pumice soils, *N. Z. J. Agric. Res.*, *9*:93 (1966)
226. A. G. Sinclair, C. C. Boswell, I. S. Cornforth, A. Lee, C. Morgan, J. Morton, M. L. Nguyen, W. M. H. Saunders, P. Shannon, R. G. Smith, and G. Whelan, "Agronomic requirements of sulphur in New Zealand Pastures," Proc. 20th Tech. Conf. NZFMRA, New Zealand, pp. 538–572 (1985).
227. A. G. Sinclair, Sulphur fertilizer recommendation scheme in New Zealand, *Sulphur Agric.*, *10*:12 (1986).
228. C. C. Boswell, "Sulphur fertilizer requirements," Proc. Towards the More Efficient Use of Soil and Fertilizer Sulphur (R. E. White and L. D. Currie, eds.), Massey University, Palmerston North, New Zealand, pp. 253–256 (1988).
229. M. R. McCaskill and G. J. Blair, Development of a simulation model of sulphur cycling in grazed pastures, *Biogeochem.*, *5*:165 (1988).
230. M. R. McCaskill and G. J. Blair, A model for S, P and N uptake by perennial pasture, 1. Model construction, *Fert. Res.*, *22*:161 (1990).
231. M. R. McCaskill and G. J. Blair, A model for S, P and N uptake by a perennial pasture, 2. Calibration and prediction, *Fert. Res.*, *22*:173 (1990).
232. L. K. Heng, Measurement and Modelling of Chloride and Sulphate Leaching from a Mole-Drained Soil. Ph.D. Diss., Massey Univ., Palmerston North, New Zealand (1991).
233. S. Phimsarn, A Comparison of the Fate of Elemental Sulphur and Sulphate Sulphur Based Fertilizers in Pasture Soils. Ph.D. Diss., Massey Univ., Palmerston North, New Zealand (1991).
234. N. D. Grace, Animal requirements for sulphur. Proc. Sulphur in New Zealand Agriculture (P.E.H. Gregg and J. K. Syres, eds.), Massey University, Palmerston North, New Zealand, pp. 85–94 (1983).

4

Sulfur Fertilizers for Grazed Pasture Systems

Colin C. Boswell
New Zealand Pastoral Agricultural Research Institute, Invermay Agricultural Centre, Mosgiel, New Zealand

Paul E. H. Gregg
Massey University, Palmerston North, New Zealand

I. INTRODUCTION

Fertilizer sulfur (S) is required to be added to improved pastures because of an imbalance of natural inputs and outflows of S from the systems. It is needed where soils are naturally S deficient (e.g., most New Zealand soils, upon which much of our experience is based), and where deficiencies develop or are exacerbated because losses from the systems are greater than natural inputs.

Natural inputs in most pastures are relatively low and result either from atmospheric S or through the process of weathering of minerals in soil parent material. Atmospheric S is usually deposited in rainfall as sulfate or dry as sulfur dioxide on the surface of soil and/or plants where the atmosphere is more polluted. The release of sulfate by weathering is often offset by its rapid loss by leaching, and the process has negligible net effect in New Zealand soils at least. Leaching is the major process of loss from pasture systems [1,2]. Other soil losses may include the retention of sulfate by adsorption to soil particles and the accumulation of S in organic-S form in animal feces for extended periods of time [3,4], especially in dry environments. The accumulation of S in feces is often part of a nutrient transfer process that occurs through grazing of pasture by animals [1,2]. The net result of nutrient transfer is movement of nutrients such as S from some parts of a pasture (e.g., a hillside) and accumulation in other smaller areas where stock camp, such as on the brow of a hill. Like other nutrients, S is lost from the grazed pasture system through the sale of animal products such as wool, meat, and milk, off the property.

Sulfur fertilizer requirements for pastures differ fundamentally from those for crops. Grazed pastures are typically expected to persist for from 5 to 50 years, whereas many crops are typically short-lived and in tropical environments there may be one, two or even three crops per year.

However, one crop per year is the norm in soils likely to be deficient in S. Generally, cropping is continued on the same soil area for several to many years. For rapidly growing crops to respond to any fertilizer, the nutrients must be available in sufficient quantities during growth. With most crops, fertilizer S is applied only at sowing.

With perennial pastures the range of possible S fertilizers and application options is much broader than with annual crops. Pastures vary from rangeland to intensively farmed ryegrass/clover pastures, or intensively managed grass pastures maintained for cut-and-carry managements (e.g., for feedlot or housed animal feeding). However, in this chapter we rely heavily on our experience of New Zealand pastures, which are invariably composed of mixtures of grasses and legumes (with the role of the legume being to supply sufficient nitrogen to the pasture to meet its needs), and which are grazed by sheep or cattle.

II. PATTERNS IN THE DISTRIBUTION OF S-DEFICIENT PASTURES

Key features in the distribution of S-deficient pastures are as follows: they occur (1) where there are low inputs of S from the atmosphere, (2) on coarse-textured, low S-retention soils, (3) on soils with low S reserves, (4) on soils that may have had a prolonged cropping history in absence of S fertilizer, and (5) on highly leached soils formed under a high rainfall [5].

If pastures on soils receiving low atmospheric inputs of S also receive such low annual rainfall to be classified semi-arid (300–450 mm per year, [6]), then they are likely to be even more deficient—e.g., valley floor soils of Central Otago in the South Island of New Zealand, where sulfur is often the only plant deficiency [7]; S-deficient semi-arid soils frequently support savanna grassland that is/has been regularly burned, presumably volatilizing organic-S [8,9]. Worldwide, semi-arid soils are sandy or coarse-textured, low in organic matter, highly susceptible to loss of S by leaching, and would normally be expected to support crops with moderately high S requirements—such as sorghum, millet, groundnuts, pulses, soybeans, oilseed, and cotton—before being used for pasture.

Coarse-textured soils also occur in wetter environments. Many of the S-deficient soils of temperate North America have a coarse (e.g., sandy) texture and have been the subject of review papers [10–12]. Under high rainfall (>2500 mm) on a gley podzol soil (coarse-textured topsoil), S deficiency results from a high rate of loss of sulfate by leaching [13]. Coarse-textured pumice soils, from volcanic eruption, are S-deficient by virtue of rapid loss of sulfate by leaching from a parent material low in S [14]. Coarse-textured soils are also likely to have limited mineralization of S due to lower moisture contents and lower organic matter contents [15], and low sulfate adsorption capacity [16]. Tropical soils forming in rapidly weathering volcanic ash that are accumulating organic matter may be S-deficient if S-mineralization is too slow to meet plant needs [8].

In cool-temperate Canada, the pattern of S-deficient pastures have been related to soil-formation factors and processes [15]. Sulfur deficiency was particularly associated with Luvisol soils (gray-wooded soils formed under forest) at the wetter end of an environmental gradient of moisture effectiveness. The increasing deficiency in the Luvisols was attributed to greater leaching losses of S, higher producing crops and higher S-demanding crops (e.g., rapeseed/canola) causing higher removal of S, and especially the utilization of labile organic matter reserves from soils with a low mineralization potential.

A large proportion of improved pastures in New Zealand have a consistent requirement for fertilizer S as associated soils have low levels of S. Some pastoral soils have higher S levels due to the accumulation of organic matter in the soil. Where this is the case, there is still a need to replace the nutrient lost as a consequence of the grazing or pastoral cropping (e.g., silage, hay), in order to sustain the long-term productivity of the soil.

III. NATURAL INPUTS AND LOSSES OF S FROM PASTURES

In New Zealand pastures, natural additions are restricted to low atmospheric inputs of 1–2 kg S ha^{-1} yr^{-1} [17] over most of the country, rising to 8–15 kg S ha^{-1} yr^{-1} at the coasts [18,19]; similar levels of input can be expected in environments remote from industrialized areas elsewhere in the world. Net additions of S by weathering are negligible because sulfate released by weathering is rapidly lost by leaching [17,20] and the only other additions may occur in limited areas of irrigated pastures. In New

Zealand up to 10 kg S ha^{-1} yr^{-1} may be added in this form (e.g., 1 ppm S in water; 1000 mm irrigation applied/year [21]).

In industrialized countries, atmospheric inputs from the smoke and pollution has resulted in historically higher atmospheric S inputs [22], but recent pollution control laws have reduced such inputs, especially in North America and Europe. Reduced pollution is only one of several changes in the environment or in farming practice that have been precursors to the development of sulfur deficiency. Others include decreased use of high S fuels such as coal; replacement of fertilizers containing S such as ammonium sulfate and single superphosphate, applied nominally to supply nitrogen (N) and phosphorus (P) respectively, with higher-analysis fertilizers such as urea and triple superphosphate or reactive phosphate rock (RPR), which contain little or no S; more intensive cropping (e.g., cropping previously uncropped land, multiple cropping, reduced fallowing, increasing yields by irrigation with the attendant likelihood of increased leaching losses, use of higher-yielding cultivars or alternative crops, change to high S-demanding crops, all result in the removal of greater amounts of S from the soil); use of crop residues for food or fuel in developing countries; declining reserves of soil S; and decreased use of pesticides that contain S [23]. An additional factor has been a reduction in the use of fertilizer in general as a result of poor returns for agricultural products. This downturn in the use of fertilizers was general in New Zealand pastoral farming over the period 1984–1990 [24,25]. Not all of these factors apply in all countries; however, the change to high-analysis fertilizers is widespread, as is the use of higher-producing cultivars.

In grazed pastures, most losses of S from the system are by takeoff in animal products, excretal transfer by grazing animals, runoff, and especially leaching. Some losses occur in removal of plant products, such as hay and silage; in cut-and-carry pasture harvesting systems, removal in plant produce is a major source of loss. Typical New Zealand S losses in animal product can be calculated as 0.22 kg S/sheep stock unit/year (e.g., 4.5 kg wool × 3.4%S + 30 kg lamb × 0.15%S + 60/5 kg ewe × 0.15%S)—assuming 20% replacement of ewes; for intensive dairy production similar losses of 0.24 kg S/(equivalent sheep) stock unit/year are expected. At 7.5 stock units per dairy cow, this equates to 1.8 kg S/cow/year [21]. Excretal transfer losses occur when animals transport S either away from the pasture to yards and so forth or within a paddock to stock camping sites [1,2]. Transfer losses are greater in extensively grazed hill country than on intensively grazed improved flat pastures and are greater under dairying than under sheep grazing. Much of the transfer loss occurs because S in urine is leached beyond the rooting zone of the pasture plants. Urine S losses of 70% have been recorded in New Zealand experiments

[26]. Leaching losses are higher on soils with low S-retention properties, especially when pastures are irrigated [2] or where rainfall is high [13]. The extent of losses in irrigated pastures is dependent on the proportion of the applied water that can be stored in the soil (i.e., losses = water applied − water-holding capacity of the soil; in some schemes this can be as high as 60% [21]). Saggar *et al.* in Chapter 3 identify leaching as the main source of loss of sulfur from pastures.

IV. EVALUATING FERTILIZER S REQUIREMENTS

Essentially, S fertilizer requirements of grazed pastures involve making decisions about what form, quantity, and time the fertilizer should be applied. Many factors will influence these decisions, and often the final recommendation may differ from theoretical requirements.

Several techniques are used to provide the information that assists in making decisions about fertilizer S requirements of grazed pastures. Evaluation techniques include field trials, glasshouse experiments, S balance models, and plant and soil analyses.

Field trials can be designed to assess, either separately or collectively, what form, quantity (rate), and time that S fertilizer should be applied at either a particular site or over a range of sites. Although providing useful information, field trials are expensive to conduct and are site-specific, requiring great care in extrapolating to other environments.

Glasshouse trials, although a relatively low cost method, have many limitations [27,28]. The most serious of these is that glasshouse trials often assess nutrient availability in the topsoil. In contrast, in the field, some plants may be able to utilize subsoil sulfate effectively [29]. Besides this, the low soil volume for the plant root system and the favorable moisture and temperature regimes maintained during the trial period tend to accentuate the deficiencies and thus overestimate the apparent S requirements. Alternatively, the preparation of some soils for potting can also increase the mineralization of soil organic S [30,31] and thus mask the symptoms of S deficiency. Therefore, wherever possible, it is desirable that the results from glasshouse trials be confirmed under field conditions for their reliability in assessing S deficiency or fertilizer S form differences. They are not suitable for assessing the quantity of S required for given situations.

In recent years, fertilizer S requirements (especially rates) have normally been calculated by estimating the net losses of S from the pasture system. In New Zealand grazed pastures, this is often done by using a model published by the Ministry of Agriculture and Fisheries [32]. The model allows for the natural inputs of S and the losses expected on different soils and under different farming systems and recommends the amount of fertilizer S to be applied annually to replace losses from the system. In order to calculate losses in the soil there must be data of soil type (especially to account for soil texture), drainage characteristics, slope, rainfall and irrigation (if applicable), and anion retention characteristics, to account for leaching losses, and to a lesser extent, to account for any "fixation" of S by strong adsorption. Animal losses are computed from the stocking rate, farming system and intensity, irrigation or none, and topography. The New Zealand model also indicates the most efficient fertilizer form (e.g., sulfate or elemental sulfur [S^0]) and application time. These topics are considered further later in the chapter.

A review of the model by Nguyen and Goh [33] highlights its function as a maintenance model that is reliable where a steady state exists but shows a need for better understanding of the dynamics of movement of S between different soil S pools and plant and animal pools. Knowledge of rates of transfer between these pools is increasing [e.g., 3, 26, 34], with research being directed at movement of S into and out of soil organic matter, especially [34–36]. Fertilizer requirements (amounts, forms, and timing of application) are calculated essentially from the net losses from the pastoral system; other indicators are used to provide an indication of the level of fertility in the soil or as data against which to measure changes in the soil fertility status of S.

Soil tests are used to indicate the level of fertility in a soil. In New Zealand, "available" soil sulfate, which includes soluble sulfate and readily released adsorbed sulfate, is measured (normally in the 0–75 mm soil depth) as the sulfate-S extracted by a calcium phosphate solution [37]. A New Zealand topsoil with <7 ppm sulfate is regarded as S-deficient [38]; in the majority of New Zealand pastoral soils (those derived from sedimentary rock), a topsoil with 10–12 ppm sulphate is required for biological optimum (producing at 97% of maximum yield) [39]. An additional new test that measures the ability of soils to release sulfate from soil organic matter reserves to the pool of S available for plant uptake has been developed [40]. The amounts of organic matter–S (ester-S) extracted by potassium phosphate solution has been correlated with the ability of pastoral soils to replenish the soil sulfate pool as S is lost (e.g., through plant uptake and leaching). At biological optimum, the target range for the organic-S test is 15–20 [39].

Plant total S contents below which pasture plants are regarded as deficient in S—i.e., "critical" S levels (generally the S concentration in the plant tops required to support 90–95% of maximum yield)—can provide an indication of the requirement for S fertilizer. Critical levels for temperate pasture species summarized from many sources cover a broad range (Table 1) and probably reflect differences in environmental factors, plant factors such as stage of growth, and possibly differences in analytical procedures between different data sets [41]. In improved New Zealand pastures, a standardized sample taken at grazing height from plants at a young stage of growth shows the critical level for both clovers and grasses is 0.25–0.3%S [42].

The tropical legume data and tropical grasses data shown in Table 1 are far less variable because legume data are from one pot experiment and grasses data from another. In both cases the environmental conditions were fixed within experiments, and the stage of growth at harvesting was common between species. The medicago species appeared to have a higher S requirement than other tropical legumes [27], and tropical grasses have even lower requirements. Lucerne (*Medicago sativa* L.) grown in temperate and tropical environments has similar critical levels (Table 1).

Problems with critical plant S levels for assessing the internal S status of plants have led to the adoption of other diagnostic indices. The ratio of nitrogen to sulfur (N:S) in plant proteins is sometimes used. Where S is deficient, N tends to accumulate in nonprotein form and N:S increases above that normally formed in plant proteins [27]. The normal ratio varies with plant species but the general guidelines suggested by Smith and Stregar [27] are 18 for temperate legumes and 14 for temperate grasses. The respective N:S ratios collated in reviews are 15–20 for legumes and 6–8 for grasses [41,43,44]. Recently attention has been centered on balanced pasture nutrition [45], especially relative S and phosphorus (P) contents of pasture. An S:P ratio of 0.72 in clover tops indicated nutritional balance for the two nutrients [45].

Table 1 Summary of Critical S Levels in Pasture Plants in Australia and New Zealand

Pasture types	Critical S (% of DM)	References
Temperate species		
Clovers (tops)	0.15–0.3	Collated by [41,43]
Lucerne (tops)	0.2–0.3	[41,43]
Mixed grasses (tops)	0.14–0.3	[41,43]
Tropical legumes		
9 legume spp—excluding *Medicago* spp.	0.14–0.18	[44]
3 *Medicago* spp.—including lucerne	0.2	[44]
Tropical grasses		
7 species	0.07–0.11	[27]

Both soil tests and plant analyses are useful in assessing whether S fertilizer is needed or not but do not provide information on the amounts of S required.

V. FORMS OF S FERTILIZER

Sulfur fertilizers are almost exclusively either in sulfate or elemental sulfur (S^0) forms. Sulfate (e.g., as single superphosphate, ammonium sulfate [sulfate of ammonia], potassium sulfate, and gypsum) is the most widely used form (see also review papers [46,47] for a more complete list). Single superphosphate is the main fertilizer used for pastures in Australia and New Zealand. It contains 9% P and 12% S as sulfate. Elemental S fertilizers represent a minor (but growing) part of the S fertilizer market. Elemental S fertilizers are more suitable for perennial pastures than for crops, especially for use with high-analysis fertilizers containing high concentrations of other nutrients. Elemental S fertilizers available in various parts of the world [5] include dry-screened S^0, mixtures of S^0 with superphosphate, S-urea, S-bentonite mixtures, S^0 pellets, popcorn S, pumice S, S–rock phosphate mixtures, S-anhydrite, S-coated urea, S^0 slurries and emulsions, and ammonia-S solution. An increasing number of fertilizer products consist of a mixture of sulfate and S^0—e.g., S^0-superphosphate mixtures and the Canadian "Sulgro" (S^0–ammonium sulfate) mixtures. The effectiveness of these are considered in more detail later in the chapter.

Plants can take up sulfur from soil solution only as sulfate-S.

The S in sulfate fertilizers has the advantage of being immediately available to plants following dissolution of the fertilizer. The rate of dissolution is largely influenced by the size of the fertilizer granules and the availability of soil moisture, and these factors have been incorporated into a model that predicts the rate of release of sulphate from superphosphate [48]. The disadvantage of sulfate fertilizers is that the sulfate is released to the soil as a relatively large dose over a short period of time, and the highly mobile sulfate ions are likely to be lost beyond the rooting zone of plants by leaching, especially in soils with low anion retention properties.

In contrast, S^0 has to be oxidized to sulfate before it can be taken up by plants. It is regarded as a slow-release fertilizer. The rate of oxidation is dependent on the surface area of the particles of S^0 available to be colonized by oxidizing micro-organisms in the soil. For a given mass the surface area is inversely proportional to particle diameter [49,50]. For rapid oxidation, fine S^0 particle size is required (e.g., with 1–2 μm particles in slurry, oxidation is practically instantaneous [51]). However, very fine dry particles of S^0 are difficult to handle and to apply as a fertilizer, are irritating to the eyes and lungs, and are susceptible to loss by wind or to transfer on the surface of applied irrigation water [52]. But, most seriously, fine particles of S^0 represent a fire and explosive hazard [53]. Fine dry S^0 per se is thus impractical as a commercial fertilizer [54]. So, satisfactory forms of S^0 should contain at least some finely divided S^0, but the particles should form part of a matrix of the fertilizer and not be able to be separated off from the fertilizer particles as a fine dust. The products already listed in this section meet this safety criterion. As will be discussed later, although they may be safe, not all can be recommended as pastoral fertilizers, because their rate of oxidation is likely to be too slow to be effective, at least in temperate environments.

A. Oxidation of Elemental S

The availability of sulfate from S^0 fertilizers depends on the rate of S^0 oxidation, which depends primarily on particle size of the S^0 and soil temperature in temperate New Zealand [55].

Recent research advances in New Zealand concerned with the identification of residual S^0 in the soil [56], which in turn provides a measure of S^0 oxidation over time, and modeling the rate of oxidation of S^0 [57,58], allow measurements of the efficiency of different S^0 fertilizers in different environments [5]. The model (see Ch. 5) also offers a systematic approach to the selection of S^0 fertilizers to suit different environmental and/or pasture situations; further it has created the datum on which fertilizer manufacturers can base S^0 formulations [21]. Watkinson [57,58] showed that the oxidation of S^0 particles fitted a cubic model. In the model, the amount of S^0 remaining unoxidized to sulfate after a time (t) is given by:

$$m/m_0 = [1 - at/(bc)^{0.5}]^3$$

where:

m = mass of S^0 remaining
m_0 = mass of S^0 originally applied
a = S^0 oxidation rate constant in terms of reduction in diameter of the S^0 particles per unit of time
t = time in days since application
$(bc)^{0.5}$ = geometric mean of the particle size range of S^0; b = minimum diameter, c = maximum diameter (mm); were $c/b < 2$

The rate constant is an indicator of the rate of oxidation for a given set of conditions, or better, for given sites, or climatic regions. The rate constant can be calculated by substitution:

$$a = [1 - (m/m_0)^{0.333}] \cdot (bc)^{0.5}/t$$

because m_0, b, c are known and m and t are measured.

Alternatively, an oxidation rate constant k can be defined as equal to ½ *p*a [57], where *p* is the S^0 particle density of 2.07 g cm^{-3}, and k becomes the mass of S^0 oxidized per unit area per unit time [55]. Once calculated, using S^0 with a relatively narrow range of particle size (i.e., agrees with the relationship $c/b < 2$), the rate constant applies to S^0 irrespective of particle size.

The rate constant for a region [55,59] or specific site [60] is an integrated measurement taking account of the various physical, biological, and chemical factors that affect S^0 oxidation [61]. Five regional rate constants have been calculated within New Zealand (Table 2, Fig. 1 [55]); these account for differences in soil temperatures and rainfall (and therefore different potential leaching losses).

Although Watkinson and Lee [55] found soil temperature to be the major factor influencing the rate constant for S oxidation, and that southwest Australian data fitted the pattern of New Zealand results, it is not possible to extrapolate rate constants to other situations simply on the basis of soil temperatures. For example, the New Zealand study did not involve sites where rainfall was <500 mm yr^{-1} and most sites were at least double this amount. Also it is likely that the S^0-oxidizing microorganisms (thio-

Table 2 Rate Constants for Oxidation of Elemental S in the North and South Islands of New Zealand

Region	Mean soil temperature (100 mm)°C	Mean annual rainfall (mm)	Rate constant *a*
North of North Island (warm temperate)	13–15	1500	0.21
Central and southern North Island (cool temperate)	12–13	800–1500	0.16
Western South Island (cool temperate, high rainfall)	9–11	2000–3000	0.17
Northern South Island (cool)	9–11	800–1000	0.15
Southern South Island (cool temperate)	8–10	400–1200	0.12

Source: Refs. 55 and 59.

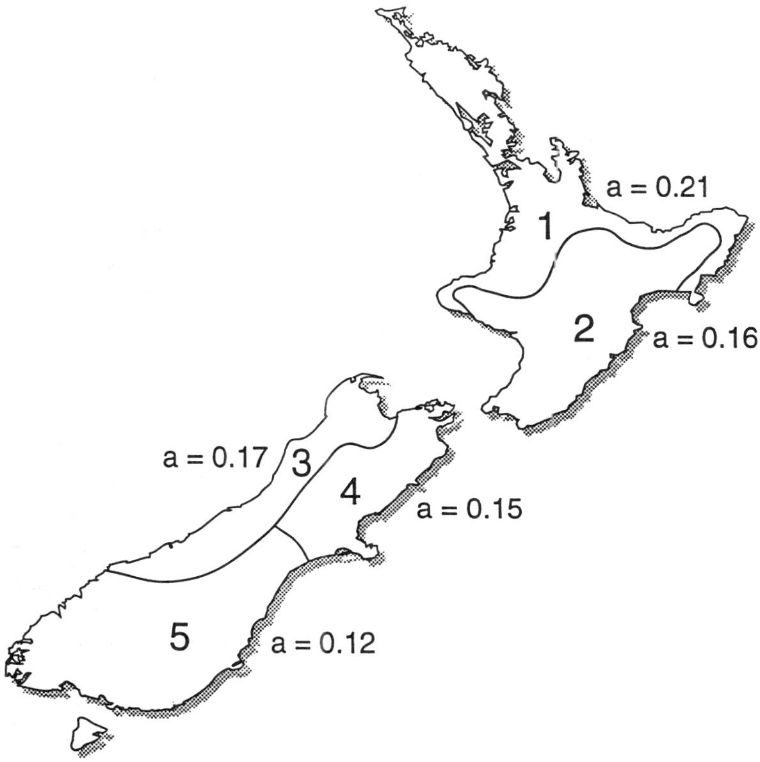

Figure 1 Map of New Zealand showing five climate regions whose rate constants of oxidation are listed in Table 2. (From Ref. 55.)

bacilli) were the same in New Zealand and western Australia, but the organisms are likely to differ in different parts of the world. In Canada, for example, heterotrophs dominate S^0 oxidation [62], and the rates of oxidation are slower [58]. An Australian S^0-oxidation model [48], which uses the same process of oxidation reducing S^0 particle size by dissolution of the outer layers, incorporates rate modifiers for soil temperature and soil moisture as they vary from optimal oxidation conditions.

Where the rate constant for a site is known, the amount of S^0 of known particle size that is oxidized in a given time can be calculated. From this information, the suitability of particle size ranges of S^0 to a location (climatic region, etc.), can be evaluated and can form the basis of specifications for an effective S^0 fertilizer for the region and/or pasture system. Alternatively, rate constants can be used to calculate the rates of fertilizers with known S^0 particle size distributions that will supply a required rate of sulfate-S over a given time [5].

Figure 2 shows the effects of two particle sizes on S oxidation at the two locations with a = 0.21 and 0.12, respectively.

The proportion (%) of S^0 of given particle size range oxidized in a given time (1, 2, 3 years, etc.) can be calculated using the rate constant equation.

$$
\begin{aligned}
m/m_0 &= [1 - at/(bc)^{0.5}]^3 \\
m &= m_0[1 - at/(bc)^{0.5}]^3 \\
\text{mass oxidized} &= m_0 - m \\
&= m_0 - m_0[1 - at/(bc)^{0.5}]^3
\end{aligned}
$$

Figure 3 summarizes the amounts of oxidation occurring in a wide range of S^0 particle sizes at sites with a = 0.21; S^0 particles <150 μm are oxidized within 1 year of application; particles <250 mm are oxidized within 2 years, and particles <500 μm are oxidized within 4 years (96% oxidized within 3 years). Particles of S^0 need to be finer in locations with a rate constant similar to the southern New Zealand region (a = 0.12); here particles <75 μm are oxidized within 1 year of application, particles

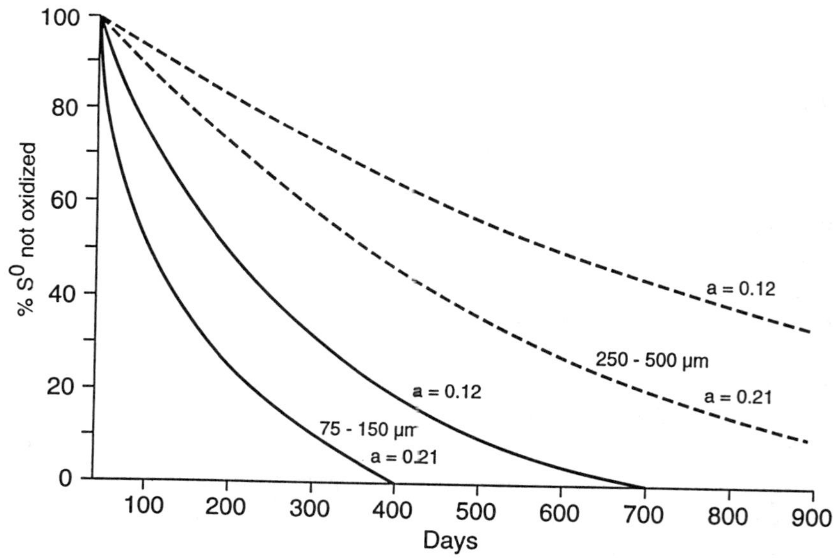

Figure 2 Modeled effects of rate constant and S^0 particle size on percentage of S^0 oxidized with time. (From Ref. 5.)

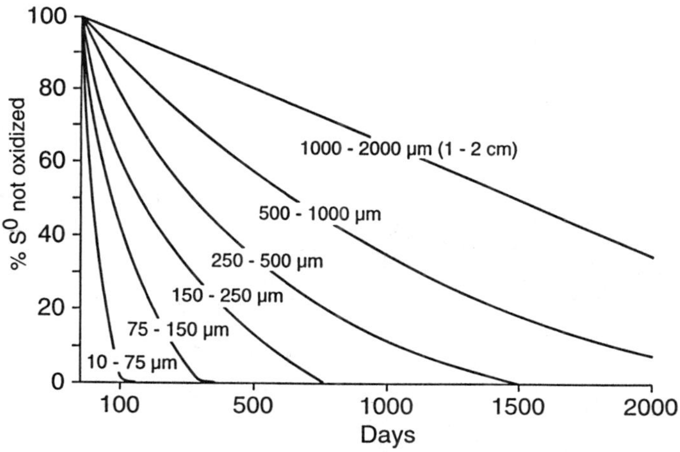

Figure 3 Modeled oxidation times for different ranges of S^0 particle size where the rate constant a = 0.21. (From Ref. 5.)

<150 μm are oxidized within 2 years, and particles <250 are oxidized within 3 years. Australian research [63] similarly showed that the sulfate requirements of annual crops from S^0 applied at the beginning of the growing season was best met with particles of 200 μm at two northern sites (Darwin and Cairns) and 100 μm at southern Australian sites (e.g., Armidale NSW). The distribution of the ranges of particle size, and efficiency of the S^0 component of a 50% S superphosphate product (44% S^0, 6% sulfate-S) expressed in terms of the oxidation of the S^0, is calculated for sites with 'a' values of 0.21 in Figure 4.

In fertilizers with very fine particle size (e.g., Thiovit 2–8 μm) the calculated time for 99% oxidation of S^0 particles at sites with rate constants of 0.21 and 0.12 is 15 and 26 days, respectively.

B. Elemental Sulfur Fertilizers

Forms of elemental sulfur fertilizers have been reviewed [46,47,64,65]. Forms produced using dry S^0 vary from screened S^0 (100% S), mined raw S forms (e.g., pumice S, Middle East ores), refined ores, to mixtures of S^0 superphosphate (S-superphosphates, sulfur-fortified [SF] superphosphates), or mixtures with rock phosphate, or with partially acidulated rock phosphate. Molten sulfur is used in the preparation of other mixtures; e.g., with superphosphate, bentonite clay (S-bentonite), urea, di-ammonium phosphate or di-ammonium phosphate–urea phosphate, calcium sulfate anhy-

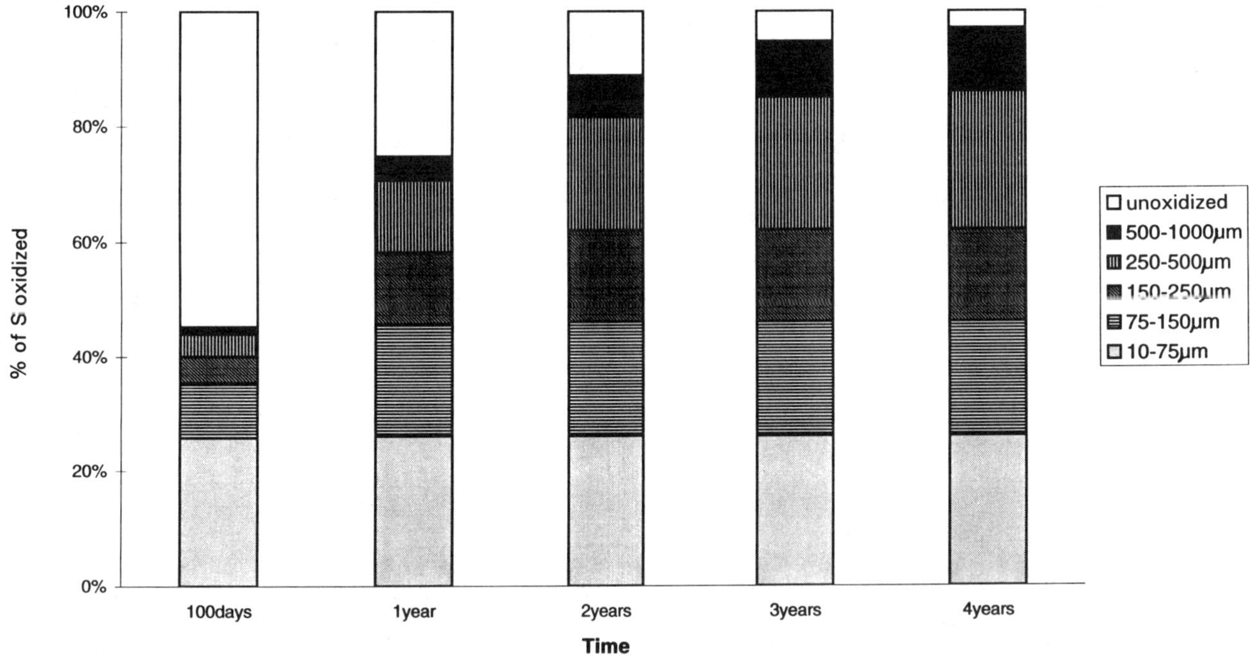

Figure 4 Efficiency of oxidation of S^0 in 50% S-superphosphate fertilizer at different times after application at a site where a = 0.21 (% S^0 oxidized). (From Ref. 5.)

drite, gypsum, and triple superphosphate. Elemental S is also applied in liquid form as S slurries. The suitability of a selection of S^0 fertilizers for pastures has been discussed by Boswell and Friesen [5] and is summarized here.

1. Dry-Screened S^0

Much of the early S^0 research was with dry-screened particles of various sizes (Tables 3 and 4). Dry S^0 has been evaluated both as a fertilizer [7,16,66–73], and to determine S^0 particle size/fertilizer effectiveness relationships [49,50,74–79].

The effectiveness of the S^0 fertilizer was a measure of how much of it was oxidized to sulfate and was a function of its particle size; the size of particles proving "effective" in turn being dependent on the local climatic conditions and on the value placed on residual (longer term = 2–4 years) oxidation of the S^0. Table 3 summarizes some of the sizes of particles recommended or used in field or glasshouse studies. We contend that the rate constant technique offers a practical method of estimating suitable S^0 particle sizes in fertilizers. In New Zealand, dry S^0 as screened agricultural S is no longer commercially available.

2. Mixtures of S^0 with Superphosphate

The best-known S^0-containing fertilizers in New Zealand and Australia are mixtures of S^0 and superphosphate known as S-superphosphates, containing 20, 28, 33, 40, or 50% S (Table 4). Thus they also contain sulfate-S. In New Zealand they are manufactured by spraying molten elemental S onto the acidulating phosphate rock during the manufacture of the superphosphate. The S^0 component is fully integrated into the matrix of the superphosphate and is relatively finely divided, and the products are safe. These S-superphosphate products differ from their predecessors, which were simply mixtures of dry-screened S^0 and dry superphosphate [21]. The screened agricultural S shown in Table 4 represents the particle size distribution that would be expected in earlier S-superphosphates prepared by dry-mix methods.

The maximum S^0 content of commercial S-super mixtures is 44% (total S content 50%) (Table 4). There is a general relationship between the S^0 content and particle size of the S^0 in the fertilizer—as the S^0 content increases,the fineness of the S^0 particles declines [21]. However, the importance of this differential in size is lost when mixtures of different (less than 50% S) S contents are simply prepared by mixing 50% S-superphosphate with single superphosphate, since all of the S^0 particles are of moderately coarse size expected in the high S concentration product.

Table 3 S^0 Particle Size Recommendations for Pastures and Crops

	Site features	S rate	S^0 size	Reference
Pasture				
Legume based	Central South Island NZ, 500 mm rainfall, establishment	50 kg ha^{-1} for 2–3 years	$S^0 < 150$ μm or preferably sulfate	[71,79,80]
”	Central South Island NZ, 500 mm rainfall, maintenance	50 kg ha^{-1} for 2–3 years	$S^0 < 250$ μm	[7]
Sub clover/grass	West Australia, sand	45 kg ha^{-1}	S^0 210–250 μm	[16]
”	West Australia, soil	Up to 120 kg ha^{-1}	S^0 150–420 μm	[81]
White clover based pasture	NSW Australia, podzol	30 kg ha^{-1} for 2–3 years	$S^0 < 250$ μm	[73]
Sub clover/grass	California, gravelly loam	20 kg ha^{-1}	$S^0 < 300$ μm	[82]
”	California, 1000 mm rain, Mediterranean climate	45 kg ha^{-1} for 2 years	$S^0 < 300–600$ μm	[68]
”	N Oregon 1000 mm +	30–40 kg ha^{-1} annually 80 kg ha^{-1} biennially	$S^0 < 250$ μm annually Up to 1–2 mm biennially	[83]
Clover/grass	Australia extensive grazing	—	S^0 177–420 μm	[65]
”	”	20 kg ha^{-1}[a]	S^0 350–850 μm[b]	[72]

[a]Using commercial screened S^0 (cumulative % S^0 particle size: 4% < 150 μm, 17% < 250 μm, 40% < 450 μm, 61% < 1.0 mm, 87% < 2.8 mm, 13% >2.8 mm).
[b]Particle size range chosen for glasshouse experiment.

Table 4 S^0 Particle Size Distributions in Molten-Mixed S-Superphosphates Manufactured in New Zealand (cumulative % of S in different size ranges)

		Molten-mixed S-superphosphate products			
Screened agricultural S	S^0 fertilizer	18% S super	27% S super	33% S super	50% S super
Total-S%	100	18	27	33	50
SO_4-S%	—	9	9	8	6
S^0%	100	9	18	25	44
Size ranges of S^0 particles					
<2.0 mm	99	97	100	100	99
<1.0 mm	67	96	100	100	98
<500 μm	42	93	98	97	90
<250 μm	26	90	89	75	54
<150 μm	17	77	65	59	22
<75 μm	9	36	29	37	7
<38 μm	3	15	12	17	1

Source: Ref. 5.

Sulfur superphosphate mixtures are the preferred forms of S^0 fertilizer in New Zealand and Australia; they fit in with the mainly single superphosphate manufacturing industry, they are safe to handle and apply, and the mix of readily available and slow-release S suits their major use as pasture fertilizers. 50% S superphosphate has superseded screened agricultural S in supplementing the S content of fertilizer mixtures (e.g., with diammonium phosphate [= DAP; 18%N, 20%P]; reactive phosphate rock [=RPR; 13%P] a slow-release P fertilizer; partially acidulated rock phosphate [=PAR; 10%P, 8%S]; "long life" super, a product containing readily available and slowly available P) and, as already mentioned, it is used in mixtures with superphosphate to produce S-superphosphates with lower S contents.

Research with S-superphosphate mixtures has not been widely reported outside of Australia. However a concentrated (triple) superphosphate-S mixture prepared by granulating triple super with S^0 to form prills has been reported [67]. Experimental triple super-S^0 mixtures were investigated in cool temperate southern New Zealand [84]; the rate of release of sulfate was dependent on the size of the S^0 particles incorporated in the mixtures. Molten S (12% S) added in the manufacture of triple super were much coarser (maximum 25% < 250 μm) than particles mixed by a similar process in single superphosphate.

In West Africa, the S^0 in both a granulated triple superphosphate-S^0 mixture and dry S^0 <250 μm was at least as effective as sulfate-S and in some trials more effective [85]. In Australia S-triple superphosphate products consist of triple superphosphate granules coated with S^0; 10% and 20% S mixtures involve different thicknesses of S. Sulfate release from the 20% product is more prolonged than from the 10% mixture (B. Bull, personal communication). A glasshouse study [86] demonstrated that choice of adhesives affected the rate of release of sulfate from the fertilizers.

3. S-Bentonite Mixtures

Mixtures of S^0 and bentonite clay offer a method of applying S^0 in a high concentration (e.g., 70–90% S) of fine to moderately fine particle size in a safe-handling (dust-free) product. As with S-superphosphate, S-bentonite mixtures can be prepared as dry mixes of ground S^0 and clay, which are then granulated; or as wet mixes of molten S^0 and bentonite clay. The latter mixes may be granulated or flaked (=slated) or prilled by one of several methods [87–89]. When wetted the bentonite takes up water and the clay disperses (i.e., the prill breaks down) to release the S^0 for oxidation.

An example of the dry-mix process are granules described as "granulated ground (100% < 425 μm, 71% < 250 μm) S" [78,84,90]. In cool temperate southern New Zealand, there was a lag time of about 2 months before plant sulfate levels were significantly increased; however, they proved effective S fertilizers for pasture both in the year of application [74,78,91] and for a further 2 years [78].

Most commercially available S-bentonites are prepared as wet mixes; such products have been variously reported in field or laboratory evaluations as "Agrisul," "Dispersul," "Degrasul," "Solusul," and "Tiger 90." Most are prepared as 90% S and about 10% bentonite. In wet mix products the S^0 forms the structure of the prill, the expanding bentonite clay is mixed within it; when wetted, the bentonite expands and fractures the S^0 surrounding it. The extent of the shattering of the S^0 depends on the bentonite used, the proportion of bentonite in the mixture, and the homogeneity of the mixing. Sodium bentonite swells to 14 times its volume when wetted [87] and calcium bentonite swells to 1–2 times its volume [88]. Some manufacturers use calcium bentonite or a mixture of calcium and sodium bentonites. Sodium bentonite can be produced by treating a montmorillonite clay with sodium carbonate [89].

The choice of 10% bentonite in the mixture appears to be historical rather than based on agronomic effectiveness of the product. Boswell et al. [91] found 10% bentonite mixtures to lack homogeneity in their mixing and recommended that mixtures should contain 15–20% bentonite to ensure consistent rapid dispersion, and have fine S^0 particle size (e.g., 15% bentonite—73% $S^0 < 150$ μm, 95% $S^0 < 250$ μm; 20% bentonite—87% $S^0 < 150$ μm, 98% $S^0 < 250$ μm [87]). At these bentonite contents the oxidation of the fine S is rapid [77], and sulfate is quickly available to plants in temperate environments [91,92].

In pasture experiments, commercial (90% S) S-bentonite prills have always exhibited a lag period before measurable release of sulfate is observed. In some cases the lag period is a matter of months [92], but in others 2–3 years [93–95]. Most found 90% S-bentonite to be less effective than gypsum in the year of application [69,92,93,95]. Although variable between experiments (and between batches of a S-bentonite product [96]), the effectiveness of the S^0 fertilizers were better sustained into the second and third years; i.e., the S-bentonites had greater residual effects than sulfate fertilizers [69,92,93,95,97]. From these results it is clear that such S-bentonites are best suited to a pasture maintenance fertilizer role.

4. Sulfur Urea

S-urea (=sulfur-impregnated urea) is manufactured by injecting molten S^0 into liquid urea and prilling the melt [98]; it contains 36% N and 20% S. The S^0 is finely divided (<100 μm [98], see also Table 6), and homogeneously distributed through the urea. In a field experiment (ryegrass/white clover pasture) in which plant sulfate and total-S contents were used to indicate the relative rates at which sulfate was oxidized from three different S sources (applied at 25 kg S ha^{-1}), rapid S oxidation from the S-urea was reflected in elevated plant sulfate levels within 58 days of application to pasture (Table 5).

Solberg and Nyborg [54], in Canada, used S-urea mixtures (pellets 1–3 mm) containing 10 and 22% S with 100% of particles <50 μm, and found the urea-S to be oxidized equally as rapidly as fine powdery sublimed S^0 (100% particles in range 50–100 μm). The fertilizers were worked into the fallow soil to a depth of 10–15 cm. Recovery of S from S-urea as sulfate 1–2 months after application was at least 33% that from sulfate. This indicated that the fine S^0 in S-urea would overcome S-deficiency in cereal grains or rapeseed/canola, and probably establish pasture; however, to be equally as effective as sulfate in the establishment of the first-year crop, the S^0 fertilizers should be applied at least double the sulfate rate.

Subsequent Canadian studies have shown that over 5 years, the mean response to S in S-urea (rates not quantified) in both forage and cereal crops was 15.9% [99].

Table 5 Plant Sulfate (%) and Sulfur Uptake (kg S ha^{-1}) from Three Alternative S Fertilizers in Cool Temperate Southern New Zealand

Fertilizer	Plant SO_4-S				S Uptake
	23 days	58 days	79 days	174 days	0–174 days
Control (nil S)	0.113	0.138	0.157	0.145	23.03
S-urea	0.124	0.179	0.194	0.164	28.07
S-bentonite (85% S)[a]	0.123	0.153	0.178	0.177	24.74
S-bentonite (90% S)[b]	0.126	0.141	0.154	0.165	24.98
Gypsum	0.154	0.197	0.181	0.127	26.28
LSD	0.036	0.056	0.047	0.070	4.35

[a]Experimental product described in Ref. 87.
[b]Granulated ground S-bentonite granules prepared by Mintech NZ Ltd.
Source: Ref. 5.

5. S–rock Phosphate Mixtures

Two types of S–rock phosphate mixtures have been evaluated. Dry-ground rock phosphate and dry-ground S^0 have been granulated together in an attempt to produce "biosuper," and ground S^0 has been granulated with partially acidulated rock phosphate (PARP).

In biosuper the oxidation of S^0 by soil microorganisms to H_2SO_4 results in dissolution of the P from the rock phosphate [100–103], as well as release of sulfate. The mixture may also be inoculated with thiobacilli [103,104]. Comparison of reactive rock phosphate plus S^0 with S^0 alone has shown that a larger amount of S^0 was oxidized in the presence of rock phosphate than when S^0 was applied alone [105]. Biosuper is probably most applicable to tropical or subtropical climates [103–107]. However, the New Zealand field studies using reactive phosphate rocks [101] were conducted in a moist temperate environment. Tandon [107] believed that physical mixes of rock P and S^0 had possibilities but they were required to be tested extensively; he saw them being applied in advance of the nutrients being required, and requiring to be intimately mixed with the soil. The potential for biosuper has probably been understated by the use of unreactive, usually local rocks [100,103,104], sometimes on soils with moderately high pH [100].

The effectiveness of biosuper is dependent on such factors as fineness of particle size of rock phosphate and S^0 [100,108] and ratio of rock phosphate to S^0 [108]. Biosuper prepared by coating pelleted rock phosphate with S^0 and inoculated with thiobacilli at application required 3 years before release of P or S was apparent [109]; the rate of S^0 oxidation was no doubt affected by the low surface area/mass ratio.

As with S–rock phosphate, the major benefit of co-granulation of S^0 (10% w/w) with PARP (using unreactive rocks as the P source) is in the increased availability of soluble phosphate [110,111], especially with PARPs acidulated with sulfuric rather than phosphoric acid [111].

6. Pumice S

Pumice containing elemental S can be mined from natural lake bed sites in New Zealand (e.g., Rotokawa pumice—Table 6), and applied as a fertilizer "as is" (12–27% S), or enriched by a froth flotation of crushed raw material (50–92% S^0) before application. The pumice S is cited as an example of natural S^0 deposits. The froth flotation technique is also used in Egypt to refine natural elemental S deposits for agricultural use [112]; surface sulfur ore is crushed to pass a 250 mm sieve, so 100% of particles are <250 μm and the S^0 (96% S) is likely to be finer than the refined Rotokawa S shown in Table 6. In field evaluations, 12% S Rotokawa pumice S, applied at 30 kg S ha^{-1}, proved similar to S^0 < 150 μm with pasture responses of 10%, 12%, and nonsignificant 3%, respectively, in years 1–3 following application [113].

Table 6 Particle Size Distributions of S^0 in Fertilizer Products Evaluated in New Zealand (cumulative % S^0 in particle size ranges)

	S^0 Fertilizer					
	Screened agricultural S	12-mesh agricultural S	Popcorn S	Rotokawa pumice S[a]	Froth flotation pumice S[b]	S Urea
Size ranges						
<2.0 mm	99	100	82	91	100	100
<1.0 mm	67	98	36	77	99	100
<500 μm	42	61	14	62	95	100
<250 μm	26	22	6	45	88	100
<150 μm	17	10	3	33	67	97
<75 μm	9	—	1	13	36	36
<38 μm	3	—	—	3	—	5

[a]27% S.
[b]92% S.
Source: Ref. 5.

7. Sulfur Slurries and Sulfur Emulsion

Hagstrom [46] reported the availability of several slurry products, suspensions of S^0 in water, containing 50–74% S, with the particle size of S^0 typically 1–2 μm. The S^0 was rapidly oxidized to sulfate in soil. Micronized S applied as an emulsion in water (e.g., Thiovit)—80% S, with S^0 particle size 2–8 μm [114,115], has been used increasingly as a foliar-applied fertilizer [115,116]. McGrath and Till [117] demonstrated that the extremely fine S^0 was not taken directly into the foliage but was probably washed onto the soil, where it was rapidly oxidized to sulfate.

8. Mixtures of S^0 and Ammonium Sulfate

Two products (Sulgro 95 and Sulgro 68) are composed of 95% S^0 <74 μm and 60% S^0 <74 μm plus 40% ammonium sulfate, respectively [118], and have been designed to meet the agronomic requirements of crops in western Canada. Surface application followed by tillage to improve soil/fertilizer contact allowed rapid oxidation. The Sulgro 68 type of product is similar to New Zealand S-superphosphates with both a sulfate and a finely divided S^0 component and therefore could be expected to be a suitable maintenance fertilizer in temperate pastures.

Among the other S^0 fertilizer products listed in [5] are several that are not likely to have a significant role in pastures either because of costs of manufacture (anhydrite), problems with application (ammonium S), or because they are impractical because of their large particle size and consequently slow rate of oxidation (e.g., sulfur pellets, popcorn S, and S-coated urea).

VI. S FERTILIZERS FOR PASTURE ESTABLISHMENT AND MAINTENANCE

Fertilizer requirements for two separate pasture situations are recognized. Most fertilizer is applied to maintain production from existing pastures—this is known as maintenance fertilizer. More specialized fertilizer requirements may be required when new pastures are established (establishment fertilizer).

A. Establishment Fertilizer

Seedling pasture plants are similar to young annual crops; they have very small root systems and can take up only nutrients that are chemically and physically accessible to them. Thus chemically available sulfate-S is often chosen for pasture establishment, and the rate of application is often higher than that required to maintain pasture to ensure an adequate supply of the nutrient is in close proximity to the roots. In New Zealand, single superphosphate is the fertilizer most frequently sown at pasture establishment. In soils where S is the major or only deficiency, a S-superphosphate mixture (e.g., 20–28% S-superphosphate) has also proven suitable [71].

Some S fertilizers may cause damage to pasture seeds at sowing—e.g., if they are in direct contact with the seed. Seeds of legumes (e.g., clover) are more susceptible to fertilizer injury than grasses. In general, fertilizers with a content of >15% (N plus K) are likely to cause seed injury: ammonium sulfate, potassium sulfate, and mixtures of potassium chloride with superphosphate are all potentially injurious to seed; superphosphate may be a minor cause for concern, especially with clover seeds, and prolonged contact should be avoided. In many cases, establishment of intensively managed pastures is accomplished by drilling the seed and broadcasting the fertilizer on the soil surface in a single pass of the machinery. In extensive rangeland situations, seed and fertilizer are often mixed and applied from a fixed-wing aircraft; in these circumstances, the seed and fertilizer are in contact for a short period between mixing and application; this time should be kept to a minimum.

B. Maintenance S Fertilizers

In established pastures, immediacy of nutrient supply is not a priority. In fact, there are many situations where slow-release sulfur fertilizers offer a definite advantage—e.g., in severely leached soils due to high rainfall [38], or in coarse-textured soils [14,16,24], or where savings can be made from fertilizer application to extensively grazed pastures at intervals of 2–3 years [119]. In many soils where S deficiency occurs, pastures are based on legume-grass mixtures, which rely on the legume to fix atmospheric N for pasture needs. There is therefore no requirement for fertilizer N in such pastures. However, requirements for fertilizer P are general and for K occasional. In countries where a cut-and-carry system of pasture management may have been practiced for centuries K, deficiency can be widespread.

Data in Table 7 [74] serve to demonstrate differential effects of forms of S fertilizer on the establishment and initial 3 years' growth of lucerne. Sulfur was applied at sowing as sulfate fertilizers, a mixture of sulfate and S^0 fertilizer–S-superphosphate prepared by the molten sulfur mixing process (MMSS), and two forms of S^0—fine S^0 (94% < 295 μm) extracted from the S-superphosphate (S^0 extracted), and screened agricultural S^0 (see Table 4 for size distribution), as single dressings, at 15, 30, 60, and 120 kg S ha^{-1}. The two higher rates have been found to be adequate for legume

Table 7 Effects of Form of S Fertilizer on Lucerne Cover (%) and Plant Vigor (scored 0–10) 3-year-old Lucerne at Flowering; Dryland; 550 mm Rainfall; South Island, New Zealand

Rates of S applied kg/ha[a]	SO_4-Sfertilizer	MMS S	S° extracted	Screened S°	LSD
Cover (%)					
15	30	12	17	6	
30	52	65	35	31	
60	72	78	12	16	
120	83	92	53	29	27.2%
Mean	59	62	29	20	14.6%
Control (nil S)	6				
Plant vigor					
15	5.8	5.8	5.7	3.2	
30	8.8	7.7	6.5	5.2	
60	8.8	8.8	8.2	7.7	
120	9.2	9.7	8.8	8.3	2.72
Mean	8.3	8.0	7.3	6.1	1.36
Control (nil S)	3.1				

[a]Single spring application.
Source: Ref. 5.

establishment in this environment [71]. The percentage cover values provide a measure of the success of plant establishment, and plant vigor provides a measure of the residual benefits of the fertilizers.

At the two high rates of fertilizer, successful establishment resulted in high percentage cover values with the sulfate fertilizers and sulfur superphosphate (mean 81%), but restricted establishment with S^0: fine S^0 = 32%, coarse S^0 = 23% (control = 6%). In contrast, residual benefits of the S fertilizer forms were measured by vigor scores of 9.0, 9.2, 8.5, and 8.0, respectively, which did not differ significantly. The effectiveness of S-superphosphate for lucerne establishment was largely attributable to its sulfate content rather than the fine S^0 component. Thus, a ready supply of sulfate-S was necessary for establishment. All forms of S fertilizer were satisfactory for maintaining the pasture in this environment.

Pasture production data from trials establishing legumes at similar locations in the dry inland South Island New Zealand S-deficient environment [70,71] similarly show that sulfate fertilizers or sulfate mixed with fine S^0 (as S-superphosphate [71]) is required for successful plant establishment.

VII. TIMING AND FREQUENCY OF FERTILIZER S APPLICATIONS

The timing and frequency of fertilizer S applications to pasture is dependent on whether the pasture is cut or grazed, the intensity of the grazing system, the form of fertilizer chosen, the method of fertilizer spreading, and weather patterns.

The key issue with the timing of fertilizer S application is the need to limit losses of the mobile sulfate anion from the soil by leaching.

In the special case of pasture establishment, fertilizer S is required in an immediately available form at sowing, usually in spring or early autumn in New Zealand.

The timing of maintenance S fertilizer applications depends on the farming systems; replacement of S losses in intensively grazed dairy pastures, irrigated pastures, or in intensive cut-and-carry systems ideally occur at relatively frequent intervals, e.g., during the growing season, following harvesting in cut-and-carry systems. In less intensive systems, application time is less critical, especially where S^0 fertilizer is used. Application may be annual, again usually in spring, but timing may depend on the wetness of soil and its ability to support ground-spreading equipment. In much of New Zealand and Australian semi-intensive and extensive pastures, especially on hill country, fertilizer is applied by fixed-wing aircraft; ground surface conditions do not affect the timing. Sulfate fertilizers are preferably applied in spring; S^0 fertilizers may be applied in spring through to early autumn. Essentially the timing reflects a desire to boost production according to the normal seasonal pattern of pasture growth; in temperate New Zealand pastures, this typically involves a period of peak production in spring, reduced production in summer, and a lesser autumn peak in production.

In the most intensive pastoral systems, especially where leaching losses may be exacerbated—e.g., irrigated pastures under intensive dairying—maintenance topdressing may be required two to three times during a growing season. In other intensively grazed pastures, maintenance fertilizer application is generally annual. On less intensive systems where drier conditions restrict pasture production and consequently stocking rate, slow-release S fertilizers such as S-superphosphate mixtures, applied biennially or even at 3-year intervals, are effective for lucerne and clover-based pastures in dry inland New Zealand environments [7,21].

VIII. S FERTILIZER FOR LUCERNE PASTURES

Plant rooting depth can affect choice of S fertilizers. Most temperate pasture species are comparatively shallow-rooted. In contrast, lucerne is a deep-rooted species that is often grown as a specialist pasture in dryland, S-deficient, situations [7,29,75]. Occasionally dryland soils have reserves of sulfate accumulated in the soil below the rooting depths of the shallow-rooted plants [15,29], usually the consequence of incomplete leaching of S caused by restricted rainfall. Lucerne has the ability to take up sulfate from these reserves where grass/clover pasture cannot [29]. In these circumstances, shallow-rooted plants would be expected to respond to surface-applied S fertilizer whereas lucerne would not [29].

The ability of lucerne to take up sulfate over a large rooting depth in a soil profile with no evident accumulation of sulfate has been shown in New Zealand [75] and in Pennsylvania [120]. In the New Zealand experiment (annual rainfall 550 mm per year), the soil profile (0–1.2 m) contained little sulfate at the start of the experiment on a 3-year old lucerne stand; responses to applied forms and rates of sulfur were initially recorded only as elevated plant sulfate levels [75]. Yield responses developed with time; initially in the nil S control treatment in the second growing season after S fertilizer was applied [75], and subsequently in other treatments, indicating that leaching of sulfate through the depth of lucerne rooting took 2–3 years. With lucerne—under these low-rainfall, relatively slow-leaching conditions—sulfate is likely to be as effective as slow-release S^0 fertilizers. Under similar dry conditions in California, sulfate was similarly effective [67].

The amount of S required for pasture establishment is 40–50 kg S ha^{-1} for 2 to 3 years if the lucerne is harvested as a hay crop [71]. Maintenance fertilizer needs for a 4-year-old lucerne-based pasture in a dry (550 mm rainfall), mountainous, South Island New Zealand environment, under a cutting management, for 2 years, were 50 kg S ha^{-1} as superphosphate or S-superphosphate, or 100 kg S ha^{-1} as S^0 [7]. Maintenance requirements for grazed pastures in the same environment would be substantially less than under the above experimental management because there would be greater returns of nutrients from plant litter and (albeit uneven) in feces and urine. Grazing may also affect the rate of oxidation of S^0 [59,66]. Elevated plant sulfate levels derived from $^{35}S^0$ (25 kg S ha^{-1}, improved grass/clover pasture, irrigated, rotationally grazed at a high stocking rate, 25 sheep ha^{-1}) occurred 2–3 months earlier than in adjacent ungrazed plots (mown and clippings returned) [66]. In addition, the amount of residual S^0 in the soil 3 months after application was significantly lower in grazed than in ungrazed pasture [66].

IX. FACTORS AFFECTING CHOICE OF S FERTILIZER FORM

The choice of S fertilizers for any given pasture situation is governed first by the effectiveness of the fertilizers, second by costs, and third by their suitability with respect to the other nutrient needs of the pasture. The effectiveness of the fertilizer depends on the immediacy of requirement of sulfate, the requirement for other nutrients, and the need to restrict leaching losses. The important cost considerations are the relative costs of the S and other nutrients, and the concentration of nutrients that affect the cost of transport and application.

A. Availability of Sulfate

Immediate availability of sulfate is required at pasture establishment. Sulfate fertilizers should therefore be used. For maintenance of established pasture, S^0 fertilizers supply sulfate at a rate better suited to the requirements of the perennial pasture plants.

B. Other Nutrients

Most New Zealand pastures require S and P fertilizer inputs. S requirements are greater than P in dry soils. Increasing requirements for P relative to S follow the pattern of increasing rainfall and consequently greater soil development, weathering, and leaching. Relative S and P requirements thus change with soil groups [21], which can similarly be related to landform (topography—e.g., over relatively small distances in mountains and valleys) [119]. For example, in the North Canterbury area (South Island, New Zealand) shown diagrammatically in Figure 5, S is the major deficiency in the subhygrous yellow-gray earth soils in the 600–700 mm rainfall valley floor zone (with the exception of an irrigated area of plains soils where P requirements approach S requirements); progressively more P is required as soils change (with altitude and rainfall) from dry hygrous yellow-gray earths (700–800 mm), to yellow-gray–yellow-brown earth intergrades, and the lowland yellow-brown earths (800–1200 mm), to upland yellow-brown earths (1200–2400 mm). Phosphorus and S requirements are similar in the lowland yellow-brown earths. A summary of S fertilizers suitable for the different soil groups and the associated land forms is shown in Table 8.

Possible exceptions to the pattern are soils formed on limestone (rendzina soils). Under low-moderate rainfall (<800 mm), S is likely to be the only deficiency in rendzinas, which usually have high natural fertility, especially P and K. Sulfur deficiencies probably result from rapid leaching

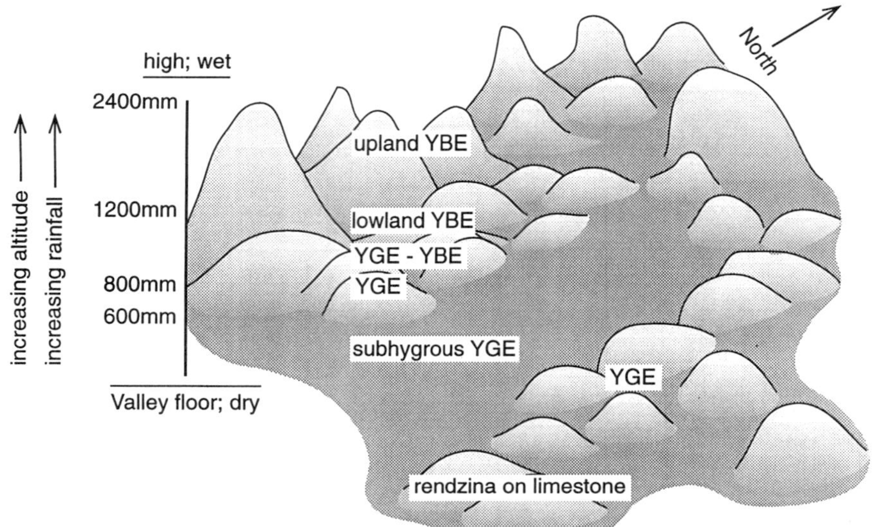

Figure 5 Stylized diagram of landscape in North Canterbury, New Zealand, showing relationship between landscape, soil groups and rainfall, used to determine choice of S fertilizers recorded in Table 8. (From Ref. 21.)

of sulfate from the often shallow (15–30 cm) topsoils into the parent rock directly beneath [21].

Recognition of the importance of P and nitrogen (N) requirements additional to S brings a need for a brief consideration of suitable P and N fertilizer forms for pastures. In New Zealand, P is mainly supplied in superphosphate or S-superphosphates. Where a high-concentration P fertilizer is required (usually in mixtures with high concentration S-superphosphate) DAP is favored for lower, drier soils and soils with high pH; reactive phosphate rock (RPR) may be practical as a maintenance fertilizer for wetter (>800 mm rainfall) soils with pH < 5.5. The acid soil conditions assist in the rate of chemical dissolution of the RPR. Similarly, in mixtures with S^0 fertilizer, the acidifying effect of oxidizing S^0 increases the rate of release of phosphate from the rock [105]. In the North Canterbury example, N is required for specific situations (e.g., dairying on irrigated soils); DAP is the preferred N fertilizer because of its high P content and its suitability for mixtures with 50% S-superphosphate.

Table 8 Fertilizer Forms for Different Land Forms and Pastures in North Canterbury, New Zealand

Soil group	Annual rainfall (mm)	Land form	S fertilizer
Plains			
Dry subhygrous YGE	600–700	Irrigated	20–28% S-superphosphate, DAP/50% S-superphosphate mixture
		Dryland pasture	20–50% S-superphosphate
		Dryland lucerne	Superphosphate, 20–50% S-superphosphate
Hills			
YGE	700–800	Low dry hills	20–50% S-superphosphate
YGE/YBE	800–1000	Rolling hills	Superphosphate, 20% S-superphosphate
Lowland YBE	1000–1200	Wetter hills	Superphosphate
Rendzina (on limestone)	600–800	Rolling hills	50% S-superphosphate
High Country			
Upland YBE	1200–2400	High country	20–28% S-superphosphate, RPR/50%S-superphosphate mixture

YGE = yellow-gray earth; DAP = diammonium phosphate; YBE = yellow-brown earth; RPR = reactive phosphate rock.

C. Restricting Leaching Losses and Application Frequency

In soils subject to leaching, slow-release S fertilizers are required to restrict sulfate losses. They are best suited for maintenance fertilizer applications. Annual applications are required on irrigated intensively farmed soils. In New Zealand high country, slow release of sulfate from S^0 over a 2–3-year period can be tailored to plant production needs while reducing the frequency of fertilizer application.

X. INFLUENCE OF FERTILIZER S ON THE AVAILABILITY OF OTHER NUTRIENTS FOR PLANTS AND ANIMALS

Fertilizer S application can alter the availability of other nutrients through soil acidification (lowering of soil pH). Acidification results from the addition of ammonium ions in ammonium sulfate and the oxidation of S^0 fertilizers in general. With S^0 fertilizers, one unit of S^0 has an acidifying effect that requires three units of lime to neutralize [13]. Acidification is generally limited at the rates of S^0 applied to most pastures. The effect on soil pH is likely to be greatest in poorly buffered soils (e.g., dryland, often coarse-textured soils with low organic matter contents), at high rates of S ($>$100 kg S ha^{-1}) and where oxidation of S^0 is rapid. The greatest degree of acidification reported in New Zealand pasture occurred in the development of a legume pasture in a rangeland (natural tussock grassland) situation where the topsoil (0–7.5 cm) contained little organic matter; the initial pH was 6.3, and the maximum reduction in soil pH over 3 years following the application of 100 kg S^0 ha^{-1} was 0.25 units [71]. However, some of the acidifying effect was due to the legume itself [71].

Reduction in soil pH is likely to decrease the availability of molybdenum, and possibly phosphorus (below pH 5.5, through increased concentration of soluble aluminum), and increase the availability of manganese, boron, zinc, and iron to plants [13]. In New Zealand pastures, the effect on molybdenum availability is likely to be of most relevance; however, it can be readily rectified by the application of sodium molybdate at 50 g ha^{-1} in the fertilizer mixture every 4–5 years [121].

S^0 is actually used to amend high pH soil conditions caused by either high salinity or, more especially, calcareous deposits. The H_2SO_4 product of S^0 oxidation dissolves calcium carbonate and may release other nutrients in the soil [66]. The amendment effects are most spectacular in semi-arid conditions such as in the high-pH Egyptian soils; application of very fine S^0 ($<$44 μm) at 200 kg S ha^{-1} increased available P levels in the soil, increased sulfate movement through the soil profile, and increased the rooting depth of lupins [52].

High plant sulfate levels and high plant molybdenum levels can induce copper deficiency in ruminants grazing pasture because copper becomes bound in insoluble complex salts such as copper thiomolybdates in the rumen and copper absorption is restricted [122]. High plant sulfate levels can be expected immediately following application of sulfate fertilizers because plants tend to take up sulfur in the sulfate form in the presence of excess of the nutrient [123]. However, small amounts of sulfate fertilizer prevent high concentrations of molybdenum in pasture [121]. In New Zealand soils, molybdenum-induced copper deficiency is likely to occur only if there is excessive or unnecessarily frequent application of molybdenum, or a combination of molybdenum application and liming [13].

Sulfur and selenium (Se) are similar chemically, and Australian studies [124,125] have indicated that the two elements can compete with each other in plant and animal systems. The trace element Se is essential for animals but not plants; Se deficiency has been shown to dramatically reduce sheep fertility and wool production in New Zealand [126]. In soils where Se deficiencies are expected, repeated S fertilizer applications may increase the likelihood of Se deficiencies in grazing animals [126]. Some of the effect is likely to be a consequence of changes in pasture either sown with the initial S fertilizer, or resulting from the fertilizer applications. For example, Se levels in browntop (*Agrostis tenuis* Sibth.) >ryegrass (*Lolium perenne* L.) >white clover (*Trifolium repens* L.) [127], and pasture improvement in much of New Zealand typically changes the species from browntop toward ryegrass and clover. Selenium deficiency can be overcome by treatment of the animals by injection or oral drench, or by including a slow-release Se prill in the fertilizer program (annually or biennially at the equivalent of 1 kg Se ha^{-1}).

XI. CONCLUSIONS

The sulfur requirements of grazed pastures are influenced strongly by the local climate and intensity of land use. Soil characteristics have a modifying effect, especially with respect to influencing leaching losses and acting as a reservoir for supplying plant-available sulfur.

Under intensive grazing, the cycling of S through the soil-pasture-animal system, although creating losses, is a crucial mechanism in mobilizing and reducing the system's dependence on high rates of fertilizer S.

To efficiently utilize S added to grazed pastures, choice of a suitable S fertilizer is of paramount importance. Fortunately a wide range of fertilizer S forms have been developed to cover a wide range of environmental conditions. In current pastoral farming practice in Australia, superphosphate is probably the most widely used fertilizer—a reflection of similar S and P nutrient maintenance requirements in developed pastures. The attention given to S^0 fertilizers in this chapter reflects an increasing interest in high S concentration fertilizers incorporating S^0, which offer the benefits of slow release and reduced potential leaching losses of sulfate while meeting plant requirements.

With regular use of S fertilizers to overcome or reduce S deficiencies in grazed pastures, both sulfate-S forms and elemental S forms may have an influence on other nutrients essential for plant and animal growth. Consultants involved in giving fertilizer advice need to be aware of these possible detrimental effects.

REFERENCES

1. S. Saggar, M. J. Hedley, A. G. Gillingham, J. S. Rowarth, S. Richardson, N. S. Bolan, and P. E. H. Gregg, Predicting the fate of fertiliser sulphur in grazed hill country pastures by modelling the transfer and accumulation of soil phosphorus, *N. Z. J. Agric. Res. 33*:129–138 (1990).
2. M. L. Nguyen and K. M. Goh, Nutrient cycling and losses based on mass-balance model in grazed pastures receiving long-term superphosphate applications in New Zealand 2. Sulphur. *J. Agric. Sci.* (*Camb.*) *119*:107–122 (1992).
3 G. J. Blair, A. R. Till, and C. C. Boswell, Rate of recycling of sulfur from urine, faeces and litter applied to the soil surface, *Aust. J. Soil Res. 32*: 543–554 (1994).
4. M. L. Nguyen and K. M. Goh, Chemical fractionation, water solubility and temporal distribution of sheep faecal sulfur fractions in grazed pastures receiving long-term sulphate fertilization, *Commun. Soil Sci. Plant Anal. 24(11 & 12)*:1087–1113 (1993).
5. C. C. Boswell and D. K. Friesen, Elemental sulfur fertilizers and their use on crops and pastures, *Fert. Res. 35*:127–149 (1993).
6. G. G. Cossens, Agriculture and climate in Central Otago, *Proc. N.Z. Grassld Assoc. 48*:15–21 (1987).
7. C. C. Boswell and B. Swaney, Comparison of the long-term effects of different sulphur fertilisers on legume pasture production on a toposequence of North Otago soils, New Zealand, *N. Z. J. Agric. Res. 44*:95–104 (1991).
8. R. L. Fox and G. J. Blair, Plant response to sulphur in tropical soils, *Sulphur in Agriculture* (M. A. Tabatabai, ed.), Agron. Monogr. 27, Am. Soc. Agron., Madison, Wisc., pp. 405–434 (1986).
9. B. T. Kang, E. Okoro, D. Acquaye, and O. A. Osiname, Sulfur status of some Nigerian soils from the savanna and forest zones, *Soil Sci. 132*: 220–227 (1981).
10. R. G. Hoeft and R. H. Fox, Plant response to sulphur in the midwest and north-eastern United States. *Sulphur in Agriculture* (M. A. Tabatabai, ed.), Agron. Monogr. 27, Am. Soc. Agron., Madison, Wisc., pp. 345–374 (1986).
11. E. J. Kamprath and U. S. Jones, Plant response to sulphur in the south-eastern United States. *Sulphur in Agriculture* (M. A. Tabatabai, ed.), Agron. Monogr. 27, Am. Soc. Agron., Madison, Wisc., pp. 323–343 (1986).
12. P. E. Rasmussen and P. O. Kresge, Plant responses to sulfur in the mid west and north-eastern United States. *Sulphur in Agriculture* (M. A. Tabatabai, ed.), Agron. Monogr. 27, Am. Soc. Agron., Madison, Wisc., pp. 357–374 (1986).

13. C. During and D. J. Martin, Sulphate nutrition, movement, and sorption, with special reference to a gley podzol, West Coast, South Island, *N. Z. J. Agric. Res. 11*:665–676 (1968).
14. M. R. J. Toxopeus, Sulphur deficiency in grassland on young soils from volcanic ash and its possible prediction by laboratory tests, *Proc. XI Int. Grassl Congr.*, pp. 345–346 (1970).
15. J. R. Bettany, H. H. Janzen, and J. W. B. Stewart, Sulphur deficiency in the prairie provinces of Canada. *Proceedings of the Sulphur 82 Conference* (A. I. More, ed.), The British Sulphur Corporation Limited London 2:787–799 (1983).
16. N. J. Barrow, Sulphate adsorption by soils as a determinant of sulphur status, and methods of supplying sulphur to non-adsorbing soils, *Proc. XI Int. Grassld Congr.*, pp. 370–373 (1970).
17. T. W. Walker and P. E. H. Gregg, The occurrence of sulphur deficiency in New Zealand, *Sulphur in Australasian Agriculture* (K. D. McLachlan, ed.), Sydney Univ. Press, Sydney, pp. 145–153 (1975).
18. S. F. Ledgard and M. P. Upsdell, Sulphur inputs in rainfall throughout New Zealand, *N. Z. J. Agric. Res. 34*:105–111 (1991).
19. C. C. Boswell, W. H. Risk, and R. J. Littlejohn, Atmospheric rainfall inputs in rainfall in southern New Zealand, *Proc. XVII Int. Grassld Congr. 1*: 667–668 (1993).
20. K. J. Syers, J. A. Adams, and T. W. Walker, Accumulation of organic matter in a chronosequence of soils developed from wind-blown sand in New Zealand, *J. Soil Sci. 21*:146–153 (1970).
21. C. C. Boswell, Fertiliser sulphur for North Canterbury pastures, *Proc. N.Z. Grassld Assoc. 56*:17–22 (1994).
22. R. A. Olson and G. W. Rehm, Sulfur in precipitation and irrigation waters and its effects on soils and plants, *Sulphur in Agriculture* (M. A. Tabatabai, ed.), Agron. Monogr. 27, Am. Soc. Agron., Madison, Wisc., pp. 279–293 (1986).
23. R. J. Morris, Sulphur—the fourth major plant nutrient, *Proceedings International Symposium on Sulphur for Korean Agriculture* (S. Jae-sung, ed.), Seoul, Korea, pp. 9–16 (1988).
24. M. G. Lambert, D. A. Clark, and A. D. Mackay, Long term effects of withholding phosphate application on North Island hill country: Ballantrae, *Proc. N.Z. Grassld Assoc. 51*:25–28 (1990).
25. M. B. O'Connor, C. E. Smart, and S. F. Ledgard, Long term effects of withholding phosphate application on North Island hill country: Te Kuiti. *Proc. N.Z. Grassld Assoc. 51*:21–24 (1990).
26. M. L. Nguyen and K. M. Goh, Factors affecting plant sulfur requirements in New Zealand soils and arable systems, *Proc. XVII Int. Grassld Congr. 2*: 1433–1436 (1993).
27. F. W. Smith and M. E. Siregar, Sulfur requirements of tropical forages, *Sulfur in SE Asian and S Pacific Agriculture* (G. J. Blair and A. R. Till, eds.), Indonesia, University of New England, pp. 76–86 (1983).

28. M. B. Jones, Sulfur availability indexes, *Sulphur in Agriculture* (M. A. Tabatabai, ed.), Agron. Monogr. 27, Am. Soc. Agron., Madison, Wisc., pp. 549–566 (1986).
29. L. C. Blakemore, T. E. Ludecke, M. L. Leamy, and A. J. Metson, A study of the pattern of sulphur responses by dryland lucerne on some brown-grey earths. *N. Z. J. Agric. Res. 12*:333–351 (1969).
30. C. H. Williams and A. Steinbergs, The evaluation of plant-available sulfur in soil. 11, The availability of adsorbed and soluble sulphates, *Plant and Soil 21*:279–294 (1964).
31. G. P. Sparling, The soil biomass, *Soil organic matter and biological activity: Development in plant and soil sciences* (D. Vaughn and R. E. Malcolm, eds.), Martinus Nijhoff/Dr W Junk Publishers 16:223–263 (1985).
32. A. G. Sinclair and W. H. M. Saunders, Sulphur, *Fertiliser and lime recommendations for pastures and crops in New Zealand, 2nd ed.* (I. S. Cornforth and A. G. Sinclair, eds.), Ministry of Agriculture and Fisheries, Wellington, pp. 15–17 (1984).
33. M. L. Nguyen and K. M. Goh, An overview of the New Zealand sulfur cycling model for recommending sulfur requirements in grazed pastures. *N. Z. J. Agric. Res. 36*:475–491 (1993).
34. M. L. Nguyen and K. M. Goh, Sulfur mineralization and release of soluble organic sulfur from camp and non-camp soils of grazed pastures receiving long-term superphosphate applications, *Biol. Fertil. Soils 14*:272–279 (1992).
35. A. Ghani, R. G. McLaren, and R. S. Swift, Sulfur mineralization and transformations in soils as influenced by additions of carbon nitrogen and sulfur, *Soil Biol. Biochem. 24*:331–341 (1992).
36. A. G. O'Donnell, J. Wu, and J. K. Syers, Sulphate-S amendments in soil and their effects on the transformation of soil sulfur, *Soil Biol. Biochem. 26*: 1507–1514 (1994).
37. A. G. Sinclair and P. D. Enright, An auto-analyser method for the determination of extractable soil sulphate using automated purification of extracts, *N. Z. J. Sci. 25*:141–146 (1982).
38. A. G. Sinclair, C. C. Boswell, I. S. Cornforth, A. Lee, J. Morgan, M. L. Nguyen, W. H. M. Saunders, P. Shannon, R. G. Smith, and G. Whelan, Agronomic requirements for sulphur in New Zealand pastures, *Proceedings of the 20th technical conference of the New Zealand Fertiliser Manufacturers' Association 2*:538–572 (1986).
39. A. H. C. Roberts, T. H. Webb, J. D. Morton, M. B. O'Connor, and D. C. Edmeades, Building a solid foundation—sulfur, phosphorus and potassium requirements for the sedimentary soils of North Canterbury, *Proc. N.Z. Grassld Assoc. 56*:7–12 (1994).
40. J. H. Watkinson and K. W. Perrott, A new soil test for sulphate and mineralisable organic sulphur. *Proceedings of the New Zealand Fertiliser Manufacturers' Research Association Conference 1990*, pp. 188–198 (1990).

41. K. Spencer, Sulphur requirements of plants, *Sulphur in Australasian Agriculture* (K. D. McLachlan, ed.), Sydney Univ. Press, Sydney, pp. 98–108 (1975).
42. W. H. M. Saunders and D. M. Cooper, Evaluation of soil and plant analysis to determine sulphur status of New Zealand soils, *Sulphur in Australasian Agriculture* (K. D. McLachlan, ed.), Sydney Univ. Press, Sydney, pp. 188–195 (1975).
43. A. J. Metson, Sulfur in forage crops. Plant analysis as a guide to the sulfur status of forage grasses and legumes, *The Sulfur Institute Technical Bulletin 20*:24 pp (1973).
44. C. S. Andrew, The effect of sulphur on the growth, sulphur and nitrogen concentrations of some tropical and temperate legumes, *Aust. J. Agric. Res. 28*:807–820 (1977).
45. A. G. Sinclair, W. H. Risk, L. C. Smith, J. D. Morrison, and K. G. Dodds, Sulphur and phosphorus in balanced pasture nutrition. *Proc. N.Z. Grassld Assoc. 56*:13–16 (1994).
46. G. R. Hagstrom, Fertilizer sources of sulphur and their use, *Sulphur in Agriculture* (M. A. Tabatabai, ed.), Agron. Monogr. 27, Am. Soc. Agron., Madison, Wisc., pp. 567–581 (1986).
47. B. Palmer, M. McCaskill, D. K. Friesen, and L. Hammond, Sulfur containing fertilizers, past, present and future, *Sulfur in SE Asian and S Pacific Agriculture.* (G. J. Blair and A. R. Till, eds.), Indonesia, University of New England, pp. 301–314 (1983).
48. M. R. McCaskill and G. J. Blair, A model for the release of sulfur from elemental S and superphosphate, *Fert. Res. 19*:77–84 (1989).
49. R. L. Fox, R. A. Olson, and H. F. Rhoades, Evaluating the sulfur status of soils by plant and soil tests, *Soil Sci. Soc. Amer. Proc. 28*:243–246 (1964).
50. P. Li and A. C. Caldwell, The oxidation of elemental sulfur in soil, *Soil Sci. Soc. Amer. Proc. 30*:370–372 (1966).
51. J. R. Bettany and H. H. Janzen, Transformations of sulphur fertilizers in prairie soils, *Proceedings of Sulphur-84,* Sulphur Development Institute of Canada, Calgary, Canada, pp. 817–822 (1984).
52. M. H. Hilal, A. Abdel-Fattah, and S. A. Korkor, Effect of fine and granular sulphur application on root depth and yield of lupinus in sandy soils, *Proceedings Middle East Sulphur Symposium* (M. H. Hilal, S. A. Korkor, and A. E. M. Selim, eds.), Cairo, Egypt, pp. 207–216 (1990).
53. H. P. Rothbaum and P. S. Groom, Fire hazards in the use of fertilisers containing elemental sulphur, *N. Z. J. Sci. 4*:476–488 (1961).
54. E. D. Solberg and M. N. Nyborg, Comparison of sulphate and elemental sulphur fertilizers, *Proceedings of the Sulphur 82 Conference* (A. I. More, ed.), The British Sulphur Corporation Limited London 2:843–852 (1983).
55. J. H. Watkinson and A. Lee, Kinetics of field oxidation of elemental sulfur in New Zealand pastoral soils and the effects of soil temperature and moisture, *Fert. Res. 37*:59–68 (1993).

56. J. H. Watkinson, A. Lee, and D. R. Lauren, Measurement of elemental sulfur in soil and sediments: Field sampling, sample storage, pretreatment, extraction and analysis by high-performance liquid chromatography, *Aust. J. Soil Res. 25*:167–178 (1987).
57. J. H. Watkinson, Measurement of the oxidation rate of elemental sulfur in soil, *Aust. J. Soil Res. 27*:365–375 (1989).
58. J. H. Watkinson and G. J. Blair, Modelling the oxidation of elemental sulphur in soils, *Fert. Res. 35*:139–150 (1993).
59. A. Ghani, A. Lee, J. H. Watkinson, C. C. Boswell, B. S. Thorrold, J. D. Morton, and M. P. Upsdell, The quantification of oxidation of elemental sulphur in pastoral soils of New Zealand, *Proc. XVII Int. Grassld Congr. 2*: 1465–1467 (1993).
60. C. C. Boswell, B. S. Thorrold, J. H. Watkinson, A. Lee, and I. L. Power, Effect of soil temperature on the rate of oxidation of elemental sulphur, *Proc. N.Z. Grassld Assoc. 54*:49–53 (1992).
61. J. J. Germida and H. H. Janzen, Factors affecting the oxidation of elemental sulphur in soils, *Fert. Res. 35*:101–114 (1993).
62. J. J. Germida, J. R. Lawrence, and V. V. S. R. Gupta, Microbial oxidation of sulphur in Saskatchewan soils, *Proceedings of Sulphur-84*, Sulphur Development Institute of Canada, Calgary, Canada, pp. 703–710 (1984).
63. G. J. Blair, R. D. B. Lefroy, M. Dana, and G. C. Anderson, Modelling of sulfur oxidation from elemental sulfur, *Plant and Soil 155/156*:379–382 (1993).
64. J. D. Beaton, Sulphur requirements of cereals, tree crops, vegetables, and other crops, *Soil Sci. 101*:267–282 (1966).
65. A. Johnson, Prescription of the types of sulphur fertilizers currently needed in agriculture, *Sulphur in Australasian Agriculture* (K. D. McLachlan, ed.), Sydney Univ. Press, Sydney, pp. 242–248 (1975).
66. K. M. Goh and M. L. Nguyen, Effects of grazing animals on the plant availability of sulphur fertilisers in grazed pastures, *Proc. N.Z. Grassld Assoc. 52*:181–185 (1990).
67. M. B. Jones, Effects of fertilizer sources on plant growth and sulphur uptake, *Proceedings of Sulphur-84*, Sulphur Development Institute of Canada, Calgary, Canada, pp. 711–716 (1984).
68. M. B. Jones and J. E. Ruckman, Effect of particle size on long-term availability of sulfur on annual type grasslands, *Agron. J. 61*:936–939 (1969).
69. M. B. Jones, W. E. Martin, and J. E. Ruckman, Effectiveness of various sulfur sources applied to annual type grasslands of California, *Proc. XI Int. Grassld Congr.*, pp. 373–376 (1970).
70. T. E. Ludecke, Further aspects of sulphur nutrition of legumes, *Proc. N.Z. Grassld Assoc. 27*:129–138 (1966).
71. P. D. McIntosh, A. G. Sinclair, and P. D. Enright, Responses of legumes to phosphorus and sulphur on 2 toposequences of North Otago soils, New Zealand, *N. Z. J. Agric. Res. 28*:505–515 (1985).

72. A. R. Till, P. W. G. Sale, and G. J. Blair, Studies of the forms of fertilizer on plant and animal responses, *Proceedings of Sulphur-84,* Sulphur Development Institute of Canada, Calgary, Canada, pp. 753–758 (1984).
73. R. G. Weir, B. Barkus, and W. T. Atkinson, The effect of particle size on the availability of brimstone sulphur to white clover, *Aust. J. Exp. Agric. Anim. Husb. 3*:314–318 (1963).
74. C. C. Boswell and B. Swanney, Alternative sulphur fertilisers in New Zealand, *Proc. N.Z. Grassld Assoc. 47*:233–242 (1986).
75. C. C. Boswell, Lucerne responses to elemental sulphur. *N. Z. J. Agric. Res. 40*:283–295 (1997).
76. F. E. Koehler and S. Roberts, An evaluation of different forms of sulphur fertilizers, *Proceedings of the Sulphur 82 Conference* (A. I. More, ed.), The British Sulphur Corporation Limited London 2:833–841 (1983).
77. M. L. Nguyen, Direct determination of elemental sulphur in soils and the application of this method in measuring oxidation potentials of elemental sulphur fertilisers, *Proceedings of the workshop towards the more efficient use of soil and fertilizer sulphur* (R. E. White and L. D. Currie, eds.), Massey University, Palmerston North, pp. 9–25 (1988).
78. B. Swanney, C. C. Boswell, P. D. Enright, and A. G. Sinclair, A comparative field evaluation of some potential sulphur fertiliser materials, *N. Z. J. Exper. Agric. 16*:183–192 (1988).
79. T. E. Ludecke, Formulation of a rational fertilizer programme in tussock country, *Proc. N.Z. Grassld Assoc. 24*:29–41 (1963).
80. P. D. McIntosh, A. G. Sinclair, and P. D. Enright, Performance of elemental sulphur, sulphur-superphosphate and superphosphate on dry inland South Island hill country, *Proc. N.Z. Grassld Assoc. 44*:172–178 (1983).
81. N. J. Barrow, Slowly available fertilizers in south-western Australia. 1. Elemental sulphur, *Aust. J. Exp. Agric. Anim. Husb. 11*:211–216 (1971).
82. M. B. Jones and J. E. Ruckman, Gypsum and elemental sulphur as fertilizers on annual grassland, *Agron. J. 58*:409–412 (1966).
83. G. Kiemnec, T. L. Jackson, and W. Mosher, Fertilizing subclover with elemental sulphur, *Sulphur in Agriculture 5*:12–16 (1981).
84. C. C. Boswell, B. Swanney, and A. C. Braithwaite, The effect of different manufacturing techniques on the availability of sulfate to pasture from mixtures of elemental sulfur with either triple superphosphate or partially acidulated reactive phosphate rock, *Fert. Res. 44*:255–263 (1996).
85. D. K. Friesen, Fate and efficiency and sulfur fertilizer applied to food crops in West Africa, *Fert. Res. 29*:45–54 (1991).
86. G. J. Blair, M. Dana, and R. Lefroy, A glasshouse evaluation of sulfur fertilizer sources for crops and pastures. 2, A comparison of sulfur coated triple superphosphates and gypsum, *Aust. J. Agric. Res. 45*:1517–1523 (1994).
87. C. C. Boswell, W. R. Owers, B. Swanney, and H. P. Rothbaum, Sulfur/sodium bentonite mixtures as sulfur fertilizers. 1. The effects of S/Na-bentonite ratios on the rate of dispersion and particle size distribution of

elemental sulfur dispersed from laboratory-produced prills, *Fert. Res. 15*: 13–31 (1988).
88. R. H. Czurak, Sulphur bentonite products, *Proceedings of Sulphur-84*, Sulphur Development Institute of Canada, Calgary, Canada, pp. 833–834 (1984).
89. H. P. Rothbaum, W. Evason, and W. R. Owers, Development of new sulphur fertilisers, *Proceedings of the technical workshop on sulphur in New Zealand agriculture*, Massey University, N.Z. (P. E. H. Gregg and J. K. Syers, eds.), pp. 139–149 (1983).
90. A. G. Sinclair and P. D. Enright, Effectiveness of elemental sulphur fertilisers, *Proceedings of the technical workshop on sulphur in New Zealand agriculture*, Massey University, N.Z. (P. E. H. Gregg and J. K. Syers, eds.), pp. 164–174 (1983).
91. C. C. Boswell, B. Swanney, and W. R. Owers, Sulfur/sodium bentonite prills as sulfur fertilizers. 2. Effect of sulfur-sodium bentonite ratios on the availability of sulfur to pasture plants in the field, *Fert. Res. 15*:33–46 (1988).
92. R. G. Smith and D. B. McDougall, Evaluation of sulphur/sodium bentonite prills as sulphur fertilisers in the Manawatu, *Proceedings of the workshop towards the more efficient use of soil and fertilizer sulphur* (R. E. White and L. D. Currie, eds.), Massey University, Palmerston North, pp. 169–178 (1988).
93. R. G. Hoeft and L. M. Walsh, Effect of carrier, rate and time of application of S on the yield, and S and N content of alfalfa, *Agron. J. 67*:427–430 (1975).
94. E. J. Kamprath and U. S. Jones. Plant response to sulphur in the southeastern United States, *Sulphur in Agriculture* (M. A. Tabatabai, ed.), Agron. Monogr. 27, Am. Soc. Agron., Madison, Wisc., pp. 323–343 (1986).
95. J. E. Matocha, Influence of sulphur sources and magnesium on forage yields of coastal bermudagrass (*Cynodon dactylon* [L] Pers), *Agron. J. 63*:493–496 (1971).
96. W. F. Nuttall, C. C. Boswell, and B. Swanney, Influence of sulphur fertilizer placement, soil moisture and temperature on yield response of rape to sulphur-bentonite, *Fert. Res. 25*:107–114 (1990).
97. B. S. Dear and K. Spencer, The Effectiveness of several fertilisers in supplying sulphur to temperate pastures, *Aust. J. Exp. Agric. Anim. Husb. 22*: 412–419 (1982).
98. E. A. Kowalenko and R. S. Baptie, Sulphur in agriculture. *Proceedings of Sulphur-84*, Sulphur Development Institute of Canada, Calgary, Canada, pp. 835–839 (1984).
99. J. D. Beaton, Desirable characteristics of sulphur containing fertilizers suitable for Asia and the Pacific region, *Proceedings of the symposium on fertilizer sulphur requirements and sources in developing countries of Asia and the Pacific. Bangkok*, pp. 101–111 (1987).
100. H. A. Kittams and O. J. Attoe, Availability of phosphorus in rock phosphate–sulfur fusions, *Agron. J. 57*:331–334 (1965).

101. S. S. S. Rajan, Phosphate rock and phosphate rock/sulphur granules as phosphate fertilizers and their dissolution in soil, *Fert. Res. 11*:43–60 (1987).
102. S. S. S. Rajan and E. A. Edge, Dissolution of granulated low-grade phosphate rocks, phosphate rocks/sulphur (Biosuper), and superphosphate in soil, *N. Z. J. Agric. Res. 23*:451–456 (1980).
103. N. M. Scott and G. Sharp, Crop responses to sulphur and sulphur research in the north of Scotland, *Proceedings of Sulphur-84*, Sulphur Development Institute of Canada, Calgary, Canada, pp. 793–798 (1984).
104. L. S. S. Pathiratna, U. P. D. Waidyanatha, and O. S. Peries, The effect of apatite and elemental sulphur mixtures on growth and P content of *Centrosema pubescens, Fert. Res. 21*:37–43 (1989).
105. A. Lee, J. H. Watkinson, G. Orbell, J. Bagyaraj, and D. R. Lauren, Factors influencing dissolution of phosphate rock and oxidation of elemental sulphur in some New Zealand soils, *N. Z. J. Agric. Res. 30*:373–385 (1987).
106. I. J. Partridge, The efficiency of biosuper and rock phosphate on forage legumes on hill soils in Fiji, *Fiji Agric. J. 44*:27–30 (1982).
107. H. L. S. Tandon, Sulphur containing fertilizers, *Proceedings of the symposium on fertilizer sulphur requirements and sources in developing countries of Asia and the Pacific. Bangkok,* pp. 95–100 (1987).
108. S. S. S. Rajan, Effect of sulphur content of phosphate rock/sulphur granules on the availability of phosphate to plants. *Fert. Res. 4*:287–296 (1983).
109. G. B. Robbins, J. J. Bushell, and J. F. Compton, Nutrient concentrations of a grass-legume pasture fertilized with single superphosphate or biosuper. *Trop. Grasslds 18*:100–102 (1984).
110. D. K. Friesen, P. W. G. Sale, and G. J. Blair, Long-term greenhouse evaluation of partially acidulated phosphate rock fertilizers. 2. Effect of cogranulation with elemental S on availability of P from two phosphate rocks, *Fert. Res. 13*:45–54 (1987).
111. P. W. G. Sale, D. K. Friesen, and G. J. Blair, Long-term greenhouse evaluation on partially acidulated phosphate rock fertilizers, 3. Agronomic effectiveness relative to single superphosphate. *Fert. Res. 29*:295–307 (1991).
112. S. F. Estefan, S. A. Korkor, and M. H. Hilal, Prefeasibility study on pilot plant flotation of sulphur ores and wastes, *Proceedings Middle East Sulphur Symposium* (M. H. Hilal, S. A. Korkor, and A. E. M. Selim, eds.), Cairo, Egypt, pp. 101–107 (1990).
113. M. Bell-Booth, R. G. Smith, W. Chatupote, and M. J. Hedley. The agronomic utilsation of hydrothermal sulphur deposits at Lake Rotokawa, *Proceedings of the workshop towards the more efficient use of soil and fertilizer sulphur* (R. E. White and L. D. Currie, eds.), Massey University, Palmerston North, pp. 149–158 (1988).
114. P. Dutruel, Evolution of sulphur presentation forms. *Proceedings of International Symposium Elemental Sulphur in Agriculture, Nice, France 1*: 383–384 (1987).

115. J. Waples and C. Sinclair, The use of 'Thiovit' on major arable crops in the UK, *Proceedings of International Symposium Elemental Sulphur in Agriculture, Nice, France 1*:355–360 (1987).
116. D. Martin, J. L. Moyart, and J. L. Gerard, Effects of fungicides on rapeseed mycoflora. Role of sulphur treatments, *Proceedings of International Symposium Elemental Sulphur in Agriculture, Nice, France 1*:371–375 (1987).
117. S. P. McGrath and A. R. Till, Uptake of sulphur following foliar applications of sulphur, *Proceedings of International Symposium Elemental Sulphur in Agriculture, Nice, France 1*:369–370 (1987).
118. E. D. Solberg, D. C. Penney, and M. N. Nyborg, Factors affecting the effective use of elemental sulphur fertilizers in Western Canada, *Proceedings of international symposium on the role of sulphur, magnesium and micronutrients in balanced plant nutrition* (S. Portch, ed.), Potash and Phosphate Institute, Hong Kong, pp. 222–233 (1992).
119. M. J. S. Floate, C. C. Boswell, B. E. Allen, and D. Scott, Towards 2000: Strategies to achieve more efficient fertiliser use in inland South Island hill and high country, *Proc. N.Z. Grassld Assoc. 48*:53–58 (1987).
120. R. G. Hoeft, Crop response to sulphur in the midwest and northeastern United States, *Sulphur in Agriculture 4*:12–15 (1980).
121. C. G. Sherrell and A. K. Metherell, Trace elements. *Fertiliser and lime recommendations for pastures and crops in New Zealand, 2nd ed.* (I. S. Cornforth and A. G. Sinclair, eds.), Ministry of Agriculture and Fisheries, Wellington, pp. 20–21 (1984).
122. N. D. Grace, Sulphur, *The mineral requirements of grazing ruminants* (N. D. Grace, ed.), New Zealand Society of Animal Production, Occasional Publication No. 9, pp. 67–75 (1983).
123. D. Bouma, The uptake and translocation of sulphur in plants, *Sulphur in Australasian Agriculture* (K. D. McLachlan, ed.), Sydney Univ. Press, Sydney, pp. 79–86 (1975).
124. J. E. Prattley, J. D. McFarlane, The effect of sulphate on the selenium content of pasture plants, *Aust. J. Exp. Agric. Anim. Husb. 14*:533–538 (1974).
125. A. L. Pope, R. J. Moir, M. Somers, E. J. Underwood, C. L. White, The effect of sulphur on ^{75}Se absorption and retention in sheep, *J. Nutr. 109*:1448–1454 (1979).
126. K. R. Millar, Selenium, *The mineral requirements of grazing ruminants* (N. D. Grace, ed.), N. Z. Soc. Anim. Prod., Occasional Publication No 9, pp. 38–47 (1983).
127. E. B. Davies and J. H. Watkinson, Uptake of native and applied selenium by pasture species: 1. Uptake of Se by browntop, ryegrass, cocksfoot and white clover from Atiamuri sand, *N. Z. J. Agric. Res. 9*:317–327 (1966).

5

Modeling the Rate of Elemental Sulfur Oxidation in Soils

John H. Watkinson
Ruakura Research Centre, Hamilton, New Zealand

Nanthi S. Bolan
Massey University, Palmerston North, New Zealand

I. INTRODUCTION

Plant roots take up sulfur (S) from the soil solution as sulfate. However, soluble and adsorbed sulfate makes up less than 10% of the total soil S. Because primary sulfide minerals are oxidized to sulfate early in soil formation, little inorganic S of lower oxidation state, such as sulfite, thiosulfate, and elemental S (S^0), remains or persists from later reactions. Further, some of this sulfate is lost through leaching, and some is retained against leaching after conversion to relatively insoluble organic compounds through biological activity. Over 90% of the total soil S, especially in humid regions, is in the organic form [1,2]. A small proportion is continually mineralized by soil microbes, which replenishes the sulfate removed by plants. Where this is insufficient for plant growth, additional S is added as fertilizer.

Traditionally S is added to soil through the application of gypsum or single superphosphate. Increases in the cost of fertilizer transportation over the past decade or two have made high-analysis forms of both phosphorus (P) and S fertilizers more attractive to manufacture and distribute. High S analysis is achieved through incorporating elemental S (S^0) in place of sulfate. Also, unlike the sulfate S in gypsum or single superphosphate, S^0 is not susceptible to losses by leaching [3].

In New Zealand, research has indicated that S^0 can be very suitable as a pastoral fertilizer, provided appropriate particle sizes are used [3]. Also S^0 has been established as an important amendment in developing desert agrosystems [4]. These uses require oxidation of S^0 to sulfate.

There is increasing research to quantify, including accurate analytical methods, and model the rate of S^0 oxidation in soils and to elucidate the factors affecting it. This will enable us to improve identification of sites, and sizes of S^0 particles suitable for S^0 application, and to improve economic fertilizer recommendations. In this chapter the mechanism of S^0 oxidation and related factors are discussed in relation to the modeling of the S^0 oxidation kinetics.

II. MODES OF S^0 OXIDATION

In soils, S^0 and other reduced forms of S are oxidized to sulfate S through both chemical (abiotic) and microbiological (biotic) processes [5,6].

A. Abiotic Oxidation

Abiotic oxidation of S^0 to sulfate is extremely slow at ambient temperature in any medium, including autoclaved soils [6–8]. In industry, oxidation of S^0 to sulfuric acid necessitates heating S^0 at high temperature (1200–1600°C) in the presence of a catalyst.

B. Microbial Oxidation

In contrast to abiotic oxidation, microbial oxidation of S^0 in soils occurs at ambient temperature and is therefore the dominant oxidative process [6,7].

Although a wide spectrum of S^0-oxidizing microorganisms have been identified, chemoautotrophic bacteria and heterotrophic fungi and actinomycetes (Table 1) play the major role in the oxidation of S^0 in most agricultural soils [6].

1. Autotrophs

The autotrophs gain energy from the oxidation of S^0 and use CO_2 as their carbon source. Among the autotrophs, the obligate chemolithotrophs of

Table 1 Some S°-Oxidizing Microorganisms

Autotrophs		
Thiobacillus thiooxidans		
T. ferrooxidans		
T. denitrificans		
Heterotrophs		
Actinomycetes	Bacteria	Fungi
Streptomyces sp	*Achromobacter* sp	*Absidia glauca*
	Arthrobacter sp	*Fusarium episphaeria*
	Bacillus sp	*Penicillium decumbens*
	Micrococcus sp	

genus *Thiobacillus* have been considered the main agents of S^0 oxidation in Australian soils [9]. More recent studies have indicated that the thiobacilli are rare in soils of western Canada compared to the heterotrophic oxidizers [10,11]. In New Zealand soils, the numbers of thiobacilli are naturally low but increase rapidly and substantially after the addition of S^0 to soil [12]. Because most thiobacilli are facultative or obligate chemolithotrophs, they are able to oxidize S independently of the supply of available carbon.

2. Heterotrophs

It has long been recognized that soil fungi and some heterotrophic bacteria including actinomycetes can oxidize S^0 [13,14]. In contrast to autotrophs, the populations of heterotrophic S-oxidizers are abundant in soil. However, rates of S^0 oxidation by heterotrophs in soils are limited by levels of available carbon present to support growth [15,16]. Using in vitro studies, Germida [17] identified two physiological functional groups of heterotrophic S^0 oxidizers: rapid and complete oxidizers (e.g., fungi) and slow and incomplete oxidizers (e.g., bacteria and actinomycetes).

There is some debate over the relative importance of autotrophs and heterotrophs in the oxidation of S^0 in soils. In the absence of methods suitable for selective inhibition of heterotrophic S^0 oxidizers in soils, it is unlikely that this question will be readily settled. It is more likely, however, that mixed populations of both microorganisms are responsible for the S^0 oxidation in most soils [6,9].

III. METHODS FOR MEASURING S^0 OXIDATION

The rate and extent of S^0 oxidation is estimated either directly, by measuring the amount of residual S^0 in soils, or indirectly, by measuring the increases in the amount of oxidation products, mainly sulfuric acid.

A. Residual S^0

Measurement of residual S^0 involves two steps: (1) extraction of S^0 from soil and (2) analysis of the S^0 in the extract.

1. Extraction of S^0

Various organic solvents have been used successfully to extract S^0 from soils and fertilizers, including acetone [18], benzene [19], carbon disulfide [20], chloroform [21,22], and toluene [23]. Because of their toxicity, benzene, carbon disulfide, and toluene are not suitable for routine analysis. The efficiency of extraction of S^0 by these solvents depends on the moisture content of the soil, the miscibility of the solvent in water, and the allotropic form of S^0. For example, because S^0 is much less soluble in water/acetone mixture than in acetone alone [24], acetone is less efficient in removing S^0 from soils with moisture contents above 75% of field capacity (FC) [18]. Chloroform is only effective as an extractant at moisture contents outside the plastic range of soil [22, 25]. Polymeric S^0 is insoluble in most solvents, including acetone (Kear and Watkinson unpublished) and chloroform (7 mg l^{-1}) [21], but is present in fertilizer S^0 at no more than 5% [22].

2. Analysis of S^0

Colorimetric, titrimetric, and chromatographic methods have been used to measure S^0 extracted from soils by various solvents.

a. Colorimetric and Titrimetric Methods. Several colorimetric methods have been developed for the measurement of S^0 over the years, the more recent being the ferric thiocyanate [20,26,27] (low sensitivity) and methylene blue [19] methods. In the latter method and the titrimetric method of Barrow [21], S^0 is reduced to H_2S by Raney Nickel and by iron powder in HCl, respectively. The H_2S is then distilled and reacted to form methylene blue [18,19] or titrated with mercuric chloride [21].

b. Chromatographic Methods. Direct gas chromatography is unsatisfactory because S^0 does not chromatograph homogeneously [23,28] and the linear range of detection is very narrow [29]. Prior reduction to H_2S

achieves a good separation [30] but removes specificity and increases the time involved.

Watkinson et al. [22] developed a high performance liquid chromatography technique that is specific for S^0, and accurate and precise down to a detection limit of 0.1 mg kg^{-1} soil. Because of its specificity and sensitivity, it can measure naturally occurring S^0, which is of particular value in relation to the persistence and numbers of natural populations of thiobacilli [31]. Also, it is significantly more accurate and precise at low concentration of added S^0, because other methods additionally measure soil organic-S compounds extracted into the solvent [22].

B. S^0 Oxidation Products

Oxidation of S^0 results in the formation of sulfuric acid, with sometimes quite stable intermediates of thiosulfate and tetrathionate [6]. The extent of S^0 oxidation can be estimated from the measurement of increases in the S anions plus microbial S and, less sensitively, protons in S^0-treated soil over the control soil.

1. Sulfate Production

The rate and extent of S^0 oxidation has traditionally been estimated by measuring the amount of sulfate extracted by dilute phosphate solution and precipitated as barium sulfate [6,32–34]. Serious positive interference can arise in turbidimetric measurements if very fine S^0 is not removed from extracts [33,35]. Other anions such as thiosulfate and tetrathionate can be measured colorimetrically [6] or, with sulfate as well, by ion chromatography [6,36].

2. Total Extractable S

Measuring the total S in soil extracts includes all S anions and the extracted organic S. Methods used have included x-ray fluorescence [7], inductively coupled argon plasma [37], and reduction to H_2S [38,39], although the latter does not measure all the organic S in the extract. The method is satisfactory for incubation studies (Fig. 1) [38], particularly short term [39], but it is quite impractical under glasshouse and field conditions, where losses to microorganisms, plants, and leaching can be extremely large (Fig. 1) [38].

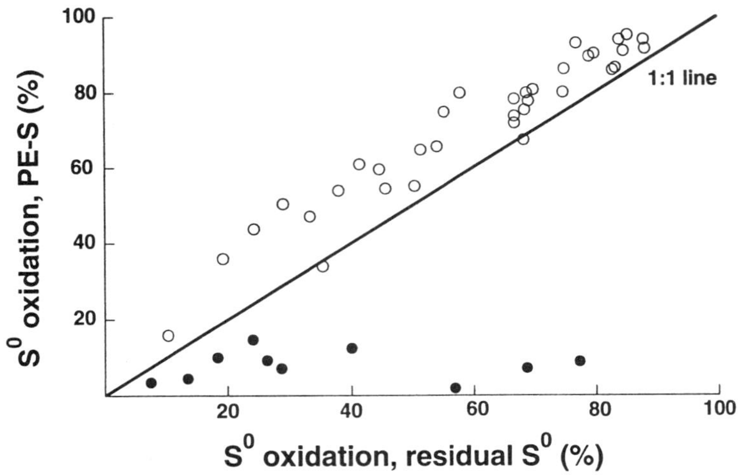

Figure 1 Relationship between S^0 oxidation measured by acetone-extractable residual S^0, and phosphate-extractable S reducible to H_2S (PE-S) in the absence (○) and presence (●) of plants and leaching. (From Ref. 38.)

3. Proton Production

The increase in titratable acidity or a decrease in soil pH has been used as a qualitative index of S^0 oxidation in some incubation [12,33,40] and field studies [38].

IV. FACTORS AFFECTING S^0 OXIDATION

Biological oxidation of S^0 is considered to require direct contact of the microbial cells with S^0 particles, and the extent of contact is important in controlling the rate of microbial oxidation [41]. From this requirement, oxidation of S^0 would involve [42,43]:

1. A film of water on an exposed surface of S^0 in contact with S^0-oxidizing microorganisms
2. Diffusion of O_2, CO_2, and nutrient ions into the water film and oxidation products (H^+ and SO_4^{2-}) away from the zone of S^0 oxidation.

Both the characteristics of the S^0 used and environmental factors affect the above system and processes and thereby influence the extent of S^0 oxidation.

A. Characteristics of S^0

The allotropic form of S^0, S^0 particle size, and the rate of S^0 application have been shown to affect the rate of oxidation.

1. S^0 Allotropes

Although S^0 exists in a large number of molecular and crystalline forms (or allotropes), orthorhombic, S_8, is the most abundant and stable form of S^0 at ambient temperature and pressure [44]. The insoluble polymeric S^0, S_∞, is formed in appreciable amounts only above 160°C and is thermodynamically unstable at ambient temperatures [44]. Under normal manufacturing conditions, S^0 fertilizer would contain no more than 5% of the sum of S_∞ and the solvent-soluble S_7 [45,46].

Headford and McSweeney [47] and Laishley et al. [48] have shown that the presence of species other than orthorhombic sulfur reduces the rate of S^0 oxidation through differences in the crystal microstructure and, consequently, binding sites for S^0-oxidizing organisms. Because fertilizer S^0 contains at least 95% of S_8 [46], this effect is of little consequence.

2. Particle Size

Because S^0 oxidation is a surface-related process, it is expected that an increase in S^0 particle size would decrease surface area and thereby decrease the rate of S^0 oxidation. Linear relationships between the amount of S^0 oxidized and the initial surface area of S^0 applied to soils have been established [49]. Use of fine S^0 particles, however, creates a dust problem and also a fire hazard during aerial spreading. To overcome these problems, granular products are obtained either by mixing finely divided S^0 particles with bentonite [50,51] or co-granulating molten S^0 with partially acidulated phosphate rock (PAPR) and single superphosphate (sulfurized super). In the latter, molten S^0 is mixed with the acid either before or after the acidulation of phosphate rock (PR) [52]. These methods lead to finely divided S^0 (75% $<150\ \mu m$) being incorporated evenly through the granules [53]. The fineness of S^0, however, depends on the total S^0 content of the final product [54]. Chatupote [38] has also shown that the fraction of S^0 particles less than 250 μm decreases with increasing amount of S^0 in superphosphate.

3. Application Rate

Being a surface reaction, the rate of S^0 oxidation is likely to be affected by the rate of S^0 application, which, in turn, would depend on the abundance of S^0-oxidizing microorganisms, the availability of substrate, and removal of products. Janzen and Bettany [39] observed a linear increase in oxidation of S^0 particles (geometric mean diameter 126 μm) with an increase in application rate up to at least 4000 mg kg^{-1}, so the effect would not be important at normal fertilizer S^0 application rates.

B. Soil and Climate Factors

1. Soil Temperature

Soil temperature influences the activity of S^0-oxidizing microbes and thereby the rate of S^0 oxidation. S^0 is oxidized over a wide range of soil temperatures (4°–40°C) with an optimum for thiobacilli of 28°–30°C [55].

An exponential increase in S^0 oxidation rate with temperature has been observed over the temperature range of 2 to 30°C [7,56] (Watkinson and Lee, unpublished). Short-term incubation gave larger exponents [56] than long-term [7] (Watkinson and Lee, unpublished), the latter being more in line with most other biological processes. Experimental error is sometimes too large to distinguish between a linear and an exponential increase [57]. Oxidation in soil cores produced a smaller, curvilinear increase [57a].

2. Soil Moisture

In short-term incubation, it has been observed that S^0 oxidation is optimum at soil moisture near to field capacity and tends to decrease at contents both below and above this point [56,57a,58,59]. This effect of moisture is probably mediated through both the activity of S^0 oxidizers and the transport of nutrients to and products of S^0 oxidation away from the zone of S^0 oxidation. Over longer periods the oxidation rate is much less sensitive to soil moisture [59,60]. In soil cores the rate is linear between wilting point and FC [57a].

3. Previous Application

Occasionally a second application of S^0 after one year on a previously untreated pastoral site in New Zealand produces a higher oxidation rate [61], but generally there is no change [62]. In a comparison on 20 sites, the mean oxidation rate for first and repeat applications were 18.7 and 18.4 $\mu g\ cm^{-2}\ day^{-1}$ (SED = 0.3), respectively [62]. These results demonstrate that the initial number of S^0-oxidizing microorganisms is not generally a limiting factor in oxidizing the S^0 in New Zealand soils under pasture. They confirm an earlier study showing that the rate of oxidation during incubation was independent of initial thiobacilli numbers [31].

V. THEORY OF MODELING THE RATE OF S^0 OXIDATION

There are basically two types of model, mechanistic and empirical. The theory of each is described separately, but they are considered together under "Experimental" and "Discussion."

A. Mechanistic Model

1. S^0 Particle Factors (shape, size, size distribution)

Janzen and Bettany [39] noted from the work of Fox et al. [49] and others that the initial oxidation rate of S^0 was linearly related to the initial surface area. They proposed an extension of this principle that as the area decreased during oxidation under standard conditions, the oxidation rate *per unit of surface area* remains constant. That is, the rate is proportional to the surface area:

$$dm/dt = -ks \tag{1}$$

where m = the mass of S^0 at time t
k = the rate constant
s = the surface area at time t

This is analogous to the approach used by Swartzendruber and Barber [63] in modeling the dissolution rate of limestone in soil.

Equation (1) would also be consistent with a mechanism involving microbes where they saturate the surface. Then, under constant conditions, the number and activity of microbes would remain constant per unit area as the particle size diminished [42].

The use of Equation (1) assumes the constancy of k under set metabolic conditions and the absence of effects from oxidation products [64]. The rate constant is a function of any rate-limiting step that would involve microbes, S^0, oxygen, and moisture and temperature, as discussed earlier. Increase in products, particularly sulfuric acid, would probably not affect k, but could require a term in Equation (1) to counteract ks. Thiobacilli are little affected by acid so would probably not require a second term. However, heterotrophs are more sensitive to acid and may need the additional term.

Equation (1) can be solved for conditions required by models of S^0 particles of increasing complexity of shape and size, starting with spheres of equal size.

a. Spheres of the Same Size. In this instance Equation (1) becomes, after substituting for s,

$$dm/dt = -k\ (3m_0^{1/3}/r_0\rho)m^{2/3} \tag{2}$$

Since

$$s = N4\pi r^2$$
$$m/m_0 = r^3/r_0^2$$
$$m_0 = N(4/3)\pi\ r_0^3\rho$$

where

N = the number of S^0 particles per unit mass of soil
m_0 = the initial mass of S^0 particles per unit mass of soil
r_0 = the initial radius of S^0 particles
r = the radius at time t
ρ = the particle density of 2.07 g cm^{-3} [44]

Integrating Equation (2) gives [39],

$$m/m_0 = (1\text{-}kt/\rho r_0)^3,\ 0 \leq t \leq \rho r_0/k \tag{3}$$

The particles have dissolved ($m/m_0 = 0$) at $t = \rho r_0/k$, hence the time restriction, otherwise the cubic would take negative values.

b. Blocky-Shaped Particles of the Same Size. The spherical model was extended by Watkinson [64] to include particles having a blocky shape. These are particles having similar dimensions along the three axes, in contrast to plates (two large dimensions and one small) and prisms (two small dimensions and one large) [64]. It was argued that the natural prismatic-shaped particles of S^0 would break down under processing and transport to a blocky shape [64]. Thus Equation (1) becomes, for N particles per unit mass of soil, each of mass M, volume V and area A,

$$dm/dt = -kJm_0^{1/3}m^{2/3}/M_0^{1/3}\rho^{2/3} \tag{4}$$

where

J = a shape factor, defined by $A = JV^{2/3}$
M_0 = the initial mass of a S^0 particle

Equation (4) integrates, providing J is constant with time, to [64]

$$(m/m_0)^{1/3} = 1 - kJt/(3\rho^{2/3}M_0^{1/3}) \tag{5}$$

Regular polyhedra are special cases of blocky-shaped particles, with calculable values of J. In particular J has a value of $(36\ \pi)^{1/3}$ for spheres, so

Equation (5) reduces to Equation (3) on substituting values of J, M_0 for spheres.

c. Porous Particles. Often S^0 particles are porous and it has been considered that the increased surface area would increase the oxidation rate over and above the geometrical surface area [43]. Highly porous particles would then be equivalent to thin plates [64], and the oxidation rate could be described by a linear equation [64],

$$m/m_0 = 1 - kst/m_0 \tag{6}$$

This view, however, is not very likely [64]. First, the microbes would colonize the outer geometrical surface and often be too large to penetrate the pores. Furthermore, the activity of any microbes that penetrated the larger internal pores would rapidly decrease as oxygen and nutrient diffusion transport became severely limiting. The outer geometrical surface is, therefore, likely to be the area involved in Equation (1).

d. Blocky-Shaped Particles of Different Sizes. Fertilizer S^0 would tend to have a range of sizes of blocky-shaped particles. These particles can be regarded as equivalent spheres of differing diameters within the range, say, of b to c. For this system the general solution to Equation (2) is a cubic of unequal roots [64,65], where

$$m/m_0 = 1 - 3B_1kt + 3B_2k^2t^2 - B_3k^3t^3 \tag{7}$$

where B_1, B_2, B_3 are functions of b, c and the mass distribution with size w(x) of the S^0 particles. The mass distribution, w(x), is defined by [64]

$$dm_0 = (m_0/\int_b^c w(x)dx)\ w(x)dx$$

Equation (7) becomes readily calculable if w(x) is a constant [64,66].

The standardization of particle shape to equivalent spheres is not unreasonable because spheres have a minimum area per unit mass of S^0. For the same reason, oxidation rate constants of other shaped particles must be correspondingly less.

Constant Mass Distribution. Wide size range (ratio of extreme diameters > 2): It has been shown [64] that the unoxidized fraction of a wide size range of equivalent spheres (diameters from b to c, $c/b > 2$) of S^0 having a constant mass distribution is given by

$$m/m_0 = (1/(c - b)) \int_b^c (1\text{-}at/x)^3\ dx \tag{8}$$

where "a" is the rate of decrease in the diameter of equivalent spheres [63,64].

The solution to Equation (8) over the time from start to when the smallest particles have just dissolved, $0 \leq t \leq b/a$, is given by [66]

$$m/m_0 = 1 - 3at((\ln(c/b))/(c - b) + 3a^2t^2/bc - a^3t^3((b + c)/2b^2c^2) \quad (9)$$

for times $t > b/a$ the function becomes increasingly less than the actual value, and a second solution is required for times up to complete dissolution of the largest particles. For $b/a \leq t \leq c/a$—i.e., time between dissolution of smallest and largest particles—the solution is [66]

$$m/m_0 = (1 + 3(at/c)(\ln(at/c) + 1/2) - 3(at/c)^2 + (at/c)^3/2)/(1 - b/c) \quad (10)$$

Thus over the period $0 \leq t \leq b/a$, Equation (9) is applicable whereas over $b/a \leq t \leq c/a$, Equation (10) is applicable.

Narrow size range ($1 < c/b < 2$): A very useful relation was developed [63,64] for the narrow size range $1 < c/b < 2$. An approximation for Equation (9) with this size restriction is

$$m/m_0 = (1 - at/(bc)^{1/2})^3 \qquad 0 \leq m/m_0 \leq 1 \quad (11)$$

Strictly, equation (11) is applicable over only the time $0 \leq t \leq b/a$; for $b/a \leq t \leq c/a$, the appropriate equation is with "at" replacing "b" [66],

$$m/m_0 = (1 - (at/c)^{1/2})^3 \quad (12)$$

However, for particles of the restricted size $1 < c/b < 2$, more than 95% of all particles have dissolved by the time the smallest particles have completely dissolved after the time b/a. At the longest time for which equation (11) is applicable—i.e., $t = \sqrt{bc}/a$, less than 0.5% of particles remain undissolved (from Equation (10)) [66]. This approximation is very useful experimentally when a restricted size range can be used [65]. Equation (11) shows that for a restricted size range of particles having a constant mass distribution, the range of sizes acts as if they were of one size, the geometric mean, of the extreme sizes. The common practice of using the arithmetic mean [39], although not strictly correct, would not introduce a serious error for particles of the restricted size range $1 < c/b < 2$.

Equation (11) is also useful in developing a rate relationship for mixtures of a wide size range and varying mass distribution as shown in the next section.

Varying Mass Distribution, Wide Size Range In sieve analysis, neighboring sieve sizes are usually chosen such that the ratio of the greater to the lesser is ≤2. This sizing for mass distribution corresponds very conveniently to the size restrictions for use of Equation (11). For each sieve fraction having this narrow size range the assumption of a constant mass distribution would be a good approximation. On this basis a mixture of S^0 particles having an indefinitely wide size range can be considered as the summation of particles of several size ranges, each having the ratio of greater to lesser sizes of ≤2, and each having its own constant mass distribution. Thus for *n* successive sieve fractions of size b_{i-1} to b_i, (b_i/b_{1-1} ≤ 2), and weight fraction w_i [66],

$$m/m_0 = \sum_{i=1}^{n} w_i \left(1 - at/(b_{i-1}\, b_i)^{1/2}\right)^3 \tag{13}$$

Watkinson [66] has shown that Equation (13) at constant w_i is equivalent to Equations (9) and (10) to within 1.5% of m_0. This is sufficiently accurate for use in estimating the oxidation rate of S^0 fertilizer mixtures.

e. Delay Time Where soils have very low numbers of S^0-oxidizing microbes, the time to reach maximum oxidation rate (defined as the delay time) [65] can be several days, even at high temperature. Watkinson and Lee [65] found a delay time of approximately 3 days with 100 μm S^0 particles at 30°C in a soil with initially <10 thiobacilli per gram. The time variable, t, in the preceding equations should therefore be replaced by ($t-t_d$) where t_d is the delay time [65].

The delay time would include time for, first, colonization of the particles [7,42] (microbes possibly following S^0 concentration gradients in the soil) and, subsequently, multiplication of the microbes on the S^0 surface until it was saturated [42]. Thereafter the oxidation rate per unit area would be constant (and maximum) under the prevailing conditions. On these assumptions the delay time would, at a given application rate (1) increase with particle size because of the lower probability of microbes contacting the fewer S^0 particles and (2) decrease with increasing temperature because of a higher microbial multiplication rate.

2. Environmental Factors (temperature, moisture)

The effects of factors such as temperature and moisture are likely to be rather complex, and no quantitative theory involving them has been developed. However, some effects can be proposed.

a. Temperature If it can be assumed that a rate-limiting step exists once a constant oxidation rate has been reached after the delay time, then an activation energy, E_a, would be involved [42]. The magnitude of E_a would be dependent on whether the step was controlled by a biochemical/metabolic, chemical reaction or diffusion process. In any event, the oxidation rate constant would be exponentially related to the Kelvin temperature, T, through the Arrhenius equation,

$$k = k' \exp(-E_a/RT) \tag{14}$$

The greater the value of E_a, the greater the effect of temperature, with values from diffusion being least (Section VI.C.1).

b. Moisture The effects of moisture would be more complex. From experimental data the oxidation rate passes through a maximum at about the FC moisture [56,57a,58,67], which represents the maximum moisture under free drainage. It is tempting to relate the effect to the supply of oxygen because it is the reactant requiring transport that is needed in greatest quantity. For example, if the basic reaction is oxidation to sulfurous or thiosulfurous acids [35,42],

$$\begin{aligned} S^0_8 + O_2 + H_2O &= H_2SO_3 + S^0_7 \\ S^0_8 + O_2 + H_2O &= H_2S_2O_3 + S^0_6 \end{aligned} \tag{15}$$

then, of the reactants, only oxygen requires transport to the S^0. Long-distance oxygen transport in soil is almost entirely in the gas phase [56,58], and only at the reaction site does it need to move across a water film. Hence filling up the pore spaces with water above FC would markedly lower the total cross-sectional area for oxygen transport without lowering the transport rate per unit area (Section VII.B.2).

The decreasing oxidation rate with decreasing moisture below FC has a less obvious explanation. Water energetics [56,58] is much less likely than, say, diffusion of nutrient to microbes [42], because diffusive transport is known to decrease markedly with decrease in moisture [68]. Alternatively, because of the hydrophobicity of the S^0 surface, lowered soil moisture could result in an increasing area becoming devoid of a water film [56], rather than simply decreasing the film thickness [42]. Since moisture is needed by both microbes and the oxidation reaction (Equation(15)) this seems a likely limiting mechanism because diffusion control is ruled out by the high activation energies found [35,42] (Section VI.C.1). However, the hydrophobic property of S^0 may well be diminished by the surface microbes and wetting agents present in soil. For example, the S^0 surface could be made hydrophilic by long chain polythionates such as produced by *T. ferrooxidans* from tetrathionate [69].

If the above mechanisms apply, then volumetric soil moisture [43,57a] would be a more appropriate moisture variable than water potential [56], perhaps as a fraction of the moisture at FC [43,57a].

c. Temperature and Moisture Equations. Janzen and Bettany [56] considered water energetics involving microbes to be important, so related moisture effects as the water potential, M. They assumed water potential and temperature (celsius), θ°C, acted through the relation,

$$k = AB^{\theta} \tag{16}$$

where

$$A = A_1[\ln(-M)]^2 + A_2\ln(-M) + C$$
$$A_1, A_2, B, C \text{ are constants}$$

Watkinson and Blair [35] noted that Equation (16) is a good approximation for the Arrhenius equation (14) at the small values of θ/T_1 that would be applicable. In this case $-E_a/(RT_1(1 + \theta/T_1))$ can be approximated by the first two terms of the binomial expansion, i.e., $(E_a\theta/RT_1^2 - E_a/RT_1)$, so that, approximately,

$$E_a = RT_1^2 \ln B, \tag{17}$$

where T_1 (273.15) on the Kelvin scale is at $\theta = 0$.

Note that increase in soil temperature can lower soil moisture under field conditions [60].

B. Empirical Model and Comparison of Models

1. Particle Factors (shape, size, size distribution)

McCaskill and Blair [43] considered that the oxidation rate of S^0 is based on an "exfoliation principle"—i.e., the particle radius is reduced by a constant amount per unit time. Using a recursive formula they were able to compute the constant reduction in radius, Δr, per unit time. This principle is identical to that of Equation (1) in Section V.A.1, as shown by Swartzendruber and Barber [63]—,i.e., a constant oxidation rate per unit area is equivalent to a constant decrease in particle radius with time. For spheres, substituting in equation (1) $m = (4/3)\pi\rho r^3$ and $s = 4\pi r^2$ gives

$$dr/dt = -k/\rho \tag{18}$$

Chatupote [38] used the approach of McCaskill and Blair, and compared their method of recursive estimation with a second in which the oxidation rate constant, k, as in Equation (18) is included:

$$r_{n+1} = (r_n^2(r_n - 3k/\rho))^{1/3}$$

This approximation is less exact than that based on Equation (18) so that smaller timesteps would be needed to give equivalent results.

McCaskill and Blair [43] also showed that the oxidation rates of spheroidal particles with relative dimensions along the x, y, z axes of 1.00, 0.75, and 0.44, respectively, gave identical oxidation rates to equivalent spheres if the oxidation rate constant for the spheres was 12% greater. Spheroidal particles are particular cases of the blocky-shaped particles of Section V.A.1.b, again showing the equivalence of the two approaches. Also, because spheres have a minimum area per unit volume or mass, their associated rate constants would be greater than for particles of a different shape. For the same reason, equivalent spheres provide a good basic standard for comparative purposes, as discussed in Section V.A.1.b.

Chatupote and coworkers [38,57a] extended the model of McCaskill and Blair to include particles of a range of particle sizes.

2. Environmental Factors (temperature, moisture)

McCaskill and Blair [43] and Chatupote [38] considered that all soils are capable of the same maximum oxidation rate expressed in terms of maximum particle radius reduction per unit time. The actual rate, Δr, is equal to the maximum, Δr_{max}, and its product with two modifying factors or scalars that take into account suboptimal temperatures ($o \leq f_T \leq 1$), and moistures ($o \leq f_\theta \leq 1$), respectively.

The temperature scalar, f_T, is a linear function of soil temperature over 3.3°C to 35.0°C, with values of zero for temperatures ≤3.3°C and unity for temperatures ≥35.0°C. This function applies to all soils.

The soil moisture scalar, f_θ is defined in terms of the fractional FC, F_θ, where

$$F_\theta = \text{(volumetric soil moisture)/(volumetric soil moisture at FC)}$$

For $o \leq F_\theta \leq 0.9$,

$$f_\theta = -0.386 + 2.37F_\theta - 0.945F_\theta^2 \text{ and } o \leq f_\theta \leq 1 \tag{19a}$$

For $0.9 \leq F_\theta \leq 1.0$,

$$f_\theta = 10 - 10F_\theta \tag{19b}$$

The foregoing treatment differs from that of Janzen and Bettany [58] (Section V.A.2.c) in the following respects:

1. All soils have the same maximum oxidation rate expressed as a radius reduction of 0.4 μm/day, and only soil moisture and tem-

perature can alter it. This is unrealistic because it neglects, for example, the effects of different microbes in different soils. In particular, Watkinson and Blair [35] have given evidence that the oxidation rate of thiobacilli in New Zealand soils is 10 times greater than that of heterotrophs in inland North American soils.
2. The linear relationship of k with temperature is an oversimplification because more recent work [7,56] (Watkinson and Lee, unpublished) has shown it to be exponential (or polynomial in cores [57a]). Also, forcing all soils to have the same effect of temperature on k is unrealistic, particularly with the known variations in microbes from soil to soil [12,15,64].
3. The use of a volumetric moisture variable [43,57a] is more appropriate than water potential [56], as discussed in Section V.A.2.b. Also, if the rate-limiting reaction is that of equation (15), then treating all soils identically could also be appropriate because the moisture coverage of S^0 is involved, which could be independent of soil properties. However, the effect at high moisture in blocking air passage for oxygen is likely to differ with soil texture [56,58], and therefore be a function of the soil.

A recent development of the model [57a] removes the restrictions of a maximum rate, and the linear relationship.

Another approach was taken by Chapman [7], who introduced the logistic equation to fit the isotherms of S^0 oxidized against time for three Scottish soils,

$$S = A/(1 + \exp(B - kt)), \tag{20}$$

where

S = sulfur oxidized per unit mass of soil at time t
A = maximum S
B = a dimensionless constant
k = a rate constant

From the variation of rate constant with temperature he calculated activation energies using the Arrhenius equation (14).

VI. EXPERIMENTAL TESTING OF MODELS

A. Methods

As discussed earlier, the most accurate data for testing the model equations are those obtained from measuring the residual S^0 [22] rather than oxidation products, mainly sulfate [39]. Rigorous testing also requires examining the conformity of data to theory over the entire range of oxidation. In this respect also residual S^0 measurements are essential because sulfate measurements may be valid for only the first week or so [39,64].

B. Particle Variables (time, shape, size, size distribution)

Because both models contain the same basic assumptions, they make the same predictions, although the mechanistic model permits a better understanding of the system and therefore allows more judicious experimentation. In particular, it is clear from Equation (3) that the decrease in mass of S^0 is related to the cubic of time. This relation is not at all obvious from the exfoliation model, but testing for constant dr/dt would be equivalent. For particles of the same blocky shape the predicted influence of time, delay time, size, and size distribution on the extent of oxidation (see under Theoretical) can be directly tested experimentally.

Although Janzen and Bettany [39] developed a relationship between the amount of S^0 oxidized and the cubic of the oxidation time, they did not test it experimentally. However, assuming the cubic and the effect of initial diameter, they calculated rate constants from data after six days for 40 soils at 23°C (mean of 48 $\mu g\ cm^{-2} day^{-1}$ S), which can be compared with other values.

The cubic relation with time was later tested by Watkinson [64] using the data of Lee et al. [12] for blocky-shaped particles of constant mass distribution over the size range 30–125 μm. Analysis of variance supported the validity of the model. Including the effects of time, phosphate rock (to examine solubilizing effects of microbes), and soil groups, it accounted for 95% of variance. Samples from the 15 soils were taken fortnightly over a 10-week period during which up to 98% S^0 was oxidized.

Further confirmation of the cubic relation of Equation (11) was given by Watkinson and Lee [65]. They followed the oxidation of S^0 of two narrow size ranges, 75–150 μm and 150–250 μm at 30°C for up to 100 days, taking samples at regular intervals on about twenty occasions. Oxidation was 90–95% complete. Allowing for a delay time, results gave excellent agreement with theory, with r^2 better than 99%. There was no measurable oxidation until after the delay time, probably because of the exponential growth of the microbes until the surface was covered. Further work on this was reported in Watkinson and Blair [35], giving the same excellent fit of data to the cubic of Equation (11) (Fig. 1). Also, the delay time increased with increasing particle size at a constant concentration of S^0, consistent with a decreased number of particles per unit volume, and the decreased probability of particle colonization within a given time. The rate constants decreased, however, with increase of particle size instead of the predicted independence with size. Possibly the closed system of the incubation at the high temperature was involved (e.g., accumulation of reaction products), because field experiments (Lee and Watkinson, unpublished) showed no significant change in rate constant over a much greater size range.

The incubation data of Li and Caldwell [33] were also shown (Watkinson, unpublished [35]) to conform to the cubic of Equation (11). This covered variations in size, application rate, and temperature. Also, over the three ranges of size investigated (geometric means 210, 420, 1300 μm) there was no significant difference in rate constant. The rate constants were much lower [35] than those found by Watkinson and Lee [65].

Over 80 field experiments in New Zealand [60,61,70] show surprisingly good conformity to Equations (11) and (13), despite varying moisture and temperature during the year. This is because simultaneous changes in these two variables partly cancel each other [35], as discussed later. Four of these experiments involved different particle sizes, covering the range of 75 to 1000 μm. The rate constants did not show any significant change with size [35].

Analysis of the field data of Barrow [71] shows a similar pattern. In one experiment the rate constants for particles of geometric means 52, 190, and 590 μm averaged 192 $\pm$ 4 μg cm^{-2} day^{-1}, with r^2 values increasing (0.69, 0.86, and 0.93) with increasing particle numbers through decreasing particle size. At a second drier site, with four sizes of geometric means 52, 88, 190, and 590 μm, the rate constants averaged 10 $\pm$ 2 μg cm^{-2} day^{-1}.

Barrow [71] also applied a mixture of six size fractions of S^0, with sizes ranging from 40 to 420 μm, at three sites. Equation (13) accounted for 89%, 92%, and 98% of the variance when fitted to the data [66], which covered more than 12 months. This supports the validity of Equation (13), and also the concept of a mean, annual rate constant (see later) because temperature and moisture were changing during the year.

Constant dr/dt is equivalent to a cubic in time (Section V.B.1), and Chatupote and coworkers [38,57a] found data from their experiments fitted in this way to be statistically significant.

C. Environmental Factors (temperature, moisture)

1. Temperature

Janzen and Bettany [56] found a good fit of their data to the empirical equation (16), after six days' incubation.

As discussed earlier, these temperature effects are predicted by the theoretical equation (14), and the corresponding activation energies, E_a, have been calculated from the data of Janzen and Bettany [56] for six soils as ranging between 77 and 102 (average 89) kJ mol^{-1} [42]. These values were independent of water potential [56]. They are too large to come from a diffusion-controlled process but are consistent with the limiting step being controlled by a biochemical or chemical reaction [42]. These results are supported by E_a calculations (Watkinson, unpublished) (Table 2) from the data of Li and Caldwell for a Minnesotan soil [33] (r^2 = 0.94), Nor and Tabatabai for five Iowan soils [8] (r^2 = 0.92), Skiba and Wainwright for an English soil [72] (r^2 = 0.99), Chapman for three Scottish soils [7] (r^2 = 0.90, using Equation (11) to obtain comparable k values, rather than from Chapman's logistic equation), Deng and Dick for four Oregon soils [73] (r^2 = 0.98), and for a New Zealand soil (Watkinson and Lee, unpublished) (r^2 = 0.99). The values for E_a averaged 80 ± 9 kJ mol^{-1} excluding the Oregon value. This is not significantly different from the average 89 ± 8 kJ mol^{-1} for the six Canadian soils. (The Oregon value of 102 kJ mol^{-1} is an overestimate because of the very short incubation time and the lack of an allowance for delay times, which increase with decrease in temperature [7,65]).

Field experiments in New Zealand also fitted Equation (14) using mean annual soil temperatures, giving an average value for E_a of 63 kJ mol^{-1} [60], which is within the lower range of incubation values.

From the variation of rate constant (logistic equation) with temperature, Chapman [7] calculated activation energies (Equation (14)) for three Scottish soils of 78, 43, and 52 kJ mol^{-1}. The value of A, the estimated maximum amount oxidizable from the logistic equation, was only about 75% of the amount of S^0 applied. This suggests some problem in either the data, such as immobilization of sulfate over the incubation period of up to 10 weeks [7], or inhibition by the sulfuric acid produced. Or it could mean the logistic equation did not fit adequately [7]. As indicated above, using the cubic plot shows no significant difference in E_a among soils, averaging 67 kJ mol^{-1}.

2. Moisture

The effect of moisture is more complex. After 1-week incubation, the relation with k is approximately parabolic, and the rate constant is maximum near the FC moisture [56,57a,58,59,73]. With increasing time, the peak flattens over a wide range of moisture [58,59]. The rate at higher moisture apparently increases more rapidly, so that after 4 weeks incubation at 25°C, the rate constant was almost unchanged over the moisture range of 0.6 to

Table 2 Summary of Results of Modeling S° Oxidation in Soils of Australasia, Great Britain, and North America, and Associated Intermediates and Chemical-Oxidizing Microorganisms

Location (use)	Delay time t_d (day)	Rate constant k (μg cm^{-2} day^{-1})	Activation energy Ea (n) (kJ mol^{-1})	Chemical intermediates		Microorganisms		
				Thiosulfate	tetrathionate	Heterotrophs	Thiobacilli Neutrophilic	Thiobacilli Acidophilic
Minnesota [33] (arable)	12.5	6	85	•	•	•	•	•
Iowa [8] (arable)	2.9 ± 6.5	(18)[a]	70 ± 11 (5)	√	√	•	•	•
Saskatchewan [15,39,56] (arable)	•	5	89 ± 8 (6)	√	√	√	√	X
N. England [72] (garden)	•	(4)[b]	79	√	√	√	√	X
NE Scotland [7,75,80] (arable)	12.3 ± 1.2 (7°C)	(12)[c]	67 ± 8 (3)	X	trace	√	√	X
New Zealand [12,64–66] (pastoral)	48	53	87	X	X	√	√	√
New Zealand [60,61,70] (pastoral)	•	(50)F	63	X	X	√	√	√
SW Australia [60,66,71] (pastoral)	•	(40)F	•	•	•	•	•	•
Oregon [73] (uncultivated)	•	51	<102	•	•	•	•	•

[a]Assumes $d_o = 50$ μm.

[b]Assumes 20-day incubation with S° of $d_o = 100$ μm.

[c]Rate enhanced by adjuvant.

Abbreviations: t_d = value at 5°C; k = values at or corrected ( ) to 25°C; n = number of soils; √ = present; X = absent; • = not measured; F = field experiment.

two times the FC [59]. A similar effect is observed for the Saskatchewan and Oregon soils incubated for only 6 or 7 days. At increasing temperature the rate constant increases very rapidly to a plateau at high moisture in the sandy soils. After 4 weeks at moisture above FC there was minimal change in rate in the sandy soil used by Nevell and Wainwright [67]. In contrast to this pattern, the sandy soil of Lee et al. [61] incubated for 2 weeks at four temperatures showed no maximum rate at 10 kPa (FC), and the rate was still increasing at the highest experimental moisture of approximately 1 kPa.

3. Temperature and Moisture

Neither Janzen and Bettany [56] nor McCaskill and Blair [43] tested their moisture–temperature relationships (Equations (16) and (19)), based on only 1-week incubations, against oxidation in field soils. However, Chatupote [38] incorporated the effects of surface area, temperature, and moisture using the procedure of McCaskill and Blair [43] to field data (two soils), and found that the significance of the fit of the data was less than with surface area alone; i.e., the model overcompensated for the effect of moisture. This is not surprising if the results of Kittams and Attoe [59], showing a decreasing effect of moisture on k with increasing time, are representative. It would seem that over the range of moisture generally encountered in field soils where S^0 is useful as a fertilizer, there would be little effect on rate constants. On present evidence, field oxidation rates would decrease only at moistures below about 50% FC [66]. Furthermore, in temperate climates at least, this would happen at higher summer temperatures, which would increase the rate, thus maintaining a relatively constant rate overall [60].

Watkinson and Lee [60] found that changes in surface area alone could largely account for the decrease in S^0 with time in over 80 field experiments [60,61,70] under the New Zealand climate. Three factors permitted the use of a mean annual rate constant [60,66].

1. Soil moisture was sufficiently close to FC as to have little effect on the rate, except in summer.
2. Even during the lower moisture of summer the higher temperature largely balanced out any diminution in rate.
3. The insular climate kept temperature changes between winter and summer to only about 10°C.

The field data of Barrow [71] in southwest Australia are consistent with this view. At the moist, higher rainfall site, the effect of the changing surface area alone (Equation (11)) gives values for r^2 of 0.9 with the two smaller particle sizes (where subsampling errors are less than with the largest size, $r^2 = 0.7$). In contrast, at the drier site (but similar temperature), the rate constant is half, and the values for r^2, which do not take into account moisture changes, are consistently lower at each size. Where the wide mixture of sizes was applied at three sites [71], the validity of equation (13) and a mean annual rate-constant [66] has been confirmed. The rate constants at sites with annual rainfall values of 560 and 970 mm differed by only 5% and had values (16 μg cm^{-2} day^{-1}) similar to those in a corresponding New Zealand climate region. The third site with the low annual rainfall of 350 mm had a rate constant value only 65% of the other two, which is very likely associated with the low moisture.

Hedley et al. [57a] found the mathematical surface relating to combined effects of temperature and moisture in undisturbed cores to be curvilinear for temperature (5° to 25°C), and linear for moisture between wilting point and FC.

VII. DISCUSSION

A. Particle Factors

The same basic assumptions are made for the two types of model discussed, so that data fit either equally well. In the mechanistic model, the basic Equation (1) is that $(1/s)dm/dt = k$; for the empirical model, it is, implicitly, Equation (18), $dr/dt = k/\rho$. The first deals with particle mass and area for reaction, the other with particle diameter. An important difference is that in the mechanistic model Equation (18) is readily derived, together with the implication of Equation (1), that the cube root of the residual mass is related to time. Such inferences, and others, are not readily made from the empirical model.

1. Shape

Particles of S^0 can be regarded, from visual evidence and light and electron microscopy [43,61], as being of blocky shape [64]. The experimental data show they conform to equations in which they can be described as "equivalent spheres" [43,64,65]. Although microscopy shows particles are often pitted and sometimes porous [61,43], the experimental data always closely conform to the cubic rather than the linear Equation (6). This is good evidence that the outside geometrical surface is involved in oxidation by microbes, and not, additionally, the internal surfaces (Section V.A.1.c).

2. Size

The fit of experimental data to the cubic show that the oxidation rate is consistent with a constant value, k, per unit area [39,64]. This fits well with approximately a monolayer coverage of the outside geometrical surface by the oxidizing microbes [42], which would therefore oxidize at a constant rate, other factors not being limiting. The surface area decreases on oxidation, so that a monolayer coverage could always be maintained.

The mechanistic model of Equation (1) implies that the value of k is independent of initial particle size, and this has been verified by numerous incubation and field experiments, with only one exception. It is not unreasonable that k is independent of initial size because all particles continuously decrease in size so that no one initial size would be expected to give rates different from others. One possible exception to this reasoning has been noted. Where particles reach dimensions approaching that of the microbes [35], a marked decrease in rate was observed.

3. Size Distribution

There are fewer data available to test Equations (9), (10), and (13) [64,66]. However, Equation (12) has been verified [35,65,66], together with the related Equation (13) [66], which is really summing terms from Equation (12). It is important to note that the appropriate mean for particles of the appropriate size range is the geometric rather than the arithmetic.

4. Delay Time

Elemental sulfur is added uninoculated to soil, so that oxidizing microbes in the soil colonize the particles, then multiply on the surface [7,42] to give approximately a monolayer coverage [42]. The apparent time to reach maximum oxidation rate has been designated the delay time and is estimated as the time intercept obtained by extrapolating the cubic to $m/m_0 = 0$ [65]. For thiobacilli, which survive at very low numbers of the order of ≤ 10 per gram soil, probably through S^0 produced in low concentrations (0.1 to 1 ppm) [12] during the natural S cycle, the delay time has been directly related to the time to reach numbers of 10^4 per gram soil (Lee and Watkinson, unpublished). It was also related to the negative exponential of the temperature ($r^2 = 0.98$), and at 5°C the delay time was 7 weeks (Watkinson and Lee, unpublished). The microbes in three Scottish soils [7] also exhibit the same relationship ($r^2 = 0.99$) but with the shorter estimated delay time at 5°C of 2 weeks (Watkinson, unpublished). The delay time can therefore be quite long, particularly at small concentrations of S^0 particles in soil at low temperatures. The colonization time is proposed to be related to the concentration of both microbes and S^0 particles, temperature, and possibly the soil moisture. Since S^0 is immobile, highly insoluble in water, but does have a measurable vapor pressure at soil temperatures [44], possibly the thiobacilli move in response to the S^0 concentration gradient in the gas phase to contact the S^0 particles. For heterotrophs, on the other hand, generally present in high concentration in soil, colonization may happen through direct initial contact of S^0 and the microbes. At a given concentration of S^0 in soil the delay time for thiobacilli decreases with decreasing particle size (i.e., larger particle numbers) [35,65]. The exponential increase in delay time with decrease in incubation temperature (Lee and Watkinson, unpublished) may reflect lower S^0 vapor concentration gradients, microbe activity, and multiplication rate on the surface.

B. Environmental Factors and Reaction Mechanisms

1. Temperature

As discussed earlier, there is a good fit of data to Equation (14) and E_a is relatively constant at 80 kJ mol^{-1} in soils of various kinds in six countries, involving different microbes and at every level of soil moisture. The effect of temperature can therefore be explained in terms of an activation energy for a chemical or biochemical reaction that limits the oxidation rate of the S^0, because E_a is too high for diffusion [42]. The linear relation between temperature and rate constant used by McCaskill and Blair [43] possibly derives from data that either cover a small temperature interval (approximating the exponential), or are insufficiently precise to distinguish the two.

The rate-limiting reaction has been suggested [42] as the oxidation of S^0 to either sulfurous or thiosulfurous acids, depending on the ratio of S^0 to O_2, e.g.,

$$S^0{}_8 + H_2O + O_2 = H_2S_2O_3 + S^0{}_6$$
$$S^0{}_8 + H_2O + O_2 = H_2SO_3 + S^0{}_7$$

With excess of O_2 the H_2SO_3 would be oxidized to sulfuric acid, which is the only product identified in New Zealand soils. Extracts from both incubated and field soils analyzed by high performance ion chromatography (HPIC) [36] have failed to detect any oxyacids of S such as thiosulfate and tetrathionate, even within a day of incubation. Only sulfate has been found. This may be related to the dominance of acidophilic thiobacilli and the fact that S-deficient soils in New Zealand are reasonably well aerated. The slower oxidizing heterotrophs seem to produce a range of intermediates containing a greater ratio of sulfur to oxygen, starting with thiosulfate and proceeding through tetrathionate, finally to sulfate [8,15,72].

2. Moisture

Because the oxidation rate shows a broad maximum with moisture at about the FC, it is likely that two different processes are involved, and there is general agreement that at moisture values greater than the FC the limitation is one of O_2 transport in the gas phase. Oxygen diffuses much faster and has a 25-fold greater concentration in air than water. Because the activation energy is high and constant over all moisture levels, it rules out any diffusion control of O_2. It is proposed that increasing moisture increasingly blocks off air passages so that the total O_2 flux to the S^0 surface is decreased but not the flux per unit area across any open air pathway. Increasing moisture reduces the effective cross section of the pathway. Hence the linear relation between apparent gas diffusion coefficient (a measure of decrease in path cross section) and pore space for moist sand, with a sharp cutoff when all the paths are blocked [74]. In addition, for clay soils, it seems possible that some pathways are not blocked entirely at intermediate moisture but produce more circuitous pathways. This implies that O_2 would take longer to reach the S^0 surface in these instances.

The net result would be that after a moisture addition, the O_2 flux would drop suddenly but gradually increase as it reached the S^0 surface from the more circuitous pathways. Such a delay in reaching maximum oxygen flux would be least in coarse-textured soils and greatest in heavy clays with a network of very fine air pores. The data of Kittams and Attoe [59] for a silt loam illustrates this point. One week after incubation, the ratio of S^0 oxidized at $2 \times$ FC compared with FC is 0.2; after 2 weeks the ratio increases to 0.4, and after 4 weeks it is 0.9. The results of Moser and Olson [58] for a silt loam at 3 and 12 weeks are consistent with the above.

The data of Janzen and Bettany [56] for six soils follow a similar pattern, but here the comparison is over four temperatures at one time (6 days). In particular, considering the sandy soil (90% sand, 7% clay) with no fine pores, at 15°C the oxidation rates steadily decreased with increasing moisture above FC, but at 23°C and 30°C there was no difference in rate. In contrast, the soil with the greatest amount of clay (60% clay, 6% sand) with many fine pores gave lower rates at moistures above FC for all temperatures, and exhibited the least oxidation (39% of maximum) at high moisture and 30°C. The few circuitous air pathways in the sand have provided maximum O_2 flux at high moisture by 23°C; the many fine pores in the clay have provided only 40% of maximum oxygen flux even at 30°C. Both soils had the same value for the maximum rate at 30°C. Deng and Dirk [73] found the same pattern with sand and clay soils.

The foregoing discussion indicates that the treatment of field moisture effects above FC in both Equations (16) [56] and (19) [43] is inadequate.

Under New Zealand conditions, S-deficient soils have a silt loam texture or coarser, so that moisture above FC would not influence the field oxidation rate over 12 months, particularly as any increase above FC would be only transitory in these free-draining soils.

There is less agreement on the explanation for the decrease in oxidation rate with decreasing moisture below FC. Moser and Olson [58] and Janzen and Bettany [56] proposed that the energy of water uptake by the microbes was the important factor, by analogy with plants. However, Janzen and Bettany [56] also thought reduced accessibility of the highly hydrophobic S^0 particles could be a factor, with which McCaskill and Blair [43] were in agreement. Watkinson [42] considered nutrient diffusion to the microbes through increasingly thinner water films was more likely than energy of water uptake, but then discounted diffusion on the grounds of the high energy of activation. The only explanation so far offered that is consistent with the high and constant energy of activation over the whole moisture range is that of diminished surface coverage of the hydrophobic S^0 surface by water at diminishing moisture contents. In other words the geometrical area of the S^0 surface available for oxidation decreases with decreasing moisture, rather than a water film over the entire surface steadily and uniformly decreasing in thickness. Evidently the hydrophobicity of S^0 is greater than the wetting effects of soil constituents or soil microbes and their products. It also presumes that a significant fraction of the surface is unaffected by microbes that would be expected to "wet" the surface. (A monolayer coverage of the surface by microbes would also not be strictly correct—see earlier.) This explanation fits with the oxidation of S^0 to sulfurous or thiosulfurous acids, mediated by microbes, as being the rate-limiting step in the process.

According to the above explanation, time would have no effect—and temperature have little or no effect—on oxidation of the dry area of S^0, in contrast to the effects at moisture levels above FC [59]. This is basically a property of the S^0, and independent of soil properties, and is consistent with the evidence. The oxidation rate decreased with the same pattern in all soils between 1.5 and 10 MPa [56,73], in contrast to the different effects at the other moisture extreme above FC.

3. Microbes

Table 2 summarizes much of the data detailed below.

Heterotrophs appear to dominate the oxidation of S^0 in central North American soils in Saskatchewan [10,15], Minnesota [33,35], and Iowa [8]. Very low values for rate constants were obtained for 40 Saskatchewan agricultural soils [39], based on sulfate produced between zero and 6 days.

Calculations for a Minnesotan soil [33] give a similar low value (Watkinson, unpublished) (Table 2). Although the average Iowan [8] rate constant (Watkinson, unpublished) is about three times greater (assuming a geometric mean d_0 of 50 μm) than the average Canadian value, it is still nearly three times less than for New Zealand thiobacilli (Table 2). The data of Li and Caldwell covered a period of up to 60 days based on sulfate analyses; there was little evidence from the cubic plots of any conversion of inorganic S to organic S forms, nor of any inhibition of the heterotrophs by the sulfuric acid produced. Nor and Tabatabai measured total inorganic S during a 74-day incubation. At 5°C there was no deviation from the cubic, but at 30°C the data deviated after 2 to 4 weeks. Only the initial rates were therefore used in calculations. Activation energies for these soils averaged 80 kJ mol^{-1} (Table 2).

Initial numbers of heterotrophs are relatively large (10^5 per gram) compared with thiobacilli and change little on incubation [12]. For this reason a shorter delay time would be expected than for thiobacilli. This is borne out by experiment. At the low temperature extreme of 5°C, the delay times for a Minnesotan soil [33], five Iowan soils [8], and three Scottish soils [7] (see later) are approximately 3 to 12 days (Watkinson, unpublished), whereas for a New Zealand soil containing thiobacilli it was 48 days (Watkinson and Lee, unpublished). This supports the validity of the values of the rate constants obtained by Janzen and Bettany [39] using the short incubation time of 6 days.

Data [72] from a fertile English garden soil give a small rate constant* and a short delay time consistent with the relatively large initial numbers of S^0-oxidizing microbes found (2×10^4 g^{-1} soil). The associated activation energy was 79 kJ mol^{-1}. Among the dune sands [72], rate constants increased from 0.8 to 3.8 $\mu g\ cm^{-2}\ day^{-1}$ with increase of C and N, from initial microbe numbers averaging 5×10^3 g^{-1} soil. These features plus the very high pH and the formation of thiosulfate and tetrathionate intermediates indicate that the organisms involved are very similar to the S^0-oxidizing heterotrophs and thiosulfate-oxidizing autotrophs found in Saskatchewan soils [15]. It is extremely unlikely that dune sands would contain sufficient natural S^0 (probably $\ll$ 0.1 ppm) to support any measurable numbers of thiobacilli such as observed in New Zealand pastoral soils [12,22].

Rate constants for three Scottish soils were calculated using Equation (11) rather than the logistic equation used by Chapman [7]. (The value for d_0 of 4.3 μm used in these calculations was obtained from Chapman's

*Assumes a 20-day incubation with flowers of sulfur having $d_0 = 10^{-2}$ cm.

value for the specific surface area (SSA) for the micronized S^0, because $d_0 = 6/(\rho.SSA)$.) The oxidizing microbes comprised heterotrophs and neutrophilic, but not acidophilic, thiobacilli [75]. They gave short delay times but rate constants several times greater than those estimated from similar microbial populations in the Canadian and English soils, but similar to those from Iowan soils. These greater values may indicate a different group of microbes, but more likely they are greater for Scottish soils because of the known enhancing effect on rate of the adjuvant present in the micronized S^0 [7]. Activation energies for the three soils did not differ significantly from each other and from the soils of North America and England. This contrasts with values calculated by Chapman [7] from the rate constants (units of reciprocal time) using the logistic equation. Two were much lower, probably because it is not easy to tell using this equation if acid products or decreasing substrate are slowing the oxidation.

Thiobacilli have been found in all pastoral soils of New Zealand [12] (Lee and Watkinson, unpublished) and appear to dominate the oxidation of S^0. Incubation at 25°C [12] produced large rate constants averaging about 50 $\mu g\ cm^{-2}\ day^{-1}$ [64] which is about 10 times that of heterotrophs at a similar temperature. On the other hand, because of low initial numbers, thiobacilli have a much longer delay time as noted above. Up to 95% oxidation has been observed without, apparently, any inhibition of thiobacilli by the large amount of sulfuric acid produced [12,64]. The activation energy in the one soil investigated is the same as for the heterotrophs of North America.

Calculations (Watkinson, unpublished) on the data from the four uncultivated soils of Deng and Dick [73] also show large rate constants, which suggests that thiobacilli may be present in the Oregon soils. However, on limited evidence the delay time appears relatively short, possibly between that for the New Zealand and the central North American soils, so that any thiobacilli are present in relatively large numbers in the soils, or a third type of microbe is involved.

Large rate constants have also been noted in Australian pastoral soils [43]. Calculations [66] on the field data of Barrow [71] give rate constants for pastoral soils in SW Australia that are only slightly lower than those found in New Zealand field experiments on pastoral soils under a similar climate [60].

It would be interesting to speculate on the absence of acidophilic thiobacilli in Saskatchewan, English, Scottish and probably Minnesotan soils and their presence in New Zealand, and probably Australian and Oregon soils, with large oxidation rate constants. Certainly thiobacilli are not inhibited by the Continental climate of Alberta, because they are naturally present in large numbers (10^5 per gram) in soils contaminated with

S^0 from nearby storage blocks from sour gas processing [76]. The one known differentiating feature of the two sets of soils (Table 2) is that the first set was cultivated at regular intervals, whereas the second consisted of undisturbed pastoral or native soils. Perhaps cultivation permits more rapid abiotic oxidation of any finely crystalline S^0 formed during the natural S cycle [22], so that even minute populations of S^0-oxidizing autotrophs cannot survive. Natural levels of 0.1 to 1 ppm S^0, along with correspondingly low numbers of thiobacilli (<10 g^{-1} soil), have been found in all New Zealand soils under permanent pasture, most of which have never been fertilized with S^0 [12] (Watkinson and Lee, unpublished).

The independence of activation energy value with microbe, country, soil, and soil moisture indicates the rate-limiting step is a basic process common to all these circumstances and conditions. The microbially mediated oxidation to sulfurous or thiosulfurous acid suggested previously is consistent with all data reviewed in this chapter.

In terms of the oxidation rate model, heterotrophs are characterized at 25°C by a negligible delay time, a small rate constant (about 5 $\mu g\ cm^{-2}\ day^{-1}$), and deviation from the cubic before full oxidation because of acid product or diminished substrate. Acidophilic thiobacilli are characterized by a long delay time (about 2 days), a large constant (about 50 $\mu g\ cm^{-2}\ day^{-1}$), and conformity to the cubic up to 95% oxidation even at 30°C (Fig. 2).

C. Fertilizer Information

1. Pastoral Soils

On the evidence presented, the important variables determining the oxidation rate constant in New Zealand pastoral soils appear to be the soil temperature and the kind of oxidizing organism (rate differences <3 at a given temperature) [60]. Soil moisture is generally at optimum near FC [60] because excess rainwater drains quickly. Watkinson and Lee [60] therefore suggested that only three particle sizes of annual fertilizer S^0 were needed in New Zealand ($<.25$, $<.15$, and $<.075$ mm), based on the mean annual soil temperatures (14, 11.5, and 9°C) of three temperature zones. This was in agreement with earlier results from agronomic experiments [60]. If moisture and temperature both fluctuate very widely, the model of Hedley et al. [57a] could be more useful.

The delay time for thiobacilli can be of the order of months at low temperature (or moisture) and low application rate of S° (Section VII.A.4). For example, there is evidence [66] for a delay time of 2 months at the Darkan site [71] in western Australia. This, therefore, has implications for

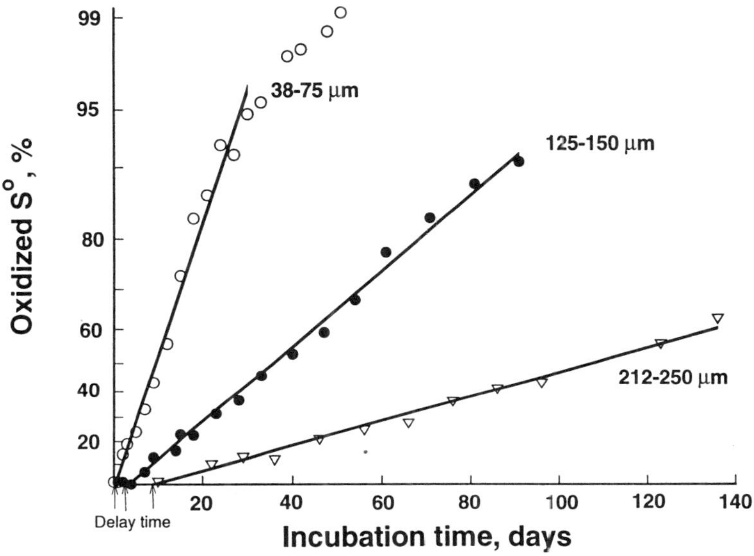

Figure 2 Linearity test of Equation (11) from m_0 = 450 mg kg^{-1} at three particle sizes: ○ 38–75 μm, ● 125–150 μm, ∇ 212–250 μm and illustration of delay time. Amount of S^0 oxidized (cubic scale) versus incubation time at 30°C and field capacity moisture. (From Watkinson and Lee, unpublished; Ref. 35.)

the timing of S^0 fertilizer applications. For example, late-autumn S^0 application could be better for supplying the peak S demand of spring growth than an early spring application [77,78] because of colonization/multiplication of S^0 during winter.

The severity of the deficiency determines the amount of S^0 fertilizer. This can be estimated from standard soil tests for sulfate and mineralizable organic-S and a fertilizer recommendation model (New Zealand AgResearch, Soil Fertility Service) [79].

A constancy of k during the year is also observed at four sites in southwest Australia [66,71], as noted earlier. At the two drier sites in southwest Australia (350 and 500 mm rainfall) the effect of very low moisture apparently exceeded the temperature effect and gave a lower average rate constant. Optimum particle size for fertilizer S^0 could possibly be estimated from soil temperature and moisture as for New Zealand soils [57a,60].

2. Cultivated Soils

An order of magnitude difference in oxidation rate exists between cultivated and uncultivated soils (Table 2), possibly related to the absence of thiobacilli and natural S^0 in the cultivated soils (previous section). It would therefore be worthwhile testing this, as indicated in the next section.

D. Further Work

1. Reasons for the Presence/Absence of S^0-Oxidizing Autotrophs in Field Soils

Because of the far greater rate of oxidation by the S^0-oxidizing autotrophs, it would be important to determine the reason for their absence in agricultural soils of, for example, Saskatchewan and Minnesota, with a view to the possible inoculation of S^0 fertilizer applied to these soils. The effects of cultivation, and very low temperatures, on natural levels of S^0 and S^0-oxidizing autotrophs (as discussed earlier) could be readily investigated. If the suggested effects of cultivation were verified, then the procedure of inoculating S^0 fertilizer with thiobacilli, followed by regular additions of S^0 to ensure their survival, could be examined. Differences in oxidation rate at different field sites among different strains of the same organism, particularly thiobacilli, need investigation [60].

2. Effects of Soil Temperature and Moisture on the Rate Constants under Field Conditions

The model relating oxidation rate to soil temperature and moisture [57a] needs to be tested further, particularly under field conditions. The question of self-regulation, because of the opposing effects of temperature and moisture, particularly in summer [60], is simple and useful for annual fertilizer S^0, but needs wider verification. These matters could be investigated by measuring residual S^0, average soil temperature, and moisture over short intervals at several sites during at least one year.

VIII. CONCLUSIONS

The theoretical equations based on the assumptions of a constant oxidation rate per unit area of S^0, and an activation energy for the rate-limiting step, have been verified in various laboratory and field experiments. Experiments have involved both heterotrophs and autotrophs. Other variables involved were residual mass of S^0, particle size, shape, time, soil temperature, and delay time.

The empirical soil moisture relation observed in incubations showing a maximum rate constant at approximately the FC is consistent with involvement of atmospheric oxygen in two ways:

1. A decreased cross-sectional path for diffusion in soil air at moisture above FC.
2. Decreased reaction sites on the S^0 surface at moisture below FC. The information available from field experiments suggests that moisture is of greater importance at values much less than FC and near the wilting point.

Soils divide basically into two groups according to the magnitude of their rate constant:

1. Those with a small constant (5 μg cm^{-2} day^{-1} at 25°C) are probably associated with S^0-oxidizing heterotrophs and thiosulfate-oxidizing heterotrophs and autotrophs, and intermediates of thiosulfate and tetrathionate present during the early part of the oxidation to sulfate.
2. Those with a large constant (50 μg cm^{-2} day^{-1}) are associated with S^0-oxidizing autotrophs and the absence of intermediates.

Both groups have the same activation energy of approximately 80 kJ mol^{-1}, which, with other evidence, is consistent with the controlling mechanism being the oxidation of S^0 to sulfurous or thiosulfurous acid.

REFERENCES

1. J. R. Freney, Sulphur-containing organics, *Soil Biochemistry* (A. D. McLaren and G. H. Peterson, eds.), Marcel Dekker, New York, p. 229 (1967).
2. J. K. Syers, R. J. Skinner, and D. Curtin, Soil and fertilizer sulphur in U.K. agriculture, *The Fertilizer Society of London*, Proceedings No. 264, p. 43 (1987).
3. A. G. Sinclair, C. C. Boswell, I. S. Cornforth, A. Lee, C. Morgan, M. L. Nguygen, W. M. H., Saunders, P. Shannon, R. G. Smith, and G. Whelan, "Agronomic requirements for sulphur in New Zealand pastures," Proceedings New Zealand Fertilizer Manufacturers' Research Association, 2: 538–72 (1985).
4. M. H. Hilal, "Sulphur in desert agrosystems," Proceedings of the Middle East Sulphur Symposium 1990, Cairo, Egypt, pp. 19–50 (1992).
5. A. E. Konopka, R. H. Miller, and L. E. Sommers, Microbiology of the sulphur cycle, *Sulphur in Agriculture* (M. A. Tabatabai, ed.) *Agronomy Society of America*, p. 23 (1986).
6. M. Wainwright, Sulfur oxidation in soils, *Adv. Agron., 37:* 349–96 (1984).
7. S. J. Chapman, Oxidation of micronized elemental sulphur in soil, *Plant and Soil, 116:* 69–76 (1989).
8. Y. M. Nor and M. A. Tabatabai, Oxidation of elemental sulfur in soils, *Soil Sci. Soc. Am. J., 41:* 736–41 (1977).

9. M. I. Vitolins and R. J. Swaby, Activity of sulphur-oxidizing microorganisms in some Australian soils, *Aust. J. Soil Res., 7:* 171–83 (1969).
10. J. J. Germida, J. R. Lawrence, and V. V. S. R. Gupta, "Microbial oxidation of sulphur in Saskatchewan soils," Proceedings of International Sulphur 1984 Conference, pp. 703–10 (1984).
11. J. R. Lawrence and J. J. Germida, Relationship between microbial biomass and elemental sulfur oxidation in agricultural soils, *Soil Sci. Soc. Am. J., 52:* 673–7 (1988).
12. A. Lee, J. H. Watkinson, G. Orbell, J. Bagyaraj, and D. R. Lauren, Factors influencing dissolution of phosphate rock and oxidation of elemental sulphur in some New Zealand soils, *New Zealand J. Agric. Res., 30:* 373–85 (1987).
13. S. Yagi, S. Kitai, and M. Kimuru, Oxidation of elemental sulphur to thiosulphate by Streptomyces, *Appl. Microbiol., 22:* 157–9 (1971).
14. K. Killham, N. D. Lindley, and M. Wainwright, Inorganic sulphur oxidation by *Aureobasidium pullulons, Appl. and Environ. Microbiol., 42:* 629–32 (1981).
15. J. R. Lawrence and J. J. Germida, Enumeration of sulfur-oxidizing populations in Saskatchewan agricultural soils, *Can J. Soil. Sci., 71:* 127–36 (1991).
16. I. L. Pepper and R. H. Miller, Comparison of the oxidation of thiosulfate and elemental sulfur by two heterotrophic bacteria and *Thiobacillus thiooxidans, Soil Sci., 126:* 9–14 (1978).
17. J. J. Germida, Modified sulfur-containing media for studying sulfur oxidizing microorganisms, *Planetary Ecology* (D. E. Caldwell et al, eds.), Van Nostrand Reinhold, New York, p. 333 (1985).
18. M. L. Nguyen, "Direct determination of elemental sulphur in soils and the application of this method in measuring oxidation potentials of elemental sulphur fertilizers," Massey Fertilizer and Lime Research Centre Occasional Report No. 2: 9–25 (1988).
19. S. L. Chopra, Determination of elemental sulphur in soils, *J. Indian Soc. Soil. Sci., 11:* 33–8 (1963).
20. C. B. Fliermans and T. D. Brock, Assay of elemental sulfur in soil, *Soil Sci., 115:* 120–2 (1973).
21. N. J. Barrow, Determination of elemental sulphur in soils, *J. Sci. Food. Agric., 19:* 454–6 (1968).
22. J. H. Watkinson, A. Lee, and D. R. Lauren, Measurement of elemental sulfur in soil and sediments: Field sampling, sample storage, pretreatment, extraction and analysis by high-performance liquid chromatography, *Aust. J. Soil Res., 25:* 167–78 (1987).
23. K. Y. Chen, M. Moussavi, and A. Sycip, Solvent extraction of sulfur from marine sediments and its determination by gas chromatography, *Environ. Sci. Technol., 7:* 948–51 (1973).
24. M. G. R. Hart, A turbidimetric method of determining elemental sulphur, *Analyst, 86:* 472–5 (1961).
25. N. J. Barrow, Note on incomplete extraction of elemental sulphur from wet soil by chloroform, *J. Sci. Food. Agric., 21:* 439–40 (1970).

26. J. K. Bartlett and D. A. Skoog, Colorimetric determination of elemental sulphur in hydrocarbons, *Anal. Chem., 26:* 1008–11 (1954).
27. D. G. Maynard and P. A. Addison, Extraction and colometric determination of elemental sulfur in organic horizons of forest soils, *Can. J. Soil Sci., 65:* 811–3 (1985).
28. D. L. Struble, Quantitative determination of elemental sulfur by GLC with an electron capture or a flame photometric detector. *J. Chromatogr. Sci., 10:* 57–9 (1972).
29. D. R. Lauren and J. H. Watkinson, Elemental sulphur analysis using HPLC on 10 μm rigid polymer particles, *J. Chromatogr., 348:* 317–20 (1985).
30. C. Heim, I. Devai, and J. Harangi, Gas chromatographic method for the determination of elemental sulphur in sediments, *J. Chromatogr., 295:* 259–63 (1984).
31. A. Lee, J. H. Watkinson, and M. L. Nguyen, "Oxidation of elemental sulphur by thiobacilli in soils from New Zealand," Proceedings of the Middle East Sulphur Symposium 1990, Cairo, Egypt, pp 109–24 (1992).
32. C. Bloomfield, Effects of some phosphate fertilisers on the oxidation of elemental sulphur in soil, *Soil Sci., 103:* 219–23 (1967)).
33. P. Li and A. C. Caldwell, The oxidation of elemental sulfur in soil, *Proc. Soil. Sci. Am., 30:* 370–2 (1966).
34. M. Wainwright, Microbial sulphur oxidation in soil, *Sci. Prog. Oxford, 65:* 459–75 (1978).
35. J. H. Watkinson and G. J. Blair, Modelling the oxidation of elemental sulfur in soils, *Fert. Res., 35:* 115–26 (1993).
36. J. H. Watkinson and M. J. Kear, "High performance ion chromatography measurement of sulphate in calcium phosphate extracts of soil," Massey Fertilizer and Lime Research Centre Occasional Report No. 5: 102–8 (1991).
37. M. J. Kear and M. M. Sutton, "Reduction of calcium interference in the measurement of sulphur by ICAP-Atomic emission spectrometry using the 182.03 nm line," Massey Fertilizer and Lime Research Centre Occasional Report No. 5: 96–101 (1991).
38. W. Chatupote, An investigation of some factors influencing the rate of oxidation of elemental sulphur fertilizers, PhD Thesis, Palmerston North, New Zealand: Massey University (1990).
39. H. H. Janzen and J. R. Bettany, Measurement of sulfur oxidation in soils, *Soil Sci., 143:* 444–52 (1987).
40. S. S. S. Rajan and E. A. Edge, Dissolution of granulated low-grade phosphate rocks, phosphate rocks/sulphur (Biosuper) and superphosphate in soil, *New Zealand J. Agric. Res., 23:* 451–6 (1980).
41. S. Takakuwa, T. Fujimori, and H. Iwasaki, Some properties of cell-sulfur adhesion in *Thiobacillus thiooxidans, J. Gen. Appl. Microbiol., 25:* 21–9 (1979).
42. J. H. Watkinson, "Rate equations for the oxidation of elemental sulphur in soil," Massey Fertilizer and Lime Research Centre Occasional Report No. 2: 26–31 (1988).

43. M. R. McCaskill and G. J. Blair, A model for the release of sulfur from elemental S and superphosphate, *Fert. Res., 19:* 77–84 (1989).
44. B. Meyer, Sulfur, energy and the environment, Amsterdam: Elsevier (1977).
45. A. C. Ludwig and J. M. Dale, The effect of S_μ on the mechanical properties of sulphur, *Sulphur Inst., 9:* 10–3 (1973).
46. R. Steudel, "Recent developments in the chemistry of elemental sulfur and sulfur-rich compounds," Proceedings International Symposium, Elemental Sulphur in Agriculture, *1:* 7–14 (1987).
47. C. E. I. Headford and G. McSweeney, Allotropes of sulfur—some preparations and soil oxidations, *New Zealand Fertilizer Manufacturers' Research Association* 1984 Research Reports for March, p. 373 (1984).
48. E. J. Laishley, B. W. Bryant, J. B. Kobryn, and J. B. Hyne, The effect of particle size and molecular composition of elemental sulphur on ease of microbiological oxidation. *Alberta Sulphur Research Ltd Quarterly Bulletin 20:* 35–50 (1983).
49. R. L. Fox, H. M. Atesalp, D. H. Kampbell, and H. F. Rhoades, Factors influencing the availability of sulfur fertilizers to alfalfa and corn, *Proc. Soil. Sci. Soc. Am., 28:* 406–8 (1964).
50. H. P. Rothbaum, W. Evason, and W. R. Owers, "Development of new sulphur fertilizers," *Proceedings of the Technical Workshop and Sulphur in New Zealand Agriculture* (P. E. H. Gregg and J. K. Syers, eds.), Occasional Report No. 4, Department of Soil Science, Massey University, Palmerston North, p. 139 (1983).
51. H. P. Rothbaum, J. R. Sewell, and W. Evason, Water degradable sulphur/anhydrite pellets for fertilizer use, *New Zealand J. Sci., 23:* 371–82 (1980).
52. D. A. Rogers and A. C. Braithwaite, The incorporation of molten sulphur in triple superphosphate and factors affecting the particle size of the sulphur, *New Zealand Fertilizer Manufacturers' Research Association Research Reports*, p. 361 (1984).
53. C. C. Boswell and B. Swanney, Alternative sulphur fertilizers in New Zealand, *Proc. New Zealand Grassland Assoc., 47:* 233–42 (1986).
54. R. G. Weir, Elemental sulphur particle size in sulphur-fortified superphosphate, *The Agric. Gazette, New South Wales, 78:* 486–91 (1967).
55. S. A. Waksman and J. S. Joffe, Micro-organisms concerned with oxidation of sulfur in soil, *J. Bacteriol., 7:* 239 (1992)).
56. H. H. Janzen and J. R. Bettany, The effect of temperature and water potential on sulfur oxidation in soils, *Soil Sci., 144:* 81–9 (1987).
57. C. D. Shedley, An evaluation of elemental sulfur as a pasture fertilizer, PhD Thesis, Armidale, Australia: University of New England (1982).

57a. M. J. Hedley, W. Chatupote, L. K. Heng, N. S. Bolan, P. Loganathan, and M. R. Bretherton, "An elemental sulphur oxidation model for all seasons," Massey Fertilizer and Lime Research Centre Occasional Report No. 8: 170–81 (1995).

58. U. S. Moser and R. V. Olson, Sulfur oxidation in four soils as influenced by soil moisture tension and sulfur bacteria, *Soil Sci., 76:* 251–6 (1953).

59. H. A. Kittams and O. J. Attoe, Availability of phosphorus in rock phosphate-sulfur fusions, *Agron, J., 57:* 331–4 (1965).
60. J. H. Watkinson and A. Lee, Kinetics of field oxidation of elemental sulphur in New Zealand pastoral soils and the effects of soil temperature and moisture, *Fert. Res., 37:* 59–68 (1994).
61. A. Lee, J. H. Watkinson, and D. R. Lauren, Factors affecting oxidation rates of elemental sulphur in a soil under a ryegrass dominant sward, *Soil Biol. Biochem., 20:* 809–16 (1988).
62. A. Ghani, A. Lee, J. H. Watkinson, C. C. Boswell, B. S. Thorrold, J. D. Morton, and M. P. Upsdell, "The quantification of oxidation of elemental sulphur in pastoral soils in New Zealand," Proceedings XVII International Grassland Congress II: 1465–7 (1993).
63. D. Swartzendruber and S. A. Barber, Dissolution of limestone particles in soil, *Soil Sci., 100:* 287–91 (1965).
64. J. H. Watkinson, Measurement of the oxidation rate of elemental sulfur in soil, *Aust. J. Soil Res., 27:* 365–75 (1989).
65. J. H. Watkinson and A. Lee, "A mechanistic model for the oxidation rate of elemental sulphur in soil tested on results by HPLC," Proceedings of Middle East Sulphur Symposium 1990, Cairo, Egypt, pp. 163–71 (1992).
66. J. H. Watkinson, Oxidation rate of elemental sulfur particles with a wide range of sizes *Aust. J. Soil Res., 31:* 67–72 (1993).
67. W. Nevell and M. Wainwright, Influence of soil moisture on sulphur oxidation in brown earth soils exposed to atmospheric pollution, *Biol. Fertil. Soils, 5:* 209–14 (1987).
68. P. H. Nye, Diffusion of ions and uncharged solutes in soils, *Adv. Agron., 31:* 225–72 (1979).
69. R. Steudel, G. Holdt, T. Göbel, and W. Hazeu, Chromatographic separation of higher polythionates $S_nO_6^2$ (n = 3. . . 22) and their detection in cultures of *Thiobacillus ferrooxidans*; Molecular composition of bacterial sulfur secretions, *Angew Chem. Int. Ed. Engl., 26:* 151–3 (1987).
70. A. Lee, C. Boswell, and J. H. Watkinson, Effect of particle size on the oxidation of elemental sulphur, thiobacilli numbers, soil sulphate, and its availability to pasture, *New Zealand J. Agric. Res., 31:* 179–86 (1988).
71. N. J. Barrow, Slowly available fertilizers in southwestern Australia, 1. Elemental sulfur, *Aust. J. Exp. Agric. Anim. Husb., 11:* 211–6(1971).
72. U. Skiba and M. Wainwright, Oxidation of elemental S in coastal dune sands and soils, *Pl. Soil*, 77: 87–95 (1984).
73. S. Deng and R. P. Dick, Sulfur oxidation and rhodanese activity in soils, *Soil Sci., 150:* 552–60 (1990).
74. K. Shimamura, Gas diffusion through compacted sands, *Soil Sci., 153:* 274–9 (1992).
75. S. J. Chapman, Thiobacillus populations in some agricultural soils, *Soil Biol. Biochem., 22:* 479–82 (1990).

76. D.G. Maynard, J. J. Germida, and P. A. Addison, The effect of elemental sulfur on certain chemical and biological properties of surface organic horizons of a forest soil, *Can. J. For. Res., 16:* 1050–4 (1986).
77. S. F. Ledgard, B. S. Thorrold, M. S. Sprosen, and M. F. Hawke, "Effectiveness of sulphate and elemental sulphur fertilisers applied to pastures in autumn or spring," Proceedings XVII International Grassland Congress II: 1464–5 (1993).
78. N. L. Nguyen, C. E. Wright, and C. B. Dyson, "Elemental sulphur, wet mix sulphur superphosphate and sulphur bentonite granules as sulphur fertilisers," Massey Fertilizer and Lime Research Centre Occasional Report No. 2: 214–21 (1988).
79. J. H. Watkinson and K. W. Perrott, "A new soil test for sulphate and mineralised organic sulphur," Proceedings New Zealand Fertilizer Manufacturers' Research Association Conference, pp 188–98 (1990).
80. S. J. Chapman, Microbial sulphur in some Scottish soils, *Soil Biol. Biochem., 19:* 301–5 (1987).

6
Role and Effect of Sulfur in Tree Biology

Gary D. Hogan
Natural Resources Canada, Canadian Forest Service, Sault Ste. Marie, Ontario, Canada

Heinz Rennenberg and Siegfried Fink
Institute for Forest Botany and Tree Physiology, Freiburg, Germany

I. INTRODUCTION: SULFUR ABUNDANCE AND FORMS IN NATURE

Sulfur (S) is a ubiquitous element. It is present throughout the environment in minerals, soils, plants, animals, air, and water. It is generally reported that the average concentration of S in the earth's crust is 0.05 to 0.06%. Sulfur is present in rocks, minerals, hydrocarbons, and soils in a number of forms, principally occurring as sulfide and sulphate minerals and organic matter. Sulfur is an essential element, forming part of a number of chemical compounds and cycles required for normal cell function.

Because of its ubiquitous nature and its close association with base metal minerals and with hydrocarbons, S has become a dominant form of air pollution. This activity has resulted in releases or emissions of S in a number of forms—S, hydrogen sulfide (H_2S), sulfur dioxide (SO_2), and sulfate (SO_4) as a component of acid rain and fog.

II. REQUIREMENT FOR S

A. Sulfur-Containing Compounds

As a constituent of protein, S is an essential component of all living cells. In many plant species, the requirement for S is therefore determined by protein synthesis. This is often reflected by a relatively uniform, species-specific sulfur-to-nitrogen (N) ratio of the plant biomass (Rennenberg 1984). However, plants can contain a wide range of S-containing secondary metabolites, such as complex thiols, thiazoles, cyclic disulfides, alliins, glucosinolates, thionins, and phytochelatins (Schnug 1993; Bohlmann 1993; Rauser 1993). Some of these compounds are produced in plants only under particular environmental conditions—e.g., phytochelatins in the presence of heavy metals such as cadmium in the soil. Still, S-containing secondary metabolites may constitute a significant fraction of the S present in the plant or a particular organ of the plant and will then also increase the plant's requirement for S. In addition, excess sulfate present in the environment may be taken up and stored inside the plant (Rennenberg 1984). Storage of reduced S in S-rich proteins may be essential for plant development—e.g., the synthesis of S-rich seed storage proteins by *Bertholletia excelsa* (Humb. & Bonpl.), the Brazil nut tree (Kreis 1990). Therefore, the actual requirement of a given species for S may not depend only on the amount of S used for protein synthesis; it may change during development or as a result of environmental constraints. In perennial plants such as trees, S uptake will usually not reflect S requirement in growth and development because storage and mobilization of S can dominate S nutrition, even during periods of intensive growth like bud burst. In Norway spruce (*Picea abies* (L). Karst), for example, the early growth of the new flush is dependent on the translocation of reduced S from the previous year's foliage and the sulfate stored inside the buds during fall (Schupp and Rennenberg 1992; Schupp et al. 1992).

B. Roles of S-containing Compounds Within Trees

Sulfate, the dominant S source of plants in their environment, is the ubiquitous precursor of all S compounds in plants. In evergreen and deciduous trees, sulfate is the main form of S transported long-distance from roots to the shoot (Schupp 1991; Rennenberg et al. 1994a). As sulfate is readily exchanged between xylem and phloem (Biddulph 1956), the existence of a cycling sulfate pool filled up by the roots and used up at the sites of sulfate reduction (e.g., the mature leaves) is generally assumed. However, experimental evidence for the transport of sulfate in the phloem is lacking. Sulfate is usually found in high concentrations in leaf tissue of trees (e.g., in needles of Norway spruce) (Schupp and Rennenberg 1992; Kaiser et al. 1993; Polle et al. 1994). However, most of sulfate found in mature leaves is thought to be retained within the vacuole. Trapped within this cellular compartment, where sulfate efflux is slow, it is at best only marginally available even if the growth of plants is limited by the availability of S (Cram 1990). Thus, foliar sulfate found in mature leaves must be considered as a permanent rather than a transient store that cannot be used to meet the needs of plants for S during growth and development. Recent experiments with beech, however, suggest that, unlike that held in foliage, sulfate is transiently stored in the wood during fall and is mobilized in early spring (Herschbach and Rennenberg 1996). The high concentrations of sulfate in the xylem sap of beech trees during spring may therefore at least partially reflect mobilization of sulfate stored in wood tissue (Rennenberg et al. 1994a).

The S-containing amino acids cysteine and methionine as well as the S-containing peptides glutathione and γ-glutamylcysteine constitute the predominant low-molecular-weight reduced S compounds in plants. In trees all these compounds may be present in the xylem sap but in strongly varying absolute and relative amounts (Schupp et al. 1991; Rennenberg et al. 1994a,b; Schneider et al. 1994a). Apparently, cysteine and methionine not only are essential metabolites for protein synthesis but also serve as a transport form of reduced S in the xylem of numerous tree species. Alternatively, or additionally, the S-containing tripeptide glutathione is used for the long distance transport of reduced S in the xylem of many tree species. The same compound is the predominant transport form of reduced S in the phloem of spruce (Schupp et al. 1992; Schneider et al. 1994b; Blaschke et al. 1996). In contrast to herbaceous plants and deciduous trees, where basipetal transport of glutathione is well documented (Rennenberg and Thoene 1987; Herschbach and Rennenberg 1995, 1996; Seegmüller et al. 1996), phloem transport of glutathione in spruce proceeds exclusively in acropetal direction to the youngest needle generation (Schupp et al. 1992; Schneider et al. 1994b). Additional functions of glutathione in trees include the detoxification of xenobiotics as a substrate of glutathione-S-transferase (Schroder et al. 1990a,b; Lamoureux et al. 1991) and the compensation of oxidative stress as a substrate in the Foyer-Halliwell-pathway (Polle and Rennenberg 1994). Other stress situations in which glutathione is involved in herbaceous plants (Rennenberg and Brunold 1994) have not been tested with trees.

III. UPTAKE AND RELEASE OF S

A. Uptake and Transfer Across the Root

Sulfur is available to roots mainly in the form of sulfate. Sulfate is taken up into plant roots by an active, carrier-mediated process (Cram 1990; Clarkson et al. 1993). By this process sulfate may enter the symplasm of the root at the root surface, in the root cortex, or at the endodermis. The endodermis is considered a barrier for apoplastic transport. Thus, transfer across the root to the endodermis may take place in either the symplasm or the apoplastic space, but proceeds via the symplast from the endodermis to the parenchyma cells of the xylem. To reach the transpiration stream in the xylem vessels, sulfate has to be transported from the symplasm of the parenchyma cells into the apoplastic space of the xylem. This process is usually called xylem loading and is also considered to be an active carrier-mediated process. Xylem loading of sulfate may be of particular significance as a controlling factor of the S nutrition of the shoot (Herschbach and Rennenberg 1994). Sulfate uptake and xylem loading of sulfate in the roots has been studied only recently in woody species. Sulfate carriers of beech roots exhibited similar or higher affinities and similar maximum transport velocities to those previously found in studies with roots of herbaceous species (Kreuzwieser et al. 1996). The percentage of the sulfate loaded into the xylem of beech roots versus that taken up (10%) was low compared with that of the roots of herbaceous plants (25–50%). Sulfate uptake and xylem loading of sulfate by roots of pedunculate oak were lower than that observed in herbaceous plants (Seegmüller et al. 1996). These differences in sulfate uptake and/or xylem loading of sulfates between woody species and herbs may be a consequence of different rates of growth. In beech, root sulfate uptake and xylem loading increased with decreasing sulfate supply. Nitrogen deprivation decreased, whereas incubation with ammonium increased, sulfate uptake (Kreuzweiser et al. 1996). As also found in herbaceous plants, organic S compounds inhibited S uptake of beech roots, but cysteine rather than glutathione mediated this effect in beech roots. In contrast to herbaceous plants, none of the organic S compounds tested significantly affected xylem loading of sulfate (Kreuzweiser et al. 1996). Apparently, sulfate uptake and xylem loading are regulated in beech roots in an independent manner.

Whether or not ectomycorrhizal symbiosis supports the S nutrition of trees is still a matter of debate (Morrison 1962; Harley and Smith 1983). Xylem loading of sulfate, but not sulfate uptake, is stimulated by the mycorrhizal association of *Quercus robur* (L.) with *Laccaria laccata* (Seegmüller et al. 1996). Because the fungal hyphae did not cross the endodermis, the physiological explanation for this stimulation remains obscure.

Sulfate uptake is, however, enhanced by vesicular-arbuscular mycorrhizae (e.g., Gray and Gerdemann 1973).

During degradation of organic material in the soil, S-containing amino acids may become available to plant roots in addition to sulfate. However, plants have to compete for these compounds with soil microorganisms. In trees, symbiosis with ectomycorrhizal fungi may also affect the availability of reduced S compounds generated during degradation of organic matter in the soil. Experimental evidence for the uptake of S-containing amino acids by roots in their environment is lacking. At the present time, it is also unknown whether sulfur-containing amino acids are transferred from mycorrhizal fungi to the roots. Still, the xylem sap of all tree species analyzed contains considerable amounts of reduced S in the form of glutathione, cysteine, and/or methionine (Schupp et al. 1991; Rennenberg et al. 1994a,b; Schneider et al. 1994a). The origin of these S compounds is only partially understood. In the crown of Norway spruce trees, reduced S compounds in the xylem sap originate from export out of the needles in the phloem and subsequent phloem-to-xylem exchange in the twigs (Schupp et al. 1992; Schneider et al. 1994b). Reduced S compounds in the trunk may originate from (1) uptake by the roots, (2) sulfate reduction in the roots, (3) mobilization of reduced S stored in wood and/or bark, and (4) recycling of reduced S synthesized in the leaves and translocated in the phloem in basipetal direction. The latter possibility can be excluded for Norway spruce because phloem transport of reduced S compounds apparently does not proceed in basipetal direction (Schupp et al. 1992; Schneider et al. 1994b; Blaschke et al. 1996). In deciduous trees such as beech that transport reduced S in the phloem in a basipetal direction, reduced S in the xylem usually does not originate from phloem-to-xylem exchange of reduced S produced in the leaves and allocated to roots (Hershbach and Rennenberg 1996). Reduced S found in the xylem sap of beech may originate from sulfate reduction and assimilation in the roots. The significance of ectomycorrhizal association for these processes remains to be elucidated.

B. Translocation and Uptake from Vascular System by Leaves

Translocation of sulfate and reduced S compounds in xylem and phloem is an essential feature of plant S nutrition. Sulfate taken up by the roots has to be transported to the sites of sulfate reduction (i.e., the chloroplasts of the leaf mesophyll cells). Reduced S compounds produced in the mesophyll chloroplasts have to be translocated to the growing parts of the shoot and the root in order to fulfill the S requirements of these organs during growth and development. It may therefore be assumed that sulfate is translocated in the xylem from the roots to the shoot and that reduced S compounds may be transported in phloem and xylem in an acropetal direction and in the phloem also in a basipetal direction. Several observations indicate that the allocation of S compounds in plants in general, and in trees in particular, is more complex. Sulfate can be exchanged between phloem and xylem and may therefore undergo a cyclic transport inside the plant. A cycling sulfate pool may prevent accumulation of excessive sulfate in a particular organ and constitute a buffer in the presence of excess S in the environment. It may also be a controlling device adjusting the influx of sulfate to the S requirement of the plant (Clarkson et al. 1993). In Norway spruce, reduced S is exported from mature needles and transported in xylem and phloem exclusively in an acropetal direction to the new needle generation in the form of glutathione (Schupp et al. 1992; Schneider et al. 1994b). Therefore, the question has to be addressed as to how the growing tissues of the roots and the trunk are supplied with reduced S. The presence of reduced S in the xylem sap of the trunk of Norway spruce trees (Rennenberg, unpublished results) indicates that the roots may reduce sulfate in amounts that fulfill their own needs and the needs of the growing tissues of the trunk for reduced S. The finding that the enzymes of assimilatory sulfate reduction are present not only in spruce needles (Bosma et al. 1991; Schupp et al. 1992), but also in the roots of spruce seedlings (Brunold, personal communication) supports this assumption. In spring, immediately before bud break or flowering of catkins, a large increase in xylem sap reduced-S content has been observed (Schupp 1991; Schupp et al. 1991; Schneider et al. 1994a). Apparently storage pools of reduced S are mobilized and translocated in spring to support the growth of new developing tissues. Both sulfate and reduced S compounds translocated in the xylem have to be taken up from the vascular system into the symplasm of the leaves and other plant organs to be used in growth and development. This process, generally called xylem unloading, has not received much attention. From experiments with green cultured tobacco cells it is known that sulfate and glutathione are taken up from the apoplastic space by active carrier-mediated processes (Rennenberg et al. 1988; Schneider et al. 1992), but similar studies with leaf tissue of herbaceous or woody plants have not been reported.

C. Foliar Uptake

Foliar uptake of S, both gaseous SO_2 and SO_4^{2-} in precipitation, has been shown to occur through paths associated with stomata or the leaf cuticle. Uptake of SO_2 via the stomata is generally considered to be the most important path for entry of SO_2 into plants. Stomatal conductance is generally assumed to limit the uptake of SO_2 in the same way that it limits water loss. This leads to a common assumption that under "normal" conditions, the rate of SO_2 flux to the foliage can be calculated from the stomatal conductance for water vapor efflux and the SO_2 concentration gradient from the atmosphere to the substomatal cavity of the leaf. Because of the high solubility of SO_2 in water, including water held in intercellular spaces, the effective concentration of SO_2 at the cell wall is considered to be close to zero. Actual measurements of SO_2 fluxes to foliage can be at variance with modeled values, depending on the experimental conditions applied (De Kok 1990). Higher-than-predicted rates of SO_2 deposition can be expected when the cuticular surfaces are wet; greater cuticular absorption and lower-than-predicted rates of SO_2 absorption may be explained by internal resistances—i.e., physical and metabolic factors counteracting SO_2 influx (Fowler 1985; Baldocchi 1993).

D. Release of H_2S

Sulfate reduced in the leaves may not be incorporated entirely into organic S compounds but can also be emitted into the atmosphere as hydrogen sulfide (H_2S). H_2S emission may also take place as a consequence of the degradation of S-rich proteins in plants (Rennenberg 1991). Emission of volatile S from plants has first been observed with Norway spruce (Materna 1966). Until today, H_2S emission has been demonstrated for many plant species, including numerous coniferous and deciduous trees (Rennenberg 1991). Emission is observed not only in the presence of excess S in atmosphere or pedosphere, but also in the absence of significant atmospheric S and low sulfate concentrations in the soil. In Norway spruce, emission of H_2S is highly dependent on stomatal aperture. At plant-available S contents of the soil between 10 and 18 mg per 100 g soil, about 1 nmole H_2S was found to be emitted per mole water (Rennenberg et al. 1990). Apparently emission of H_2S is controlled by environmental factors that influence stomatal aperture. The processes involved in the production and emission of H_2S have been studied only in herbaceous plants. They include the diversion of sulfide during sulfate reduction, the conversion of sulfite to sulfide, as well as the enzymatic degradation of cysteine and carbonylsulfide (COS) (Rennenberg 1991). These processes may operate when plants are exposed to excess S in the atmosphere or pedosphere, but also at normal S supply when more sulfate is reduced than is actually required for growth and development. In this way sulfate assimilation is rapidly adapted to a constantly changing environment (Rennenberg 1991). A controlling device of sulfate assimilation in the leaves may be of particular significance in trees, where any signal sent from the leaves to reduce sulfate uptake by the roots will operate with a considerable delay owing to the distance between leaves and roots.

IV. EFFECTS OF INADEQUATE S SUPPLY

A. Morphology—Whole Plant and Organ

In general, S deficiency leads to destruction of chlorophyll and is generally indicated by yellowish discoloration starting adjacent to the leaf veins; in severe cases reddish or purple hues may develop finally. In contrast to symptoms of N deficiency, however, these S deficiency symptoms first appear in young leaves and the older leaves remain unaffected. Inhibited cell divisions and protein synthesis causes small, thin plants with stubborn growth habit and sometimes succulent appearance. Root growth is less affected than shoot growth (Bergmann 1988).

B. Anatomy—Organ and Cellular

No data are available for trees, but in herbaceous species the amount of mechanical tissue is increased, which leads to a higher dry weight (Bergmann 1988). At the ultrastructural level, studies carried out on beans have shown that S deficiency induces the appearance of swollen and distorted chloroplasts (Whatley 1971).

C. Physiology and Biochemistry

The physiological and biochemical consequences of S deficiency on S metabolism have not been intensively studied. The capacity for the uptake of sulfate is strongly enhanced at low sulfate concentrations in the environment (Cram 1990; Clarkson et al. 1993). The activity of key enzymes of assimilatory sulfate reduction becomes derepressed (Brunold 1990, 1993). Because there is a regulatory coupling of S and nitrogen metabolism, S deficiency reduces nitrate reductase and to a smaller extent also glutamine synthetase activity (Brunold 1993). Still, the pool of free amino acids in S-starved plants becomes greatly expanded, especially in the arginine and asparagine, but also in the *O*-acetyl-serine content (Brunold 1993). In trees, mobilization of storage pools of S may be expected as a consequence of S deficiency. However, none of the studies on the physiological and biochemical consequences of S deficiency were performed with woody plants.

V. EFFECTS OF EXCESS S

A. Acid Precipitation

1. Morphology

In acid precipitation (rain, mist) sulfuric acid is usually the main component, but nitric acid contributes to acidity, too, whereas chloride is of only minor importance (Wellburn 1988). These components may have different effects because nitric acid can also have a beneficial effect in acting as a leaf-applied fertilizer and may even stimulate growth, leading to larger leaves (Maurice and Crang 1986). sulfuric acid is more destructive and usually acts as a damaging factor only, causing lesions on leaves (Adams et al. 1990). Visible foliar injury by acid precipitation is characterized by irregular necrotic spots on the leaf surface. Such type of damage has been experimentally induced in sensitive plants such as radishes, beets, beans, or poplars at pH values below 3.4 (Evans et al. 1978; Evans 1984, Wellburn 1988). The conifers seem to be quite resistant to such damages, though it is claimed in a few cases that necrotic spots could also be induced experimentally in needles by application of acidity at pH values of 2.5 or below (Hahn 1990; Whitney and Ip 1991; Zobel and Nighswander 1991). In the field, however, such direct damage by acidity has so far not been observed without doubt, neither in conifers nor in broad-leaved trees. In nature a number of other natural factors (insects, fungi, climatic extremes) can cause very similar necrotic flecks and are more likely to be responsible for field symptoms; the same is true for some supposed ozone symptoms (Hibben 1969; Bucher and Landolt 1985; Brennan 1988; Fink 1990a).

2. Anatomy and Cytology

The first point of attack of acid precipitation, of course, is the leaf/needle surface with its epidermis, cuticle, and epicuticular waxes. As a consequence, the amorphous and fibrillar, crystalline waxes (especially in the epistomatal chambers of conifer needles) tend to become partially fragmented and eroded, and partially baked together into amorphous masses. This is especially pronounced on long-lived leaves such as conifer needles. This natural process is hastened in polluted atmospheres in the field (where gaseous and aqueous phases affect leaves). In polluted environments the formation of these waxes on young leaves is retarded and the chemical composition of the waxes is altered. In addition, their erosion is greatly accelerated to 2–5 times the rate as seen under unpolluted conditions, and the regeneration capability is reduced. This effect is especially pronounced in the epistomatal chambers of the sunken conifer stomata. Because of the hydrophobic nature of the needle surface, drops of water tend to accumulate in the pits formed by sunken stomata, where they are further concentrated during evaporation (Percy and Riding 1978; Huttunen and Laine 1983; Riding and Percy 1985; Crossley and Fowler 1986; Karhu and Huttunen 1986; Rinallo et al. 1986; Grill et al. 1987; Schmitt et al. 1987; Nebe et al. 1988; Tuomisto 1988; Rinallo and Raddi 1989; Adams et al. 1990; Turunen and Huttunen 1991; Percy et al. 1992). The fibrillar epistomatal waxes in the conifer needles get baked together within 1–2 years to form massive, plate-like amorphous plugs forming epistomatal occlusions (which may secondarily rupture again), thus increasing the diffusion resistance of the stomata, which leads to a reduction of CO_2 exchange and of stomatal transpiration, thereby reducing photosynthesis. In rare cases transpiration may also become enhanced if the plugs block stomatal closure. The enhanced erosion of the amorphous waxes on the cuticle between the stomata, on the other hand, may increase the cuticular transpiration and the wettability of the leaves, which then results in an enhanced leaching of nutrients out of the leaves. Controlled experiments could show that these erosion and degradation effects as well as the lower wax production and a thinner differentiated cuticle are due mainly to the impact of acid precipitation (i.e., aqueous phase of pollutants), being the more pronounced the lower the pH; the effect may be further enhanced by gaseous SO_2, but this alone (in the dry phase) or ozone alone had much less effect upon the waxes (Haines et al. 1985; Rinallo et al. 1986; Mengel et al. 1987, 1989, 1990; Percy and Baker 1987, 1988). The penetration of aqueous depositions through the hydrophobic stomata is normally prevented by the high surface tension of water; thus, the damage by acidic droplets may be enhanced by addition of surfactants (Rinallo and Raddi 1989).

Attempts have been made to use structural degradation of epicuticular waxes in conifer needles as a diagnostic tool in the field for the proof of the harmful impact of acid precipitation. In recent studies of "forest decline" there have been several examinations of possible changes in wax structure as indicators of the direct action of acid rain. Not surprisingly, heavy degradation of the needle surface, together with contamination by particles, were found in needles from urban or industrial areas (Crossley and Fowler 1986; Hafner et al. 1989; Fiedler et al. 1990). Under much less polluted conditions, however, fused stomatal waxes and degraded surface waxes were reported from declining trees in various areas in central Europe (Sauter and Voss 1986; Badot et al. 1988). A close correlation of wax structure was found to exist with the vitality of the trees, but not necessarily with type or amount of emissions. Some conflicting data exist—for example, for the Black Forest, where other researchers were unable to find a significant difference in wax structure between healthy and declining trees. This was largely attributed to the large intra-tree variation in wax structure. Puzzling results came from a supra-regional inventory of needle wettability with the contact angle-method (a rapid quantitative method to estimate intactness of the surface waxes). A comparison of trees from different sites in Europe showed that the healthiest trees in Scotland had the least intact waxes (highest wettability) and the most diseased trees in the Black Forest had the most intact ones and the slowest weathering rates (Cape et al. 1988, 1989). This strongly indicates that outside "classical" heavily polluted industrial areas, the structure of surface waxes may be mainly determined by natural (climatic etc.) influences, water stress, mineral nutrition, and indirectly by the general vitality of the trees. For example, internal physiological disturbances may increase synthesis of monoterpenes, which then would accumulate in the waxes, thus depressing their melting point and facilitating degradation by natural factors, without any additional external environmental stress (Güth and Frenzel 1989). Even nutrient deficiencies can lead to altered wax structures (Ylimartino et al. 1992). Abrasive damage by strong winds alone may also lead to a premature ageing of the waxes (van Gardingen et al. 1991).

Recent research has pointed to the fact that the distribution of calcium oxalate in conifer needles may be a sensitive indicator of the influence of acid precipitation (Fink 1991 a,b,c). In the epidermis, minute crystals are normally distributed within the epidermal cell walls and especially in the cuticular layer between the cuticle proper and the cell walls (Fig. 1). Under the impact of acid precipitation, this pattern of distribution now becomes altered and only very small amounts of crystals occur in the cuticular layer and the outer epidermal cell walls, whereas most of them remain restricted to the inner epidermal cell walls (Fig. 2). This probably reflects the effect of "leaching" of Ca^{2+} from the outer cell walls by the impact of acid precipitation, whereas the inner parts of the walls are not affected by the acidity. It is unlikely, however, that already-existing crystals of calcium oxalate become dissolved by the acid, but rather the cations are leached out of the apoplast before they are precipitated as oxalate. From the physiological point of view, it does not make a difference if Ca^{2+} is leached away from the apoplast and is thus lost to the needle. In the normal course of events, it would become precipitated as insoluble calcium oxalate and would become unavailable physiologically. This observation would be in accordance with the microscopic detection of crystals of calcium sulfate upon the surface of needles subjected to acid rain (Bosch et al. 1983; Nebe et al. 1988; Fiedler et al. 1990; Huttunen et al. 1990).

In addition to these effects on the structure of the cuticle and its waxes, acid precipitation may cause further damage and lead in the next step to punctual necrosis of epidermal cells on the adaxial surface, fre-

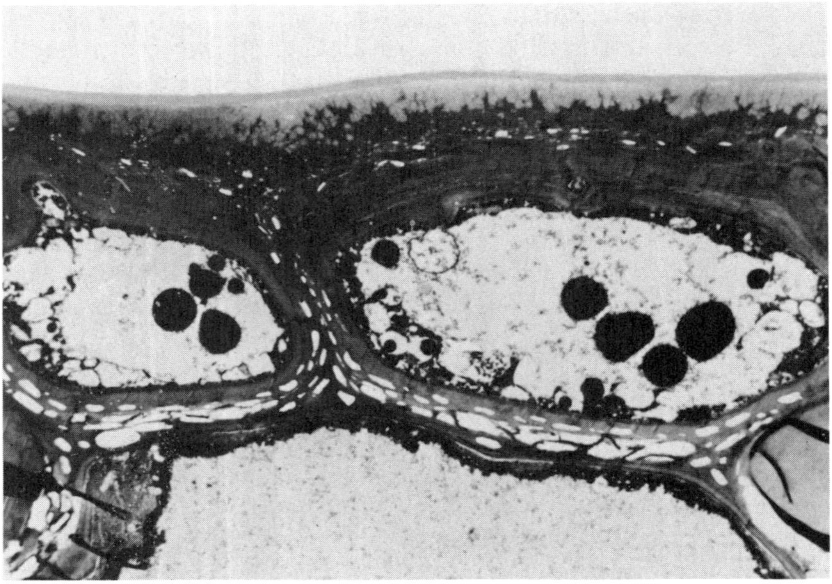

Figure 1 Electron micrograph of the normal distribution of minute crystals of calcium oxalate around the epidermal cell of a 1-year-old spruce needle well supplied with calcium (control). (× 3.300)

quently close to leaf veins or stomata, and at the bases of trichomes or glandular hairs (where the cuticle is very thin as a consequence of the secretory function of these structures). If cells adjacent to stomata collapse, the stomata may be unable to close. At pH values below about 3.5, larger epidermal necroses may proceed to form, which then may spread to adjoining cells of the palisade parenchyma and eventually spongy mesophyll. The neighboring undamaged mesophyll cells then react with hypertrophy, the cells divide again, and by hyperplasia may finally lead to circular or ringlike (with necrotic tissue in the center) proliferations of the leaves, somehow resembling insect galls. These tissues form a sort of delineation, resembling early stages of wound periderm formation, and may thus limit further progress of damage. In extreme cases, the damage may also proceed to the vascular bundles after most of the mesophyll has already collapsed. In general, developing young leaves are more affected by acid rain than older leaves (Evans et al. 1978; Evans and Curry 1979; Tung et al. 1982; Paparozzi and Tukey 1983; Adams et al. 1984; Evans 1984, Crang and McQuattie 1986; Lorenzini et al. 1988; Rinallo and Radii 1989).

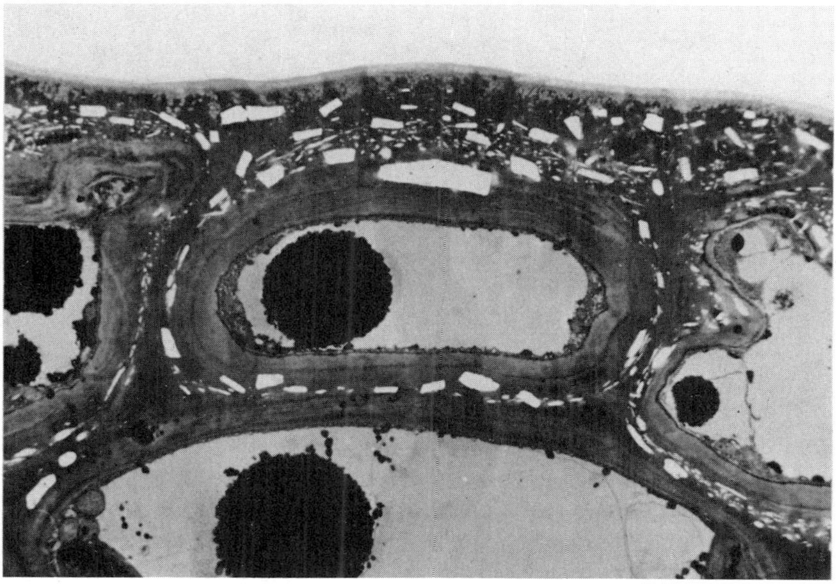

Figure 2 The corresponding pattern in the needle from an acid rain treatment (one vegetation period with pH 2.5); deposits of crystals occur only in the inner epidermal walls, whereas the outer part remains nearly free. (× 3.100)

Nonlethal changes following acid rain treatment have been observed in chloroplast structure, too. Chloroplasts in outer mesophyll cells tend to be more irregular in form with fewer thylakoids, thicker granal packing, larger starch grains, and slower starch dissolution in autumn. Sometimes acid-treated needles show more severe changes in chloroplast structure after they have been subjected sequentially to cold treatments (Holopainen and Nygren 1989; Reinikainen and Huttunen 1989; Bäck and Huttunen 1992). Needles differentiating under acid precipitation treatment may show deformations and occlusions of stomatal complexes (Turunen and Huttunen 1991; Fink 1992).

Roots of tree seedlings exposed to artificial acid rain show the development of irregular vacuoles in cortical cells of root tips, abnormal tannin deposits, and early disintegration of cytoplasm (Saastamoinen and Holopainen 1989).

From the discussion on "New-type Forest Decline," there is evidence that the symptoms of mineral deficiency (especially concerning Mg in the "Montane Yellowing" syndrome) are the result of increased leaching of cations from the soil in response to deposition of excess S and nitrogen. Macroscopically, the symptoms are characterized by a yellowing of the older leaves or needles, starting from the margins or the needle tips. Microscopically, the symptoms are characterized, especially in conifer needles, by a premature collapse of phloem cells with hypertrophy and hyperplasia of adjoining cambial and parenchyma cells, whereas the mesophyll stay intact for a long time. Only the chloroplasts show alterations, with swelling, thylakoid separation, and accumulation of starch and plastoglobuli (Fig. 3; Fink 1988, 1989, 1990b, 1992; Hüttl and Fink 1988, Holopainen and Nygren 1989; Holopainien et al. 1992).

3. Physiology

a. Effect of Simulated Acid Rain (SAR) on Growth and Component Phytomass. Numerous studies have examined the effects of acidic deposition on tree growth and component phytomass distribution. There is considerable variation among the experiments that have been carried out with respect to species examined, chemical composition of the simulated rain,

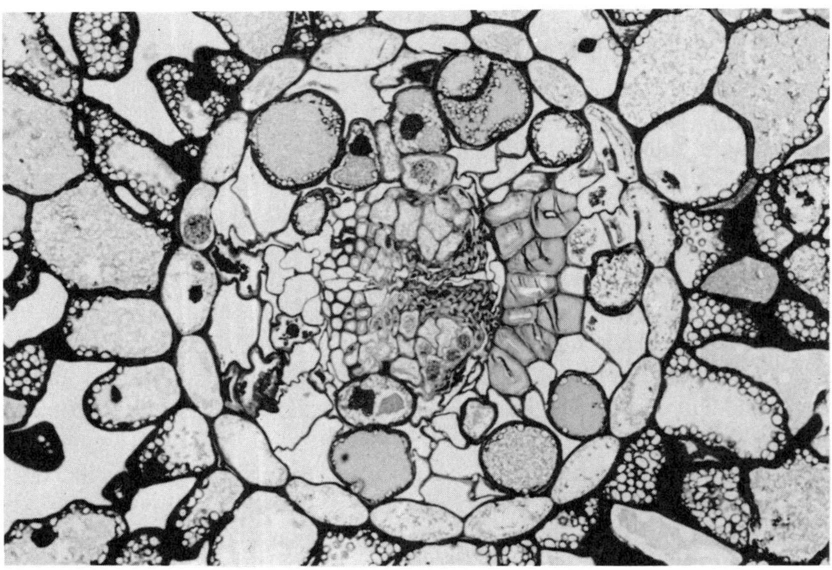

Figure 3 Spruce needle suffering from Mg deficiency as a consequence of acid rain and nitrogen inputs; mesophyll cells are intact whereas phloem shows collapse with hypertrophy and hyperplasia of adjoining living cells. (x 530)

range of acidities examined, and a host of other factors. It should come as no surprise, then, that a range of outcomes have arisen from these studies.

Studies of a number of deciduous species (*Liriodendron tulipifera* L., *Platanus* sp., *Liquidambar* sp. and *Robinia pseudoacacia* L.) (Neufeld et al. 1985), red oak (*Quercus rubra* L.), sugar maple (*Acer saccharum* Marsh.) (Reich et al. 1986), and hybrid poplar (Hogan and Taylor 1995) found little reduction in height growth except at pH 2.0 (Neufeld et al. 1985) or stem, leaf, or total phytomass, specific leaf area, number of leaves, height, stem diameter, leaf area, on root foliar or total dry weight, suggesting that these parameters are insensitive to acidic deposition except for acidities in the extreme range. In spite of a general lack of observed effect on tree growth or phytomass at pH >3.0, studies on deciduous species have often revealed that they have a greater sensitivity to acidic deposition than do conifers.

Results from studies with conifers are only slightly less illuminating than those obtained from the study of deciduous species. Studies with jack pine found that average diameter and shoot weight increased as pH of SAR decreased to pH 2.5 but that there was no significant effect on height growth (MacDonald et al.1986). Studies on *Pinus taeda* L. (Adams and O'Neill 1991; Qiu et al. 1992) over the range pH 5.2 to 3.3 found no significant effect on phytomass, root to shoot ratio, or carbon allocation. Investigations of the effects of acidic deposition on red spruce (*Picea rubens* Sarg.) have also produced conflicting results, ranging from no significant effect on height, phytomass, or root to shoot ratio (pH> 3.5) (Deans et al. 1990; Patton et al. 1991), to increases in root and shoot phytomass of red spruce at pH 4.1 (Taylor et al. 1986).

Studies on the effect of acidic deposition on allocation to belowground phytomass are not well represented in the literature, and available information is strongly weighted toward laboratory investigations carried out under controlled investigations. Few field growth experiments or natural forest studies across acidic deposition gradients have examined these criteria. Reductions in root phytomass and root relative growth rate have been observed in white ash (*Fraxinus americana* L.) following acid rain treatment at pH 3.0 (Chappelka and Chevone 1986). A similar response was seen in jack pine where root weight decreased significantly when pH of SAR fell below pH 2.5 (MacDonald et al. 1986). In addition, acidic deposition has been observed to cause shifts in carbon allocation from structural support (coarse roots) to a nutrient absorption function (fine roots). Red spruce seedlings receiving the pH 3.0 acid treatment were shown to have a greater frequency of fine roots along the coarse roots (Deans et al. 1990), an indication of a response to increased N or S availability. Studies conducted with *Pinus taeda*, however, at pH 5.2, 4.5, and 3.3 found no significant effect on fine or coarse root phytomass, or on carbon-14 (^{14}C) allocation to all components except for fine roots (Adams and O'Neill 1991).

Considered together, these results support the general conclusion, advanced by Amthor (1986), that growth inhibition occurred only when the acidity of acidic deposition was less than pH 3.0. A number of more recent studies indicate that shifts in allocation, where observed, seem to occur at a similar level of acidity.

b. Effect on Foliar Leaching, Foliar Nutrient Concentrations and Buffering Capacity. Early reports and reviews (Tamm and Cowling 1977; Morrison 1984) suggested that acidic deposition might increase foliar leaching of nutrients, resulting in reductions in foliar concentrations. This was thought to be particularly true for base cations and led to speculation that foliar cation loss would eventually lead to reductions in growth or forest health. Experiments with intact trees, both coniferous and deciduous, however, have not clearly demonstrated significant reductions of cation levels in leaves of trees exposed to SAR less acidic than pH 3.0 (Mengel et al. 1989; Patton et al. 1991; Hogan and Taylor 1995). This suggests that if accelerated leaching of cations is occurring at these levels of acidity, then there must be compensation through increased uptake or through internal redistribution.

Nutrient loss from foliar surfaces in response to changes in external acidity has been examined intensively and remains a readily identifiable mode of action of acidic deposition. Discussion related to this subject area has largely focused on the relative ease with which base cations are lost from foliage in response to increases in "acid accelerated leaching," the pH at which this occurs, the degree to which base cations are affected, and the long-term implications.

Base cations have been shown to vary in the their response to changes in the external acidity of precipitation. Efflux of calcium (Ca) from sugar maple foliage increased over the range from pH 5.0 to 3.0 but that there was no marked effect on the efflux of potassium (K) or magnesium (Mg) (Wood and Bormann 1975). A similar response for K over the pH range 5.7 and 4.3 was observed for yellow birch (*Betula allegheniensis*) (Scherbatskoy and Klein 1983), for sugar maple between pH 5.5 and 3.0 (Hogan 1992), and over the pH range 5.0 to 3.8 for sugar maple and white pine (Lovett and Hubbell 1991). Potassium is known to be quite mobile within the plant, a characteristic that would lead one to suspect a significant loss from foliage in response to external increases in acidity, but K has not always responded in a predictable fashion over the range pH 5.0 to 3.0.

Fairfax and Lepp (1975) found a fourfold increase in Ca efflux from tobacco, but no increase in K loss as pH decreased from 6.7 to 3.0.

Differences in the response of base cations could be related to their site of localisation within the leaf (Hogan 1992). Because Ca is primary localized within the cell wall "free space," it easily comes into contact with external acidic solutions. This should translate into higher leaching losses that are more directly related to changes in external acidity (Hogan 1992). K and Mg, on the other hand, are associated primarily with the symplasmic fraction of the leaf, an added barrier to cation leaching; as a result, their efflux is considerably less affected by changes in external acidity than that of Ca. In-depth studies with sugar maple leaves (Scherbatskoy and Tyree 1990) have suggested that cation loss from leaves is dominated by rapid surface exchange mechanisms, and that the time required for transcuticular exchange of these cations is too great for their loss to be of long-term significance.

A number of studies have shown that vegetation also has the capacity to resist acid injury through external and internal processes. External neutralization capacity (ENC) provides a measure of the ability of the leaf to buffer the external solution, the buffering capacity index (BCI) provides a measure of the ability of the leaf to resist internal acidification. Extensive studies of these variables have not been made in native species, but the work to date suggests there are large differences and that these characters might be used to assess species at risk. Current indications are that ENC is related to foliar nutrient levels in a general way and that BCI is correlated with Mg, N, and Ca (Liu and Cote 1993). Sugar maple (*Acer saccharum* Marsh.) and balsam fir (*Abies balsamea* (L.) Mill.) were shown to have low a ENC and BCI, characteristics that would make them more be susceptible to acidic deposition than species like paper birch (*Betula papyrifera* Marsh.) and largetooth aspen (*Populus grandidentata* Michx.). There is a suggestion that early successional species like birch and largetooth aspen, with higher foliar concentrations of elements, would be less sensitive to acidic deposition than late successional species, such as sugar maple. This could establish a clear physiological basis for a range in sensitivity to acidic deposition.

The mobilization of aluminum (Al) through soil acidification and its subsequent effect on the uptake and transport of base cations is a central issue in much of the current acidic deposition literature. Red spruce decline at high elevation in the Appalachian Mountains has been tightly linked to increases in Al/Ca molar ratio of woody tissues and to increases in respiration (McLaughlin et al. 1991). Higher Al/Ca ratios were also found for red spruce root tips at higher elevation sites, confirming that increases in Al availability are associated with the tree decline (Smith et al. 1995).

c. Effect of SAR on Chlorophyll a and b Concentration and Fluorescence. A number of studies of agricultural species (Banwart et al. 1988; Norby et al. 1985) examined chlorophyll concentrations relative to levels of acidic deposition. These studies generally confirmed that SAR did not reduce plant chlorophyll levels unless visible injury was present. Evidence from published studies on trees, however, is not as plentiful, but does exist. Barnes et al. (1990) examined the effect of acid mist at pH 3.6 on Norway spruce, observing slight, but not significantly so, increase in chlorophyll a concentration. Chlorophyll b concentration was not significantly affected, nor was total chlorophyll or chlorophyll a/b ratio. However, not all studies with conifers have reached a similar conclusion. Studies with red spruce (McLaughlin et al. 1992) have indicated that rainfall/mist acidities in the order of pH 3.0 were able to bring about an 18% reduction in foliar chlorophyll levels.

Unlike those reported for conifers, studies with deciduous species appear to have reached consensus. Chlorophyll a and b concentrations of 2-year-old sugar maple seedlings were not affected by increases in acidity of rain (pH 5.6 to 3.0) (Reich et al. 1986); neither were those of hybrid poplar over a similar range of acidities (Hogan and Taylor 1995). Investigations of chlorophyll concentrations in a number of deciduous species at pH levels as low as 2.0 (Neufeld et al. 1985) suggest that this is not a principal mode of action of this pollutant.

d. Effect of SAR on Carbon Exchange and Storage. Studies of gas exchange in deciduous trees following exposure to acidic deposition have produced a number of outcomes. Studies with *Liquidambar styraciflua* L. and *Robinia pseudoacacia* L. (Neufeld et al.1985) found significant effects on net photosynthesis (A), but only in trees exposed to pH 2.0. Similarly, investigation of the effect of SAR on photosynthesis in sugar maple and red oak (*Quercus rubra* L.) found no significant reduction after treatment at pH 3.0 (Reich et al. 1986). Other authors have reported stimulation of net photosynthesis or no effect in the range pH 3.0 to 3.5 for a number of species, both deciduous and coniferous (Norby et al. 1985; Trites and Bidwell 1985; Taylor et al. 1986; Reich et al. 1987; Sasek et al. 1991; Seiler and Paganelli 1987; Barnes et al. 1990; Hogan and Taylor 1995; Hogan 1997). An exception to this general pattern, a reduction in net photosynthesis after treatment with SAR at pH 3.0, has been reported in yellow poplar (*Liriodendron tulipifera* L.) (Roberts 1990).

Net photosynthesis in conifers appears to mirror the results found for deciduous species. A range of responses have been observed. Studies with red spruce have revealed increases of net photosynthesis at rain acidities of pH 4.1 and no apparent effect at a mist pH of 3.6 (Taylor et al. 1986). Another study by McLaughlin et al. (1992), however, reported significant effects of acidic mist at pH 3.0 on net photosynthesis and dark respiration. Net photosynthesis was decreased by 19% and dark respiration was increased by 51% after 16 weeks of acid mist treatment. Increased exposure to acidic deposition, particularly at high elevation sites, has been identified with increased dark respiration rates at southern spruce fir forest sites (McLaughlin and Kohut 1992; McLaughlin et al. 1993), resulting in carbon balances that are less conducive to sustained growth.

Carbon storage and storage products have been shown to be reduced in a number of species. Jensen and Patton (1990) found decreased starch in roots of yellow poplar treated with rains of pH 3.5 to 4.5, a result that was corroborated by Mengel et al. (1990) at pH 3.5. Although most of this work has focused on root levels of carbohydrates, levels of ethanol-soluble carbohydrates were also found to be reduced by acidic mists in current and 2-year-old conifer needles (Barnes et al. 1990).

e. Effect of SAR on Water Loss, Water Relations, and Response to Drought. Acidic deposition effects on foliar water loss, water relations, and response to drought are of considerable interest. There is a great potential for acidic deposition to influence function and structure at the leaf surface, based largely on accessibility of tree tissues to the pollutant. In addition, acidic rains are free to concentrate on the leaf surface, adjacent to structures where effects owing to increased acidity or ionic strength can occur (Evans et al. 1978)

Does acidic deposition, at ambient levels, influence plant–water relations by increasing water loss and thereby predisposing trees to drought stress? Theoretically, this could occur through (1) physicochemical changes at the leaf–air interface that lead to increased or uncontrolled water loss (i.e., altered cuticular membranes, decreased stomatal response); (2) physicochemical changes at bulk tissue level that influence the ability of leaf tissue to respond to imposed drought (i.e., changes in osmotic potential or cell wall elasticity); or (3) an interaction of the two.

A number of acid-induced structural changes have been reported (Rinallo et al. 1986) and have already been discussed herein. In addition, there has been a substantial research effort aimed at establishing a link between acidic deposition, change in the cuticular surface, stomatal apparatus function, and changes in foliar water loss. Despite of the considerable amount of effort in this area, however, studies with conifers have produced conflicting results, making it difficult to reach consensus. Red spruce has been studied extensively but this does not appear to have reduced the level of uncertainty. Data obtained with red spruce, however, suggest that SAR treatment has little or no effect on leaf conductance (Eamus et al. 1989). It has been clearly demonstrated that the control of water loss by either the stomata or the cuticle was not affected by mist acidities as strong as pH 2.5. This result is surprising because many workers consider acid mist to be more damaging than rain. Other workers (Roberts and Cannon 1989) have reported that stomatal conductance was decreased by 51% following treatment with SAR at pH 3.0.

Examinations of stomatal, or leaf, conductance in a number of broad-leaved crop species have found that leaf diffusive resistance was lowered after exposure to SAR over the range pH 2.7 to 3.4 (Evans et al. 1984). Studies on deciduous tree species have reported the opposite effect—a 33% decrease in stomatal conductance after exposure to SAR of pH 3.0 (Roberts 1990), or no apparent effect (Hogan and Taylor 1995). The reasons for this contradiction are not readily apparent. They may be related to differences in herbaceous versus woody plant leaf structure or physiology, and therefore resistance, but also serve to demonstrate that the relationship is not a simple one and that further studies are required.

Acidic deposition can alter the morphology and composition of epicuticular waxes (Rinallo et al. 1986; Barnes and Davison 1988), but the role of these changes in influencing cuticular water loss is less well defined. Eamus et al. (1989) found no significant differences in cuticular resistance to water loss in *Picea rubens* foliage treated with SAR (pH 5.0 to 2.5). Either this lack of a response of cuticular water loss to SAR treatment is specific to red spruce, or changes in the cuticular structure of conifers reported in other studies occur without any significant effect on cuticular resistance. Another study, conducted on Norway spruce, raised the possibility that stomatal conductance, and hence transpiration, was unaffected by acidic deposition under conditions of normal water supply. Mengel et al. (1989) found that trees treated with acidic fog at pH 3.0 had higher rates of transpiration but only when subsequently exposed to drought conditions. An examination of cuticular water loss under controlled conditions suggested that the site of the effect was probably the cuticular surface and not the stomata.

Studies of the effect of acidic deposition on cuticular transpiration in deciduous species are not as numerous. Hogan and Taylor (1995) found no significant effect of SAR over the range pH 5.5 to 3.0 on cuticular transpiration in hybrid poplar, a result that was supported by studies with *Acer saccharum* Marsh. (Hogan, unpublished results). Although the picture is not perfectly clear, there does not appear to be sufficient evidence to support the assumption that ambient levels of acidic deposition accelerate foliar water loss in deciduous species.

An examination of the effect of SAR on xylem water potential (Q_x) in red spruce found that it was less negative in pH 3.0–treated trees than in the controls—an effect that was attributed to stomatal closure (Roberts and Cannon 1989). Their trees showed a tendency toward a lower Q_s (osmotic adjustment), indicating that trees were responding to acid-induced stress by increasing their osmotically active constituents. This led to the conclusion that SAR was unlikely to cause major disruptions in the tree–water relations. A later examination of SAR effects on plant–water relationships by these same authors found no significant effect on Q_x, osmotic or turgor potential under non-drought conditions. Xylem water potential (Q_x) in red spruce was shown, by others, to decrease as SAR acidity increased from pH 5.0 to 2.5. This result suggested that in the predawn condition, acid mist–treated seedlings were mildly drought-stressed (Eamus et al. 1989). At pH 2.5, turgor potential was lower than in the pH 5.0 seedlings, a reflection of reduced carbon uptake, but modulus of elasticity was lower, indicating a greater elasticity and a greater potential to tolerate water stress.

Deciduous species have not been studied as intensively, but there are indications that the response is the same. SAR at pH 3.6 has been shown to decrease Q_x (more negative), indicating a significant increase in the degree of drought stress shown by white oak. (Walker and McLaughlin 1991). This response was attributed more to adverse effects of SAR on root function than to an effect at the foliar surface. Other studies with deciduous species have failed to show a clear relationship between SAR acidity and drought stress as measured by Q_x (Roberts 1990; Hogan and Taylor 1995).

4. Biochemistry—Metabolism, Cycling, and Storage

Trees exposed to sulfate-containing acid mist often have enhanced foliar sulfate concentrations, with the increase in sulfate content depending on the treatment dose (Sheppard 1994). This finding is surprising because (1) stomata are thought not to provide a pathway to aqueous sulfate uptake as a consequence of surface tension (Schonherr and Bucovac 1972) and (2) cuticles provide an effective, almost impermeable barrier for anions (Lendzian and Kerstiens 1991).

Even though low pH and increasing ion concentrations from drying acid mist on leaf surfaces may enhance the flux of anions through cuticles, sulfate movement through cuticles is far too slow to explain the increase in foliar sulfate content observed. However, sulfate may enter young leaves via incompletely formed cuticles. This assumption is supported by the finding that young leaves are generally more sensitive to damage by acid mist than older leaves (Sheppard 1994). In addition, cracks in the cuticular layer of the leaves and the periderm of the twig axis may enable additional sulfate influx. Recently, it has been suggested that ions on leaf surfaces may move into the substomatal cavity via ultrathin water layers connecting the leaf surface with the leaf interior (Burkhardt and Eiden 1994). Whether this pathway of ion influx is also operating in conifer needles, where wax tubes may occlude the epistomatal cavity, remains to be elucidated.

Sulfate from acid mist that has entered the leaf will first come into contact with the apoplastic space and, hence, is considered to enter the sulfate pool translocated from the roots to the shoot with the transpiration stream. However, it has been reported that foliar-absorbed sulfate, unlike root-absorbed sulfate, can accumulate in the apoplasm (Van Stegen and Myttenaere 1991). In addition, a significant correlation between the increase in leaf sulfate content due to foliar absorption and the loss of frost hardiness was found upon exposure of trees to sulfate-containing acid mist. High sulfate concentrations in foliage from soil derived sulfate were only poorly correlated with changes in frost hardiness (Sheppard 1994). These findings suggest that sulfate provided by the transpiration stream and sulfate from foliar absorption remain functionally separated in the leaves. The question, how this functional separation is achieved, cannot at present be addressed, as xylem unloading of sulfate and the subsequent translocation of sulfate inside the leaf are only poorly understood (Rennenberg 1984).

In red spruce seedlings, a 0.1% increase in foliar S was correlated with a 2.7°C decline in frost hardiness compared to controls (Sheppard 1994). Does this mean that the reduction in frost hardiness, or even the foliar necrosis observed upon exposure of trees to sulfate-containing acid mist, is caused by the enhanced foliar sulfate content? To address this question several observations have to be considered: (1) Although measurements of apoplastic sulfate concentrations have not been performed with leaves exposed to acid mist, foliar absorbed sulfate was found to accumulate in the apoplastic space (Van Stegen and Myttenaere 1991); (2) enhanced nitrogen supply was shown to mitigate the toxic effects of exposure to sulfate-containing acid mist (Sheppard 1994); and (3) seedlings were found to be more sensitive to sulfate-containing acid mist than older trees (Sheppard 1994). These and other observations were introduced by Sheppard (1994) into a general hypothesis on the causal relationship between the influx of sulfate from acid mist into leaves and reduced frost hardiness. In this hypothesis, an uncontrolled buildup of sulfate ions in the apoplastic space—that will be more effective in seedlings than in older trees—may cause protein denaturation in the plasma membrane and, as a consequence, a loss of membrane integrity. This direct, toxic effect of sulfate will only take place when other nutrients such as nitrogen are present in amounts insufficient to direct the extra sulfate into protein synthesis for tree growth and development. The loss of membrane integrity will induce repair processes that consume photosynthate. Consequently, the amount of sugars available for the synthesis of cryoprotectants will be reduced and frost hardiness will decrease. Even if the influx of sulfate from exposure to sulfate-containing acid mist is too low to cause an accumulation of sulfate in the apoplastic space, sulfate entering the symplasm may also become toxic. Provided (1) the sulfate entering the cytoplasmic compartment cannot readily be translocated into the vacuole and (2) nitrogen and/or carbon assimilation are insufficient to mediate rapid sulfate assimilation, sulfate accumulating in the cytoplasm may reduce the membrane integrity of the chloroplast by protein denaturation. This will result in reduced photosynthate production, a reduced availability of sugars for cryoprotectant synthesis leading to decreased frost hardiness. Although this hypothesis appears to be conclusive, detailed experimental analysis of its individual parts is required to establish a causal relationship between the accumulation of sulfate and a reduced frost hardiness. It has previously been suggested that increasing levels of thiols such as glutathione are an important factor to achieve frost hardiness in conifer needles (Esterbauer and Grill 1978). However, exposure to acid mist enhances not only the sulfate but also the thiol content of the foliage, but frost hardiness is decreased (Sheppard 1994). Enhanced thiol contents alone are therefore insufficient to improve the frost hardiness of conifer needles.

B. Sulfur Dioxide

1. Morphology

In general, SO_2 enters the leaf primarily through the stomata and only a small percentage may diffuse through the cuticle. Thus, the amount of damage strongly depends on the stomatal regulation. Young leaves are usually not sensitive, because stomata are still closed and no gas exchange occurs. Then there is an increase of susceptibility to acute injury with start of gas exchange. Though older leaves (especially several years old in evergreen species like conifers) are less sensitive to acute injury (less gas exchange), they are more sensitive to chronic influences because of accumulation of damage over the years. Young leaves/needles, however, may be influenced in their morphology/anatomy if development occurred under the influence of pollutants (e.g., thinner and smaller lamina, fewer intercellular spaces, more stomata, thinner cuticle, fewer mesophyll cells, less vascular tissue; Masuch et al. 1985, 1988). Under high air humidity, SO_2

causes the guard cells to become even more turgid and therefore the pores open wider still, allowing more pollutant to enter the leaf than would normally be the case, probably by interaction of the sulfite ion with the cations that regulate the stomatal response (Wellburn 1988).

As an acute injury symptom, SO_2 induces necroses from the tips (tip burn in needles) or margins (leaves), and is sharply delineated from healthy tissue. In broad-leaved trees, confined irregular spots in the intercostal areas or along the leaf margins show first a water-soaked appearance and subsequently dry out and turn necrotic, with a tan color. Chronic damage, especially in long-lived conifer needles, is characterized mainly by premature needle and leaf fall.

2. Anatomy and Cytology

If conifer needles differentiate under the chronic impact of gaseous pollutants, their differentiation pattern is changed from that of normal development. Under realistic SO_2 fumigation regimes, a decrease in the amount of intercellular spaces was observed, whereas ozone caused an increase in intercellular spaces. Long-term continuous or intermittent SO_2 and ozone fumigation lead to a reduction in the amount of vascular tissue formed and number of mesophyll cells, ozone having a stronger reducing effect and discontinuous fumigation causing a stronger reduction than continuous fumigation. This reduction of vascular tissue may cause water deficits and problems in assimilate translocation (Vogels et al. 1986; Masuch et al. 1988).

In general, within an already differentiated leaf or needle exposed to SO_2, the primary point of attack is the mesophyll, especially directly adjacent to the substomatal cavities, and sometimes directly the guard cells where the pollutants contact the stomata. In contrast, the vascular bundle is mainly unaffected in the beginning. This type of primary necrosis of guard cells and adjacent mesophyll has already been described in the first anatomical studies that were realized with regard to the impact of gaseous pollutants, namely SO_2 damage to needles of Norway spruce, *Picea abies* (Hartig 1896; Haselhoff and Lindau 1903; Sorauer 1908; Haselhoff et al. 1932).

This fundamental observation of primary damage and necrosis to the mesophyll cells, starting from the substomatal chambers, affecting the directly adjacent cells, and eventually spreading throughout the mesophyll, has been confirmed for many conifer needles and deciduous leaves subjected to SO_2, O_3, NO_x, HF, Cl_2, singly or in combinations; in long-living conifer needles or evergreen leaves accumulative damage proceeds through additional necrosis in subsequent years (Figs. 4, 5, 6; Solberg and Adams

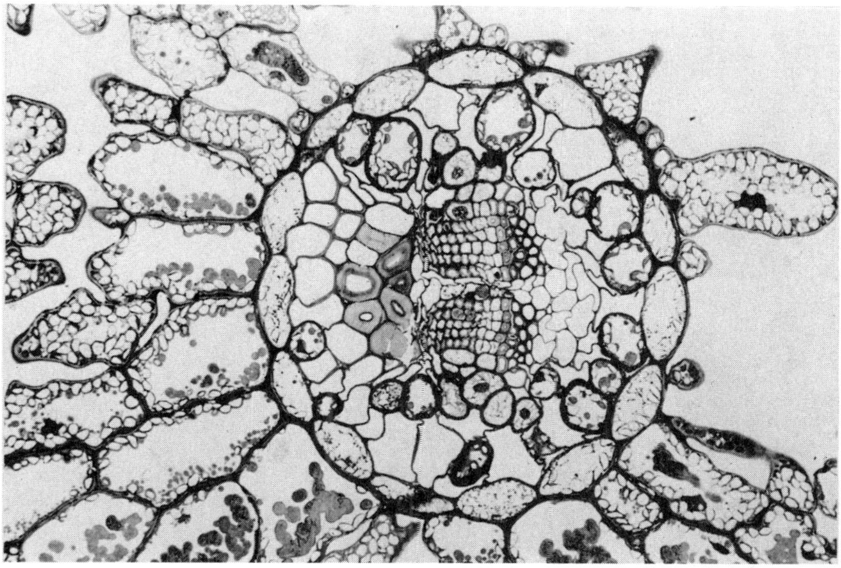

Figure 4 Cross section through a 2-year-old healthy spruce needle with intact mesophyll cells and functional vascular bundle. (× 520)

1956; Costonis 1970; Stewart et al. 1973; Evans and Miller 1975; Ito 1975; Smith and Davis 1978; Percy and Riding 1981; Werner 1982; Soikkeli 1981b; Krol et al. 1982; Kozlowski and Constandinidou 1986; Reinikainen and Kärenlampi 1987; Fink 1988, 1989, 1992; Ruetze et al. 1988; Hafner et al. 1989; Holopainen et al. 1992). The damaged mesophyll cells show plasmolysis and shrinkage, breakdown of internal compartmentalization, and finally necrosis and discoloration; the mesophyll cells directly bordering necrotic cells may react with slight discoloration, thickening of their cell walls, or hypertrophy; a special reaction in conifer needles is the hypertrophy of epithelial cells of the resin sacs, leading to their occlusion. Collapse of mesophyll may lead to depressions in the overlying epidermis. The vascular bundles generally remain unaffected during the initial phase, and only after extensive damage to the mesophyll do the sieve elements show some necrosis, accompanied by a hypertrophy of adjacent cells of the phloem and transfusion parenchyma. SO_2 induces a more general damage to all mesophyll cells, whereas ozone, for example, seems to preferentially damage the palisade parenchyma, and HF the spongy parenchyma.

Some changes also occur in the epidermis though it generally stays intact to the very end of a acute injury episode. Gaseous pollutants may

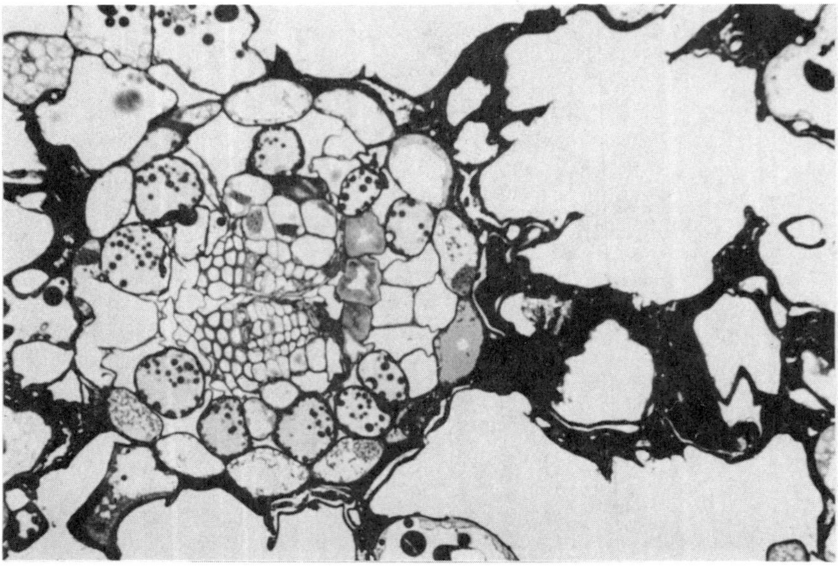

Figure 5 Cross section through a 2-year-old spruce needle damaged by acute exposure to SO_2. Note dead and collapsed mesophyll cells, vascular bundle remains unaffected. (× 550)

cause less wrinkling of the epidermis, and can provoke epidermal depressions around the stomata, which may induce either a permanent closure or a permanent opening of the stomata (thus enhancing water loss); these depressions arise through collapse of adjacent epidermal cells, which eventually exhibit preferential injury by SO_2, while guard cells themselves mostly stay alive; only in certain cases can guard cells also accessory cells also collapse (Godzik and Sassen 1978; Krause and Jensen 1979; Black and Black 1979; Godzik and Halbwachs 1986; Neighbour et al. 1988).

In addition to the impact of SO_2 on leaves, there may also be some effect upon stems. In some deciduous trees damaged by air pollution, Neger (1924) found entrance of the pollutants through lenticels in the bark, leading to necrosis of the cortical cells below and subsequently delineation of this dead zone from the adjacent healthy tissue by formation of an arched wound periderm.

Besides these direct effects of atmospheric pollutants to plants, there are also indirect effects on growth in other organs, such as reduction of root growth or, in trees, reduction of radial xylem growth as a consequence of the decreased supply of the cambium with assimilates and growth hor-

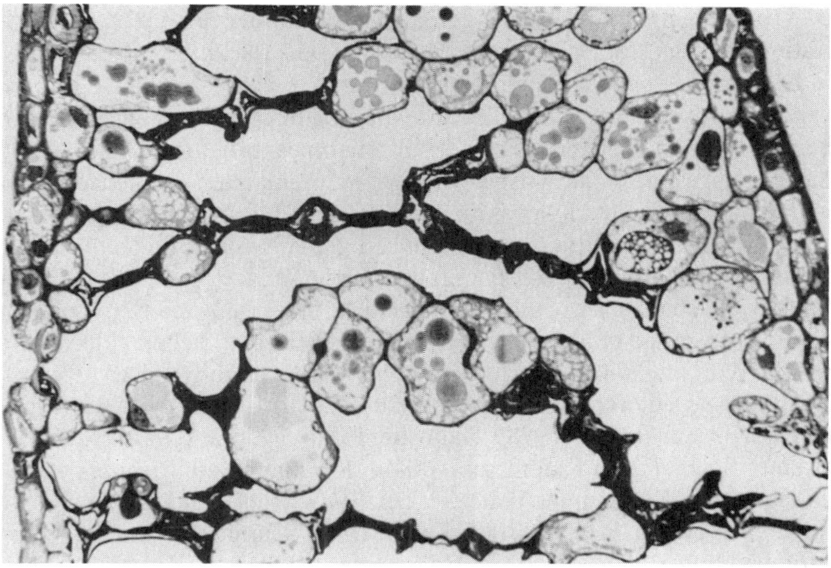

Figure 6 Tangential section through a 2-year-old spruce needle damaged by acute exposure to SO_2. Note dead and collapsed mesophyll cells (× 510).

mones (Liese et al. 1975; Grill et al. 1979; Kartusch and Halbwachs 1985; Krapfenbauer et al. 1985). In a range of species growing in different environments, a similar reaction is described: The ring widths in the wood decline, which affects especially the earlywood in conifers, and the transition between early- and latewood is more abrupt. All cells (tracheids, fibers, vessels) are shortened and have a smaller diameter, and in tracheids the number of pits frequently increase. In diffuse-porous hardwoods, the area of conducting vessels is more severely reduced than in ring-porous ones, which continue to differentiate their large earlywood-vessels at about the same rate. In hardwoods, furthermore, the relative frequency of rays and vessels increases with increasing damage.

At the cellular level, one of the first acute effects of SO_2 is enhanced membrane permeability, leading to loss of K^+, then loss of osmotic water, which enters into intercellular spaces (water soaking of tissue leads to an oily cast and watery flecks). The early symptoms are thus frequently a response to changes in the osmotic condition within a cell. Next step can be a reaction of vacuolar contents with cytoplasm along the more permeable tonoplast.

The most pronounced changes under the influence of SO_2 refer to alterations in the structure of chloroplasts. In general, different pollutants have been found to induce granulation of stroma, curling and swelling of thylakoids, reduction of grana, accumulation of plastoglobuli, and production of small, roundish chloroplasts with sometimes irregular outlines. Most of these changes are rather unspecific signs of premature senescence, however, but a few pollutant-specific changes may be speculatively distinguished. SO_2 seems to lead more to reduced grana, swollen thylakoids, and light plastoglobuli, whereas, ozone, for example, induces generally smaller and flattened chloroplasts with granulated stroma, NO_2 leads to tubular protrusions from the chloroplast envelope, and HF to stretched chloroplast envelopes and swollen or curled thylakoids; SO_2 mainly reduces sizes of starch grains in chloroplasts whereas ozone has the opposite effect (Figs. 7, 8; Godzik and Sassen 1978; Thomson 1975; Malhotra 1976; Soikkeli 1981a,b,c; Soikkeli and Kärenlampi 1984a; Kärenlampi and Houpis 1986; Reinikainen and Kärenlampi 1987; Sutinen 1987; Ruetze et al. 1988; Saastamoinen and Holopainen 1989; Ebel et al. 1990; Schmitt and Ruetze 1990; Schiffgens-Gruber and Lütz 1992; Holopainen et al. 1992). However, most changes seem to be very variable and depend more on duration and concentration of fumigation, plant species, and so on than on the nature of the pollutant. In most cases, these structural effects are the consequences of secondary or tertiary influences because the primary effect of the pollutant is most probably an induction of enhanced permeability of the membranes, which leads to water losses. Besides these changes of the chloroplasts,

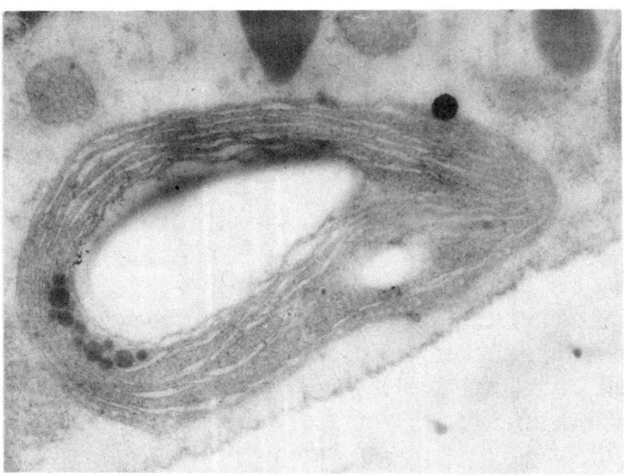

Figure 7 Chloroplast from a 2-year-old healthy spruce needle. (× 15,400)

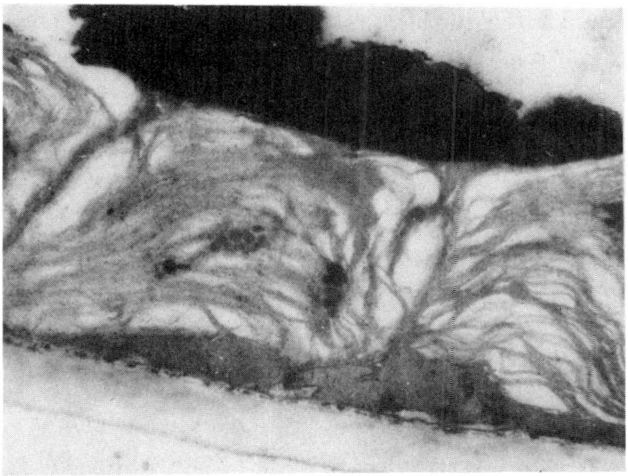

Figure 8 Chloroplasts from an SO_2-damaged spruce needle. (× 13,600)

other structural alterations such as accumulation of large lipid bodies, vacuoles with darkly staining deposits, ribbonlike aggregation of tannins in the vacuole along the tonoplast, vesicularization of cytoplasm, and lipid accumulations have been observed. Depositions of crystals of calcium oxalate within or at the outside of mesophyll cells (especially adjacent to vascular bundles) increase in spruce needles suffering from acute SO_2 injury and in leaves of *Gingko biloba* L. growing under heavy pollution from car exhausts, but partially decrease in long-term, low-concentration fumigated spruce needles (Hartig 1896; Sorauer 1908; Umemoto and Hozumi 1972; Ruetze et al. 1988).

In addition to pronounced effects on the mesophyll cells, changes in the epidermis have been reported under the influence of gaseous pollutants. Some studies claim to have found a partial secondary delignification of the walls of subsidiary and guard cells and of bundle sheath cells in spruce needles after fumigation by SO_2 or ozone, thus causing turgor loss of subsidiary cells, which leads to passive opening of the stomata and loss of the ability of these needles to reduce transpiration under water shortage (Maier-Maercker 1989; Maier-Maercker and Koch 1986). Less formation and certain fusion, but no enhanced erosion, of epicuticular waxes by dry-phase SO_2 occurs; these changes are more subtle when compared to those caused by the wet phase. With time, the dry phase, however, may become dissolved in moisture upon the cuticle and finally act in the aqueous phase as acid (Riding and Percy 1985). Long-term fumigation with realistic concentrations of ozone or SO_2 barely change wax structure, though the total quantities of waxes formed are somewhat reduced (Günthardt-Goerg 1986; Günthardt-Goerg and Keller 1987; Schmitt et al. 1987; Bytnerowicz et al. 1989).

In lichens a decrease of cortex thickness and an increase of algae with more roundish chloroplasts and smaller storage bodies occurs; the hyphae of the fungal partner show pronounced vacuoles and dark inclusions (Holopainen 1981). Accumulation of starch, swelling of thylakoids, plasmolysis of algal cells, and coalescing lipid droplets have been observed in laboratory fumigations with SO_2 and O_3 singly and in combination (Eversman and Sigal 1987).

3. Physiology and Biochemistry

The diffusive flux of SO_2 via the stomata is the most important path of SO_2 entry into plants. Therefore the rate of SO_2 absorption may be calculated from the stomatal conductance for water vapor efflux. Actual measurements of SO_2 deposition rates show lower, equal to, or higher than calculated rates of SO_2 absorption, depending on the experimental conditions applied (De Kok 1990). The finding of higher than predicted rates of SO_2 deposition indicates SO_2 flux to the cuticle, especially when leaf surfaces are wet; lower than predicted rates of SO_2 absorption may be explained by internal resistances: physical and metabolic factors counteracting SO_2 influx (Fowler 1985; Baldocchi 1993). The diffusive flux of SO_2 from the atmosphere through the stomata into the leaf is driven by the concentration gradient of SO_2 between the atmosphere and the substomatal cavity. In the gas phase of the substomatal cavity, SO_2 concentrations are low because it rapidly dissolves in the aqueous phase of the cell wall and reacts with water to form bisulfite. At least part of the bisulfite may be converted to sulfate in the apoplastic space. Alternatively, SO_2 may diffuse through the plasmalemma and may be converted to sulfate in the symplasm (Rennenberg and Polle 1994). As a consequence, accumulation of sulfate, presumably mainly in the vacuoles, is generally observed in the foliage of both herbaceous and woody plants in response to SO_2 fumigation (De Kok 1990; Ernst 1993; Polle et al. 1994). The acid produced during the processes of oxidative conversion of SO_2 is thought to be a major source of SO_2 mediated injury to the leaves (Rennenberg and Polle 1994). However, part of the SO_2 absorbed by plants undergoes reductive conversion inside the symplasm that is not connected with acid formation. Therefore, an increase in thiol content that is small as compared to the increase in sulfate content, and reemission of S from SO_2 as H_2S are observed on SO_2 exposure of both herbaceous plants and trees (De Kok 1990; Rennenberg 1991; Rennenberg and Polle 1994; Polle et al. 1994). An increased capacity for reductive rather than oxidative conversion of SO_2 may be one possible factor of increased SO_2 tolerance of plants. Symplastic SO_2 can readily pass the chloroplast envelope either by diffusion (Spedding et al. 1980) or via the phosphate translocator (Hampp and Ziegler 1977) and may be trapped efficiently at the pH of the chloroplast stroma (pH 8.0; Kurkdjian and Guern 1989) by its reaction with water to bisulfite and sulfite. Therefore, the chloroplast is a sensitive target for primary damage of SO_2 or its reaction products in aqueous solution. Damage by SO_2 is thought to be the result of free radical formation in the chloroplast, mainly by the conversion of bisulfite to sulfate (Peiser and Yang 1977; Asada and Takahashi 1987). The free radicals produced in this series of reactions may impair metabolic functions of the chloroplast by inhibition of photosystem II, lipid peroxidation, and chlorophyll destruction (Wellburn 1985; Peiser and Yang 1985). Although most of the studies in this area of research have been performed with herbaceous plants, the existing body of evidence suggests similar physiological and biochemical consequences of SO_2 absorption by trees.

C. Hydrogen Sulfide

1. H_2S Injury

H_2S may originate in soil as a consequence of reduction of sulfate under anaerobic conditions. This leads to inhibition of root growth, browning and dieback of roots, and in the shoot iron-deficiency chlorosis may be induced (Bergmann 1988). Though gaseous H_2S in the atmosphere is very toxic and unpleasant for humans, it is only slightly phytotoxic. At levels above 100 ppb, necrotic lesions and tip burn have been detected on a variety of plants (Wellburn 1988).

2. Physiology and Biochemistry

Absorption of atmospheric H_2S has been studied with only few plant species, including Norway spruce (Cope and Spedding 1982; Taylor et al. 1983; De Kok et al. 1989). Up to atmospheric levels of about 300 ppb it is mainly dependent on the atmospheric concentration and stomatal aperture. For H_2S, in contrast to SO_2, a low internal resistance is generally observed. This may be the consequence of rapid H_2S metabolism and, hence, leaf internal H_2S vapor pressures close to zero. Fumigation of plants with H_2S, as with SO_2, enhances the S content of the shoot, at least partially, but often mainly by increasing the level of sulfate (De Kok 1990). However, at similar concentrations, accumulation of sulfate proceeds more rapidly with SO_2 than with H_2S. The origin of the accumulated sulfate is obscure. It either may be produced directly from the absorbed H_2S or may originate from sulfate uptake by the roots. The latter origin appears to be possible because (1) H_2S is directly used as a source of cysteine biosynthesis and (2) reduction of sulfate uptake by the roots on H_2S exposure may be insufficient to completely prevent influx of excess S (Herschbach et al. 1995). The direct use of H_2S for the synthesis of cysteine and other organic reduced S compounds results in a rapid accumulation of water-soluble sulfhydryl compounds, mainly glutathione and cysteine (De Kok 1990). Maximum accumulation at 250 ppb H_2S is in the range of three- to four-fold and thus much higher than that observed upon SO_2 exposure (De Kok 1990).

H_2S is a phytotoxic gas. It can reduce plant growth and can cause visible injury if applied at a certain threshold level, which is different for different species (De Kok 1990). From the present literature it appears unlikely that the accumulated sulfate or thiol is a significant factor in H_2S phytotoxicity. It appears more likely that H_2S impairs metabolic reactions directly due to its high reactivity with cellular constituents such as enzymes and components of plant membranes (De Kok 1990). Still, a target directly attacked by H_2S that can be a significant factor in H_2S phytotoxicity has not been identified so far.

D. Elemental S

1. Elemental S Excess in Soil

Excess S in soil may also decrease plant growth. Leaves can show yellowing and necrosis ("burning") developing from the margins to the center and are finally abscised prematurely (Bergmann 1988)

Soil application of elemental S to cotton plants leads to marked acid injury in the roots, characterized by swelling and proliferation of the uppermost part of the main root. This is provoked by enhanced activities of cambium, pericycle, and phloem rays, which primarily segregate abnormal parenchyma-rich phloem and thick cork layers, whereas in the xylem the percentage of rays rises and the vessel members remain short and distorted (Gore and Taubenhaus 1931; Taubenhaus and Ezekiel 1931).

REFERENCES

Adams C. M., N. G. Dengler, and T. C. Hutchinson, Acid rain effects on foliar histology of *Artemisia tilesii*, *Can. J. Bot.*, *62*: 463 (1984).

Adams C. M., S. J. M. Caporn, and T. C. Hutchinson, Crystal occurrence and wax disruption on leaf surfaces of cabbage treated with simulated acid rain, *New Phytol.*, *114*: 147 (1990).

Adams M. B., and E.G. O'Neill, Effects of ozone and acidic deposition on carbon allocation and mycorrhizal colonization of Pinus taeda L.seedlings, *For. Sci.*, *37*: 5 (1991).

Amthor J. S., An estimate of the "cost" of nutrient leaching from forest canopies by rain, *New Phytol.*, *102*: 359 (1986).

Asada K., and M. Takahashi, Production and scavenging of active oxygen in photosynthesis, *Photoinhibition* (D. Kyle, C. Osmond, and C. Arntzen, eds.), Elsevier Publ., The Hague: pp. 227 (1987).

Bäck J., and S. Huttunen, Structural responses of needles of conifer seedlings to acid rain treatment, *New Phytol.*, *120*: 77 (1992).

Badot P. M., J. P. Garrec, B. Millet, M. J. Badot, and J. Mercier, Dépérissement et état hydrique des aiguilles chez le *Picea abies*. *Can. J. Bot.*, *66*: 1693 (1988).

Baldocchi D. D., Deposition of gaseous S compounds. *S Nutrition and S Assimilation in Higher Plants* (L. J. De Kok, I. Stulen, H. Rennenberg, C. Brunold, W. E. Rauser, eds.), SPB Acad. Publ., The Hague, p. 271 (1993).

Banwart W. L., P. M Porter, E. L. Ziegler, J. J. Hassett, Growth parameter and yield component response of field corn to simulated acid rain, *Environ. Exp. Bot.*, *28*: 43 (1988).

Barnes J. D., and A.W. Davison, The influence of ozone on the winter hardiness of Norway spruce (Picea abies (L.) Karst.), *New-Phytol.*, *108*: 159 (1988).

Barnes J. D., D. Eamus, and K. A. Brown, The influence of ozone, acid mist and soil nutrient status on Norway spruce *Picea abies* (L.) Karst. II. Photosynthesis, dark respiration and soluble carbohydrates of trees during late autumn, *New Phytol. 115*: 149 (1990).

Bergmann W., *Ernährungsstörungen bei Kulturpflanzen*, Gustav Fischer Verlag, Stuttgart, New York, p. 762. (1988).

Biddulph S. F., Visual indications of ^{35}S and ^{32}P translocation in the phloem, *Am. J. Bot.*, *43*: 143 (1956).

Black C. R., and V. J. Black, Light and scanning electron microscopy of SO_2-induced injury to leaf surfaces of field bean (Vicia faba L.), *Plant, Cell & Environ.*, *2*: 329 (1979).

Blaschke L., A. Schneider, C. Herschbach, and H. Rennenberg, Reduced sulphur allocation from three-year-old needles of Norway spruce (Picea abies [Karst.] L.), *J. Exp. Bot.*, *47*: 1025 (1996).

Bohlmann H., Significance of sulfur-rich proteins in seeds and leaves, (L. J. DeKok, L. J., I. Stulen, H. Rennenberg, C. Brunold, W. E. Rauser, eds.), *Sulfur Nutrition and Sulfur Assimilation in Higher Plants.* SPB Acad. Publ., The Hague. p. 276 (1993).

Bosch C., E. Pfannkuch, U. Baum and K.E. Rehfuess, Über die Erkrankung der Fichte (*Picea abies* Karst.) in den Hochlagen des Bayerischen Waldes, *Forstw. Cbl.*, *102*: 167 (1983).

Bosma W., R. Schupp, L. J. De Kok, and H. Rennenberg, Effect of selenate on assimilatory sulfate reduction and thiol content of spruce needles, *Plant Physiol. Biochem.*, *29*: 131 (1991).

Brennan E., Winter spot: a common symptom on coniferous evergreens in New Jersey, *Shade Tree*, *61*: 44 (1988).

Brunold C., Reduction of sulfate to sulfide, *Sulfur Nutrition and Sulfur Assimilation in Higher Plants* (H. Rennenberg, C. Brunold, L.J. De Kok, and I. Stulen, eds.) SPB Acad. Publ., The Hague; p. 13 (1990)

Brunold C., Regulatory interactions between sulfate and nitrate assimilation, *Sulfur Nutrition and Sulfur Assimilation in Higher Plants.* (L. J. De Kok, I. Stulen, H. Rennenberg, C. Brunold, W. E. Rauser, eds.), SPB Acad. Publ., The Hague; p. 61 (1993).

Bucher J. B., and W. Landolt, Zur Diagnose von Ozonsymptomen auf Waldbäumen. *Schweiz. Z. Forstwes.*, *136*: 863 (1985).

Burkhardt J., and R. Eiden, Thin water films on conifer needles, *Atmos. Environ.*, *28*: 2001 (1994).

Bytnerowicz A., D. M. Olszyk, S. Huttunen and B. Takemoto, Effects of photochemical smog on growth, injury, and gas exchange of pine seedlings. *Can. J. Bot.*, *67*:2175 (1989).

Cape J. N., I. S. Paterson, A. R. Wellburn, J. Wolfenden, H. Mehlhorn, P. H. Freer-Smith, and S. Fink, Early diagnosis of forest decline. Instit. of Terr. Ecology, Edinburgh, pp. 68 (1988).

Cape J. N., I. S. Paterson, and J. Wolfenden, Regional variation in surface properties of Norway spruce and Scots pine needles in relation to forest decline, *Environ. Poll.*, *58*: 325 (1989).

Chappelka III A. H., and B. I. Chevone, White ash seedling growth response to ozone and simulated acid rain, *Can. J. For. Res.*, *16*: 786 (1986).

Clarkson D. T., M. J. Hawkesford, and J.-C. Davidian, Membrane and long-distance transport of sulfate, *Sulfur Nutrition and Sulfur Assimilation in Higher Plants* (L. J. De Kok, I. Stulen, H. Rennenberg, C. Brunold, and W. E. Rauser, eds.), SPB Acad. Publ., The Hague; p. 3 (1993).

Cope D. M., and D. J. Spedding, Hydrogen sulphide uptake by vegetation, *Atmos. Environ.*, *16*: 349 (1982).

Costonis A. C., Acute foliar injury of Eastern White Pine induced by sulfur dioxide and ozone, *Phytopathology*, *60*: 994 (1970).

Cram W. J., Uptake and transport of sulfate, *Sulfur Nutrition and Sulfur Assimilation in Higher Plants.* (H. Rennenberg, C. Brunold, L. J. De Kok, I. Stulen, eds.): SPB Acad. Publ., The Hague; p. 13 (1990).

Crang R. E., and C. J. McQuattie, Qualitative and quantitative effects of acid misting and two air pollutants on foliar structures of *Liriodendron tulipifera*, *Can. J. Bot.*, *64*: 1237 (1986).

Crossley A., and D. Fowler, The weathering of Scots pine epicuticular wax in polluted and clean air, *New Phytol.*, *103*: 207 (1986).

Deans J. D., I. D. Leith, L. J. Sheppard, J. N. Cape, D. Fowler, M. B. Murray, and P. A. Mason, The influence of acid mists on growth, dry matter partitioning, nutrient concentrations and mycorrihizal fruiting bodies in red spruce seedlings, *New Phytol.*, *115*: 459 (1990).

De Kok L. J., Sulfur metabolism in plants exposed to atmospheric sulfur, *Sulfur Nutrition and Sulfur Assimilation in Higher Plants*, (H. Rennenberg, C. Brunold, L. J. De Kok, and I. Stulen, eds.), SPB Acad. Publ., The Hague; p. 111 (1990).

De Kok L. J., K. Stahl, and H. Rennenberg, Fluxes of atmospheric hydrogen sulphide to plant shoots, *New Phytol.*, *112*: 533 (1989).

Eamus, D., I. Leith, and D. Fowler, Water relations of red spruce seedling treated with acid mist. *Tree Physiol.*, *5*: 387 (1989).

Ebel B., J. Rosenkranz, A. Schiffgens, and C. Lütz, Cytological observations on spruce needles after prolonged treatment with ozone and acid mist, *Environ. Poll.*, *64*: 323 (1990).

Ernst W. H. O., Ecological aspects of sulfur in higher plants: The impact of SO_2 and the evolution of the biosynthesis of organic sulfur compounds on populations and ecosystems, *Sulfur Nutrition and Sulfur Assimilation in Higher Plants* (L. J. De Kok, I. Stulen, H. Rennenberg, C. Brunold, W. E. Rauser, eds.), SPB Acad. Publ., The Hague; p. 295 (1993).

Esterbauer H., and D. Grill, Seasonal variation of glutathione and glutathione reductase in needles of Picea abies, *Plant Physiol.*, *61*: 119 (1978).

Evans L. S., Botanical aspects of acidic precipitation, *Bot. Rev.*, *50*: 449 (1984).

Evans L. S., and T. M. Curry, Differential responses of plant foliage to simulated acid rain, *Amer. J. Bot.*, *66*: 953 (1979).

Evans L. S., and P. R. Miller, Histological comparison of single and additive O_3 and SO_2 injuries to elongating pine needles, *Amer. J. Bot.*, *62*: 416 (1975).

Evans L. S., N. F. Gmur, and F. Da Costa, Foliar response of six clones of hybrid poplar, *Phytopathology*, *68*: 847 (1978).

Eversman S., and L. L. Sigal, Effects of SO_2, O_3, and SO_2 and O_3 in combination on photosynthesis and ultrastructure of two lichen species, *Can. J. Bot.*, *65*: 1806 (1987).

Fairfax J. A. W., and N. W. Lepp, The effect of simulated acid rain on cation losses from leaves, *Nature*, *255*: 324 (1975).

Fiedler H. J., G. Baronius, and F. Ehrig, Rasterelektronenmikroskopische Untersuchungen grüner und chlorotischer Nadeln eines immissionsgeschädigten Kiefernbestandes, *Flora*, *184*: 91 (1990).

Fink S., Histological and cytological changes caused by air pollutants and other abiotic factors, *Air Pollution and Plant Metabolism* (S. Schulte-Hostede, N. M. Darrall, L. W. Blank, and A. R. Wellburn, eds.), Elsevier Appl. Sci., London, New York, p. 36 (1988).

Fink S., Pathological anatomy of conifer needles subjected to gaseous pollutants or mineral deficiencies, *Aquilo Ser. Bot.*, *27*: 1 (1989).

Fink S., Tödliche Säureflecken auf Fichtennadeln?, *Allg. Forstzeitschr.*, *45*: 989 (1990a).

Fink S., Structural changes in conifer needles due to Mg and K deficiency, *Fertilizer Res.*, *27*: 23 (1990b).

Fink S., The micromorphological distribution of bound calcium in the needles of Norway spruce, *Picea abies* (L.) Karst., *New Phytol.*, *119*: 33 (1991a).

Fink S., Unusual patterns in the distribution of calcium oxalate in spruce needles and their possible relationships to the impact of pollutants, *New Phytol.*, *119*: 41 (1991b).

Fink S., Comparative microscopical studies on the patterns of calcium oxalate distribution in the needles of various conifer species, *Botanica Acta*, *104*: 306 (1991c).

Fink S., Microscopical criteria for the diagnosis of abiotic injuries to conifer needles, *Forest Decline in the Atlantic and Pacific Region* (R. F. Huettl and D. Mueller-Dombois, eds.), Springer-Verlag, Berlin, Heidelberg, New York, p. 174 (1992).

Fowler D., Deposition of SO_2 onto plant canopies, *Sulfur Dioxide and Vegetation* (W. E. Winner, H. A. Mooney and R. A. Goldstein, eds.), Stanford University Press, Stanford, p. 389 (1985).

Gardingen P. R. van, J. Grace, and C. E. Jeffree, Abrasive damage by wind to the needle surfaces of *Picea sitchensis* (Bong.) Carr. and *Pinus sylvestris* L. *Plant, Cell & Environ.*, *14*: 185 (1991).

Godzik S., and M. M. A. Sassen, Scanning electron microscope examination of *Aesculus hippocastanum* L. leaves from control and air-polluted areas, *Environ. Pollut.*, *17*: 13 (1978).

Godzik S., and G. Halbwachs, Structural alterations of Aesculus hippocastanum leaf surface by air pollutants, *J. Plant Dis.*, *93*: 590 (1986).

Gore U. R., and J. J. Taubenhaus, Anatomy of normal and acid-injured cotton roots, *Bot. Gaz.*, *92*: 436 (1931).

Gray L. E., and J. W. Gerdemann, Uptake of S-35 by vesicular-arbuscular mycorrhizae. *Plant Soil*, *39*: 687 (1973).

Grill D., E. Liegl, and E. Windisch, Holzanatomische Untersuchungen an abgasbelasteten Bäumen, *Phytopath.*, Z. *94*: 335 (1979).

Grill D., H. Pfeifhofer, G. Halbwachs, and H. Waltinger, Investigations on epicuticular waxes of differently damaged spruce needles, *Eur. J. For. Pathol.*, *17*: 246 (1987).

Günthardt-Goerg M. S., Epicuticular wax of needles of Pinus cembra, Pinus sylvestris and Picea abies, *Eur. J. For. Pathol.*, *16*: 400 (1986).

Günthardt-Goerg M. S., and T. Keller, Some effects of long-term ozone fumigation on Norway spruce. II. Epicuticular wax and stomata, *Trees Struct. Funct.*, *1*: 145 (1987).

Güth S., and B. Frenzel, Epicuticularwachs der Tanne (*Abies alba* Mill.) und Walderkrankung. I. Die Wachsstruktur, *Angew. Botanik*, *63*: 241 (1989).

Hafner L.,W. Endler, R. Wendering, and G. Weese, Ultrastructural studies on the needles of damaged Scotch pine trees from forest dieback areas in West Berlin, *Aquilo Ser. Bot. 27*: 7 (1989).

Hahn K., Korrosion der Nadeloberfläche durch Schwefelsäureeinwirkung im direkten Sonnenlicht, *Allg. Forstzeitschr.*, *45*: 781 (1990).

Haines B. L., J. A. Jernstedt, and H. S. Neufeld, Direct foliar effects of simulated acid rain. II. Leaf surface characteristics. *New Phytol.*, *99*: 407 (1985).

Hampp R., and I. Ziegler, Sulfate and sulfite translocation via the phosphate translocator of the inner envelope membrane of chloroplasts. *Planta, 137*: 309 (1977).

Harley L. L., and S. E. Smith, *Mycorrhizal Symbiosis.* Academic Press, London, p. 483 (1983).

Hartig R., Über die Einwirkung des Hhtten- und Steinkohlenrauches auf die Gesundheit der Nadelwaldbäume. *Forstl.-naturwiss. Z. 5*: 245 (1896).

Haselhoff E., and G. Lindau, *Die Beschädigung der Vegetation durch Rauch*, Verlag Gebr. Bronträger, Leipzig, VIII + 412S (1903).

Haselhoff E., G. Bredemann, and W. Haselhoff, *Entstehung, Erklennung und Beurteilung von Rauchschäden*, Verlag Gebr. Borntraeger, Berlin, X11+472S (1932).

Herschbach C., L. J. De Kok, and H. Rennenberg, Net uptake of sulfate and its transport to the shoot in spinach plants fumigated with H_2S or SO_2: does atmospheric sulfur affect the inter-organ regulation of sulfur nutrition? *Bot. Acta, 108*: 41 (1995).

Herschbach C., and H. Rennenberg, Influence of glutathione (GSH) on net uptake of sulphate and sulphate transport in tobacco plants, *J. Exp. Bot., 45*: 1069 (1994).

Herschbach C., and H. Rennenberg, Long-distance transport of ^{35}S-S in three-year-old beech trees (Fagus sylvatica), *Physiol. Plant., 95*: 379 (1995).

Herschbach C., and H. Rennenberg, Storage and remobilization in beech trees (Fagus sylvatica), *Physiol. Plant., 98*: 125 (1996).

Hibben C. R., The distinction between injury to tree leaves by ozone and mesophyll-feeding leafhoppers, *For. Sci., 15*: 154 (1969).

Hogan G. D., Physiological effects of direct impact of acidic deposition on foliage, *Agric., Ecosystems and Environ., 42*: 307 (1992).

Hogan G. D., Effect of simulated acid rain on physiology, growth and nutrient relations of sugar maple. *Chemosphere* (in press) (1997).

Hogan G. D., and N. W. Foster, "Precipitation acidity and foliar cation losses from sugar maple," Proceedings of L'atelier sur le dépérissment des érablieres, Ste-Hyacinthe, Que., pp. 24–28 (1989).

Hogan G. D., and S. J. Taylor, Effect of acidic deposition on hybrid poplar —primary or predisposing stress? *Water Air Soil Pollution, 85*: 1419 (1995).

Holopainen T., Alterations in the ultrastructure of epiphytic lichens Hypogymnia physodes and Alectoria capillaris caused by air pollution, *Silva Fennica, 15*: 469 (1981).

Holopainen T., and P. Nygren, Effects of potassium deficiency and simulated acid rain, alone and in combination, on the ultrastructure of Scots pine needles, *Can. J. For. Res., 19*: 1402 (1989).

Holopainen T., S. Anttonen, A. Wulff, V. Palomäki, and L. Kärenlampi, Comparative evaluation of the effects of gaseous pollutants, acidic deposition and mineral deficiencies: structural changes in the cells of forest plants, *Agric., Ecosystems and Environ., 42*: 365 (1992).

Hüttl R. F., and S. Fink, Diagnostische Dhngungsversuche zur Revitalisierung geschädigter Fichtenbestände (*Picea abies* Karst.) in Südwestdeutschland, *Forstw. Cbl.*, *107*: 173 (1988).

Huttunen S., and U. K. Laine, Effects of air-borne pollutants on the surface wax structure of *Pinus sylvestris* needles, *Ann. Bot. Fenn.*, *20*: 79 (1983).

Huttunen S., M. Turunen and J. Reinikainen, Scattered $CaSO_4$-crystallites on needle surfaces after simulated acid rain as an indicator of nutrient leaching, *Water, Air and Soil Poll.*, *54*: 169 (1990).

Ito S., Histological responses of Japanese red pine needles exposed to S dioxide, *J. Jap. For. Soc.*, *57*: 300 (1975).

Jensen K. F., and R. L. Patton, Response of yellow-poplar (Liriodendron tulipifera L.) seedlings to simulated acid rain and ozone—1. Growth modifications, *Env. Exp. Bot.*, *30*: 59 (1990).

Kaiser W., A. Dittrich, and U. Heber, Sulfate concentrations in Norway spruce needles in relation to atmospheric SO_2: a comparison of trees from various forests in Germany with trees fumigated with SO_2 in growth chambers, *Tree Physiol.*, *12*: 1 (1993).

Kärenlampi L., and J. L. J. Houpis, Structural conditions of mesophyll cells of *Pinus ponderosa* var. *scopulorum* after SO_2 fumigation. *Can. J. For. Res.*, 16: 1381 (1986).

Karhu M., and S. Huttunen, Erosion effects of air pollution on needle surfaces, *Water, Air and Soil Poll.*, *31*: 417 (1986).

Kartusch B., and G. Halbwachs, Holzanatomische Untersuchungen an unterschiedlich immissionsgestressten Exemplaren von *Alnus glutinos*a L., *Angew. Bot.*, *59*: 249 (1985).

Kozlowski T. T., and H. A. Constantinidou, Responses of woody plants to environmental pollution. *For. Abstr. 47*: 5 (1986).

Krapfenbauer A., J. Bodner, and T. Knoll, Anomalien bei der Ausbildung der Jahrringe, der Hofthpfel und Tracheiden bei immissionsgeschädigen Fichten. *Cbl. Ges. Forstwes.*, *102*: 150 (1985).

Krause C. R., and K. F. Jensen, Surface changes on hybrid leaves exposed to ozone and S dioxide. *Scann. Electron Microsc.*, *3*: 77 (1979).

Kreis M., Synthesis, structure and evolutionary relationships of S-rich seed proteins in higher plants. *Sulfur Nutrition and Sulfur Assimilation in Higher Plants* (H. Rennenberg, C. Brunold, L.J. De Kok, I. Stulen, eds.). SPB Acad. Publ., The Hague; pp. 67–76 (1990).

Kreuzwieser J., C. Herschbach, and H. Rennenberg, Sulfate uptake and xylem loading of non-mycorrhizal excised roots of young Fagus sylvatica trees, *Plant Physiol. Biochem.*, *34*: 409 (1996).

Krol P. J., L. Steubing, H. G. Wolting, and A. C. Posthumus, Histologische und zytologische Untersuchungen an Trifolium repens L. und Plantago major L. nach Begasung mit einem Immissionsgemisch aus NO2, O3 und SO2. *Angew. 56*: 295 (1982).

Kurkdjian A., and J. Guern, Intracellular pH measurement and importance in cell activity, *Annu. Rev. Plant Physiol. Mol. Biol.*, *40*: 271 (1989).

Lamoureux G. L., D. G. Rusness, P. Schroder, and H. Rennenberg, Diphenyl ether herbicide metabolism in a spruce cell suspension culture: the identification of two novel metabolites derived from a glutathione conjugate, *Pest. Biochem. Physiol.*, *39*: 291 (1991).

Lendzian K., and G. Kerstiens, Sorption and transport of gases and vapors in plant cuticles. *Rev. Environ. Contamination Tox.*, *121*: 65 (1991).

Liese W., M. Schneider, and D. Eckstein, Histometrische Untersuchungen am Holzeiner rauchgeschädigten Fichte, *Eur. J. For. Path.*, 5: 152 (1975).

Liu G. E., and B. Cote, Neutralization and buffering capacity of leaves of sugar maple, largetooth aspen, paper birch and balsam fir, *Tree Physiology*, *12*: 15 (1993).

Lorenzini G., A. Panattoni, and G. Schenone, Studies on the phytotoxicity of acid smuts. *Water, Air and Soil Poll. 42*: 47 (1988).

Lovett G. M., and J. G. Hubbell, Effects of ozone and acid mist on foliar leaching from eastern white pine and sugar maple, *Can. J. For. Res.*, *21*: 794 (1991).

MacDonald N. W., J. B. Hart, and P. V. Nguyen, Simulated acid rain effects on jack pine seedling establishment and nutrition, *Soil Sci Soc.Am. J.*, *50*: 219 (1986).

McLaughlin S. B., C. P. Andersen, P. J. Hanson, M. G. Tjoelker, and W. K. Roy, Increased dark respiration and calcium deficiency of red spruce in relation to acidic deposition at high-elevation southern Appalachian Mountain sites, *Can. J. For. Res.*, *21:* 1234 (1991).

McLaughlin S. B., M. G. Tjoelker, and P. H. Freer-Smith, Growth and physiological changes in red spruce saplings associated with acidic deposition at high elevations in the southern Appalachians, USA, *Forest Ecol. Man.*, *51:* 43 (1992).

McLaughlin S. B., M. G. Tjoelker, and W. K. Roy, Acid deposition alters red spruce physiology: laboratory studies support field observations, *Can. J. For. Res.*, *23:* 380 (1993).

McLaughlin S. B., and R. J. Kohut, The effect of atmospheric deposition and ozone on carbon allocation and associated physiological processes in red spruce. *Ecol. Stud. Anal.*, *96*: 338 (1992).

Maier-Maercker U., Delignification of subsidiary and guard cell walls of *Picea abies* (L.) Karst. by fumigation with ozone, *Trees*, *3*: 57 (1989).

Maier-Maercker U., and W. Koch, Delignification of subsidiary and guard cell walls by SO_2 and probable implication on the humidity response of *Picea abies* (L.) Karst., *Eur. J. For. Pathol.*, *16*: 342 (1986).

Malhotra S. S., Effects of S dioxide on biochemical activity and ultrastructural organization of pine needle chloroplasts, *New Phytol.*, *76*: 239 (1976).

Masuch G., A. Kettrup, R. Mallant, and J. Slanina, Wirkungen von wasserstoffperoxidhaltigen saurem Nebel auf die Laubblätter junger Buchen (Fagus sylvatica L.), *VDI Ber.*, *560*: 761 (1985).

Masuch G., H. G. Kicinski, A. Kettrup, and K. S. Boos, Single and combined effects of continuous and discontinuous O3 and SO2 emission on Norway spruce needles. I. Histological and cytological changes, *Intern. J. Environ. Anal. Chem.*, *32*: 187 (1988).

Materna J., Die Ausscheidung des durch Fichtennadeln absorbierten Schwefeldioxids, *Arch. Forstwesen*, *15*: 691 (1966).

Maurice C. G., and R. E. Crang, Increase in Pinus strobus needle transectional areas in response to acid misting, *Arch. Environ. Contam. Toxicol.*, *15*: 77 (1986).

Mengel K., H.-J. Lutz, and M.-T. Breininger, Auswaschung von Nährstoffen aus jungen intakten Fichten (*Picea abies*), *Z. Pflanzenern. Bodenk.*, *150*: 61 (1987).

Mengel K., A. M. R. Hogrebe, and A. Esch, Effect of acidic fog on needle surface and water relations of *Picea abies*, *Physiol. Plant.*, *75*: 201 (1989).

Mengel K., M. T. Breininger, and H. J. Lutz, Effect of simulated acidic fog on carbohydrate leaching, CO2 assimilation and development of damage symptoms in young spruce trees (Picea abies L. Karst.), *Environ. Exp. Bot.*, *30*: 165 (1990).

Morrison I. K., Acid rain. A review of literature on acid deposition effects in forest ecosystems. *For. Abstr.*, *45*: 483 (1984).

Nebe W., E. Schierhorn, and G. Ilgen, Rasterelektronenmikroskopische und chemische Untersuchungen von immissionsgeschädigten Fichtennadeln (*Picea abies* (L.) Karst.), *Flora*, *181*: 409 (1988).

Neger F. W., *Die Krankheiten unserer Waldbäume und der wichtigsten Gartengehölze.* 2. Aufl. Verlag Ferdinand Enke, Stuttgart, VIII + 296 S (1924).

Neighbour E. A., D. A. Cottam, and T. A. Mansfield, Effects of S dioxide and nitrogen dioxide on the control of water loss by birch (Betula spp.), *New Phytol.*, *108*: 149 (1988).

Neufeld H. S., J. A. Jernstedt, and B. L. Haines, Direct foliar effects of simulated acid rain. I. Damage, growth and gas exchange. *New Phytol.*, *99*: 389 (1985).

Norby R. J., D. D. Richter, and Luxmore, R. J., Physiological processes in soybean inhibited by gaseous pollutants but not by acid rain. *New Phytol.*, *100*: 79 (1985).

Paparozzi E. T., and H. B. Tukey, Jr., Developmental and anatomical changes in leaves of yellow birch and red kidney bean exposed to simulated acid precipitation. *J. Amer. Soc. Hort. Sci.*, *108*: 890 (1983).

Patton R. L., K. F. Jensen, and G. A. Schier, Responses of red spruce seedlings to ozone and acid deposition. *Can. J. For. Res.*, *21*: 1354 (1991).

Peiser G., and S. F. Yang, Chlorophyll destruction by the bisulfite-oxygen system. *Plant Physiol.*, *60*: 277 (1977).

Peiser G., and S. F. Yang, Biochemical and physiological effects of SO_2 on nonphotosynthetic processes in plants, *S Dioxide and Vegetation* (W. E. Winner, H. A. Mooney, and R. A. Goldstein, eds.), Stanford University Press, Stanford, p. 148 (1985).

Percy K. E., and R. T. Riding, The epicuticular waxes of *Pinus strobus* subjected to air pollutants, *Can. J. For. Res.*, *8*: 474 (1978).

Percy K. E., and R. T. Riding, Histology and histochemistry of elongating needles of *Pinus strobus* subjected to a long-duration, low-concentration exposure to S dioxide, *Can. J. Bot.*, *59*: 2558 (1981).

Percy K. E., and E. A. Baker, Effects of simulated acid rain on production, morphology and composition of epicuticular wax and on cuticular membrane development, *New Phytol.*, *107*: 577 (1987).

Percy K. E., and E. A. Baker, Effects of simulated acid rain on leaf wettability, rain retention and uptake of some inorganic ions, *New Phytol.*, *108*: 75 (1988).

Percy K. E., K. F. Jensen, and C. J. McQuattie, Effects of ozone and acidic fog on red spruce needle epicuticular wax production, chemical composition, cuticular membrane ultrastructure and needle wettability. *New Phytol.*, *122*: 71 (1992).

Polle A., M. Eiblmeier, and H. Rennenberg, Sulphate and antioxidants in needles of Scots pine (Pinus sylvestris L.) from SO_2-polluted field sites in eastern Germany, *New Phytol.*, 127: 571 (1994).

Polle A., and H. Rennenberg, Photooxidative stress in trees, *Causes of Photooxidative Stress and Amelioration of Defense Systems in Plants* (C. H. Foyer and P. M. Mullineaux, eds.) CRC Press, Boca Raton, p. 199 (1994).

Qiu Z., A. H. Chappelka, G. L. Somers, B. G. Lockaby, and R. S. Meldahl, Effects of ozone and simulated acidic precipitation on above- and below-ground growth of loblolly pine (Pinus taeda), *Can. J. For. Res.*, *22*: 582 (1992).

Rauser W. E., Metal-binding peptides in plants, *Sulfur Nutrition and Sulfur Assimilation in Higher Plants* (De Kok, L. J., I. Stulen, H. Rennenberg, C. Brunold, W. E. Rauser, eds.) SPB Acad. Publ., The Hague; p. 239 (1993).

Reich P. B., A. W. Schoettle, and R. G. Amundson. Effects of ozone and acidic rain on photosynthesis and growth in sugar maple and northern red oak seedlings. *Environ. Pollut.*, *40*: 1 (1986).

Reich P. B., A. W. Schoettle, H. F. Stroo, J. Troiano, and R. G. Amundson, Effects of ozone and acid rain on white pine (*Pinus strobus*) seedlings grown in five soils. I. Net photosynthesis and growth. *Can. J. Bot.*, *65*: 977 (1987).

Rennenberg H., The fate of excess sulfur in higher plants, *Ann. Rev. Plant Physiol.*, *35*: 121 (1984).

Reinikainen J., and L. Kärenlampi, Tissue and cell injuries caused by chlorine water to Scots pine needles, *Eur. J. For. Path.*, *17*: 12 (1987).

Reinikainen J., and S. Huttunen, The level of injury and needle ultrastructure of acid rain-irrigated pine and spruce seedlings after low temperature treatment. *New Phytol.*, *112*: 29 (1989).

Rennenberg H., The significance of higher plants in the emission of sulfur compounds from terrestrial ecosystems, *Trace Gas Emissions by Plants*, (T. Sharkey, E. A. Holland and H. A. Mooney, eds.), Academic Press, San Diego, p. 217 (1991).

Rennenberg H., and C. Brunold, Significance of glutathione metabolism in plants under stress, *Progress in Bot.*, *55*: 142 (1994).

Rennenberg H., B. Huber, P. Schroder, K. Stahl, W. Haunold, H.-W. Georgii, S. Slovik, and H. Pfanz, Emission of volatile sulfur compounds from spruce trees, *Plant Physiol.*, *92*: 560 (1990).

Rennenberg H., and A. Polle, Metabolic consequences of atmospheric sulfur influx into plants, *Plant Responses to the Gaseous Environment* (R. G. Alscher and A. R. Wellburn, eds.) Chapman & Hall, London; p. 165 (1994).

Rennenberg H., A. Polle, N. Martini, and B. Thoene, Interaction of sulfate and glutathione transport in cultured tobacco cells. *Planta*, *176*: 68 (1988).

Rennenberg H., R. Schupp, V. Glavac and H. Jochheim, Xylem sap composition of beech (Fagus sylvatica L.) trees: seasonal changes in the axial distribution of S compounds, *Tree Physiol.*, *14*: 541 (1994a).

Rennenberg H., R. Schupp, and A. Schneider, Thiol composition of a xylem-tapping mistletoe and the xylem sap of its hosts, *Phytochemistry*, *37*: 975 (1994b).

Rennenberg H., and B. Thoene, "The mobility of S in higher plants," Proceedings of the International Symposium Elemental Sulfur in Agriculture 2, p. 701 (1987).

Riding R. T., and K. E. Percy, Effects of SO_2 and other air pollutants on the morphology of epicuticular waxes on needles of *Pinus strobus* and *P. banksiana*, *New Phytol.*, *99*: 555 (1985).

Rinallo C., and P. Raddi, Effects of simulated acid rain and ABS on leaf surfaces of some broadleaf seedlings, *Eur. J. For. Pathol.*, *19*: 151 (1989).

Rinallo C., P. Raddi, R. Gellini, and V. di Lonardo, Effects of simulated acid deposition on the surface structure of Norway spruce and silver fir needles, *Eur. J. For. Path.*, *16*: 440 (1986).

Roberts B. R., Physiological response of yellow poplar seedlings to simulated acid rain, ozone fumigation and drought, *For. Ecol. Manage.*, *31*: 215 (1990).

Roberts B. R., and W. N. Cannon, Changes in xylem pressure potential of red spruce seedlings treated with ozone and simulated acid rain, *Can. J. For. Res.*, *19*: 1200 (1989).

Ruetze M., U. Schmitt, W. Liese, and K. Khppers, Histologische Untersuchungen an Fichtennadeln (*Picea abies* L. Karst.) nach Begasung mit SO_2, O_3 und NO_2, *Allg. Forst- u. Jagdztg.*, *159*: 195 (1988).

Saastamoinen T., and T. Holopainen, Needle and root response of small Pinus sylvestris seedlings exposed to S dioxide and simulated acid rain, *Scand. J. For. Res.*, *4*: 273. (1989).

Sasek T. W., C. J. Richardson, E. A. Fendick, S. R. Bevington, and L. W. Kress, Carryover effects of acid rain and ozone on the physiology of multiple flushes of loblolly pine seedlings, *For Sci.*, *37*: 1078 (1991).

Sauter J. J., and J.-U. Voss, SEM-observations on the structural degradation of epistomatal waxes in *Picea abies* (L.) Karst.—and its possible role in the "Fichtensterben," *Eur. J. For. Pathol.*, *16*: 408 (1986).

Scherbatskoy T., and R. M. Klein, Responses of spruces and birch foliage to leaching by acidic mists, *J. Env. Analysis*, *12*: 189 (1983).

Scherbatskoy T., and Tyree M. T., Kinetics of exchange of ions between artificial precipitation and maple leaf surfaces, *New Phytol.*, *114*: 703 (1990).

Schiffgens-Gruber A., and C. Lütz, Ultrastructure of mesophyll cell chloroplasts of spruce needles exposed to O3, SO2 and NO2 alone and in combination, *Environ. & Exper. Bot.*, *32*: 243 (1992).

Schmitt U., and M. Ruetze, Structural changes in spruce and fir needles, *Environ. Poll.*, *68*: 345 (1990).

Schmitt U., M. Ruetze, and W. Liese, Rasterelektronenmikroskopische Untersuchungen an Stomata von Fichten- und Tannennadeln nach Begasung und saurer Beregnung, *Eur. J. For. Pathol.*, *17*: 118 (1987).

Schneider A., J. Kreuzwieser, R. Schupp, J. J. Sauter and H. Renennberg, Thiol and amino acid composition of the xylem sap of poplar trees (Populus x canadensis robusta), *Can. J. Bot.*, *72*: 347 (1994a).

Schneider A., T. Schatten, and H. Rennenberg, Exchange between phloem and xylem during long distance transport of glutathione in spruce trees (Picea abies Karst. L.), *J. Exp. Bot.*, 45: 457 (1994b).

Schnug E., Physiological functions and environmental relevance of sulfur-containing secondary metabolites, *Sulfur Nutrition and Sulfur Assimilation in Higher Plants* (L. J. De Kok, I. Stulen, H. Rennenberg, C. Brunold, W. E. Rauser, eds.), SPB Acad. Publ., The Hague, p. 179 (1993).

Schonherr J., and M. J. Bukovac, Penetration of stomata by liquids. Dependence on surface tension, wettability, and stomatal morphology, *Plant Physiol.*, *49*: 813 (1972).

Schroder P., G. L. Lamoureux, D. G. Rusness, and H. Rennenberg, Glutathione S-transferase activity in spruce needles, *Pest. Biochem. Physiol.*, *39*: 211 (1990a).

Schroder P., D. G. Rusness, and G. L. Lamoureux, Detoxification of xenobiotics in spruce trees is mediated by glutathione-S-transferases, *Sulfur Nutrition and Sulfur Assimilation in Higher Plants* (H. Rennenberg, C. Brunold, L. J. De Kok, I. Stulen eds.), SPB Acad. Publ., The Hague, p. 245 (1990b).

Schupp R., Untersuchungen zur Schwefelernärung der Fichte (Picea abies L.): Die Bedeutung der Sulfatassimilation und des Transports von Thiolen. Ph.D. Thesis. Technical University Munich, Germany (1991).

Schupp R., V. Glavac, and H. Rennenberg, Thiol composition of xylem sap of beech trees, *Phytochemistry*, *30*: 1415 (1991).

Schupp R., and H. Rennenberg, Changes in S metabolism during needle development of Norway spruce, *Bot. Acta*, *105*: 180 (1992).

Schupp R., T. Schatten, J. Willenbrink, and H. Rennenberg, Long-distance transport of reduced sulfur in spruce (Picea abies L.), *J. Exp. Bot.*, *43*: 1243 (1992).

Seegmüller S., M. Schulte, C. Herschbach, and H. Rennenberg, Interactive effects of mycorrization and elevated atmospheric CO_2 on sulphur nutrition of young pedunculate oak (Quercus robur L.) trees, *Plant Cell Environ.*, *19*: 418 (1996).

Seiler J. R., and Paganelli, D. J., Photosynthesis and growth response of red spruce and loblolly pine to soil applied lead and simulated acid rain, *For. Sci.*, *33*: 668 (1987).

Sheppard L. J., Causal mechanisms by which sulphate, nitrate and acidity influence frost hardiness in red spruce: review and hypothesis, *New Phytol.*, *127*: 69 (1994).

Smith H. J., and D. D. Davies, Histological changes induced in Scotch pine needles by S dioxide, *Phytopathology*, 68: 1711 (1978).

Smith K. T., W. C. Shortle, and W. D. Ostrofsky, Aluminum and calcium in fine root tips of red spruce collected from the forest floor, *Can J. For Res*, *25*: 1237 (1995).

Soikkeli S., A review of the structural effects of air pollution on mesophyll tissue of plants at light and transmission electron microscope level, Savonia, 4: 11 (1981a).

Soikkeli S., The types of ultrastructural injuries in conifer needles of northern industrial environments, *Silva Fennica*, *15*: 399 (1981b).

Soikkeli S., Comparison of cytological injuries in conifer needles from several polluted industrial environments in Finland, *Ann. Bot. Fenn.*, *18*: 47 (1981c).

Soikkeli S., and L. Kärenlampi, Cellular and ultrastructural effects, *Air Pollution and Plant Life* (M. Treshow, ed.), John Wiley & Sons, Chichester, New York, p. 159 (1984).

Solberg R. A., and D. F. Adams, Histological responses of some plant leaves to hydrogen fluoride and sulfur dioxide, *Amer. J. Bot.*, *43*: 755 (1956).

Sorauer P., Beitrag zur anatomischen Analyse rauchgeschädigter Pflanzen, *II. Landwirtsch. Jahrb.*, *37*: 673 (1908).

Spedding D. J., I. Ziegler, R. Hampp and H. Ziegler, Effect of pH on the uptake of ^{35}S-sulfur from sulfate, sulfite, and sulfide by isolated spinach chloroplasts, *Z. Pflanzenphysiol.*, *96*: 351 (1980).

Stewart D., M. Treshow, and F. M. Harner, Pathological anatomy of conifer needle necrosis, *Can. J. Bot.*, *51*: 983 (1973).

Sutinen S., Ultrastructure of mesophyll cells of spruce needles exposed to O_3 alone or together with SO_2, *Eur. J. For. Pathol.*, *17*: 362 (1987).

Tamm C.O., and E.B. Cowling, Acidic precipitation and forest vegetation, *Water Air Soil Pollut.*, *7*: 503 (1977).

Taubenhaus J. J., and W. N. Ezekiel, Acid injury of cotton roots, *Bot. Gaz.*, *92*: 430 (1931).

Taylor G. E., S. B. McLaughlin, D. S. Shriner, and W. J. Selvidge, The flux of sulfur containing gases to vegetation, *Atmos. Environ.*, *17*: 789 (1983).

Taylor G. E., R. J Norby, S. B. McLaughlin, A. H. Johnson, and R. S. Turner, Carbon dioxide assimilation and growth of red spruce seedlings in response to ozone, precipitation chemistry and soil type, *Oecologia*, *70*: 163 (1986).

Thomson W. W., Effects of air pollutants on plant ultrastructure, *Responses of Plants to Air Pollution* (J. B. Mudd and T. T. Kozlowski, eds.), p. 179 (1975).

Trites L. F., and R. G. S. Bidwell, Effect of acidic precipitation on bean plants, *Can. J. Bot.*, *65*: 1121 (1985).

Tung G., A. L. Kuja, and S. N. Linzon, Histopathology of plant leaf injury caused by simulated acid rain, *Trans. Microscop. Soc. Canada*, *9*: 64 (1982).

Tuomisto H., Use of *Picea abies* needles as indicators of air pollution: epicuticular wax morphology, *Ann. Bot. Fenn. 25*: 351 (1988).

Turunen M., and S. Huttunen, Effect of simulated acid rain on the epicuticular wax of Scots pine needles under northerly conditions, *Can. J. Bot.*, *69*: 412 (1991).

Umemoto K., and K. Hozumi, Correlation between the degree of air pollution and the distribution of calcium oxalate crystals in the Gingko leaf, *Microchem. J.*, *17*: 689 (1972).

Van Stegen J., and C. Myttenaere, Status of sulfur in the foliage of Norway spruce (Picea abies (L.) Karst.) in relation to the mode of contamination, *Environ. Pollut.*, 69: 327 (1991).

Vogels K., R. Guderian, and G. Masuch, "Studies on Norway spruce (*Picea abies* Karst.) in damaged forest stands and in climatic chambers experiments," Proceedings of Acidification and its Policy Implications, Amsterdam, Holland (1986).

Walker R. F., and S. B. McLaughlin, Growth and root system development of white oak and loblolly pine as affected by simulated acidic precipitation and ectomycorrhizal inoculation, *Forest Ecol Manag.*, *46*: 123 (1991).

Wellburn A. R., SO_2 effects on stromal and thylakoid function, *Sulfur Dioxide and Vegetation.* (W. E. Winner, H. A. Mooney, and R. A. Goldstein, eds.) Stanford University Press, Stanford, p. 133 (1985).

Wellburn A., Air Pollution and Acid Rain: The Biological Impact, Longman Scientific & Technical, Harlow, UK, pp. 274 (1988).

Werner A., Histological changes induced in *Weigela*, Japanese larch and Scots pine leaves by SO_2, *Arboretum Kornickie*, *26*: 189 (1982).

Whatley J. M., Ultrastructural changes in chloroplasts of *Phaseolus vulgaris* during development under conditions of nutrient deficiency, *New Phytol.*, *70*: 725 (1971).

Whitney R. D., and D. W. Ip, Necrotic spots induced by simulated acid rain on needles of *Abies balsamea* saplings, *Eur. J. For. Path.*, *21*: 36 (1991).

Wood T., and Bormann F. H., Increases in foliar leaching caused by acidification of an artificial mist. *Ambio*, *4*: 169 (1975).

Ylimartino A., E. Pääkönen, and T. Holopainen, Foliar nutrient concentrations and epicuticular wax formation of Scots pine needles. Proceedings of "15th Int. Meeting Air Pollution and Interactions between Organisms in Forest Ecosystems," Tharandt/Dresden, Germany, p. 81 (1992).

Zobel A., and J. E. Nighswander, Accumulation of phenolic compounds in the necrotic areas of Austrian and red pine needles after spraying with sulfuric acid: a possible bioindicator of air pollution. *New Phytol.*, *117*: 565 (1991).

7

Responses of Forest Ecosystems to Changing Sulfur Inputs

Dale W. Johnson
Biological Sciences Center, Desert Research Institute, Reno, Nevada

Myron J. Mitchell
State University of New York, Syracuse, New York

I. INTRODUCTION

There was little information on sulfur (S) cycling in forests compared with that of other nutrients (especially N) until the past two decades. Interest in S nutrition and cycling in forests was heightened with the discovery of deficiencies in some unpolluted regions (Kelly and Lambert, 1972; Humphreys et al., 1975; Turner et al., 1977, 1980) and excesses associated with acidic deposition in other regions of the world (Mewies and Khanna, 1981; Shriner and Henderson, 1978; Johnson et al., 1982, 1986b). Several reviews have been written about S cycling in forests, and the interested reader should refer to those papers for a general treatise on the subject (Turner and Lambert, 1980; Johnson, 1984; Mitchell et al., 1992a,b). It is our primary purpose to consider an important but often overlooked aspect of S cycling in forests—namely, the response to changing S inputs.

Several studies have addressed the effects of increased S inputs in forest soils and forest ecosystems (Johnson and Cole, 1977; Khanna and Beese, 1978; Singh et al., 1980; Lee and Weber, 1982; Khanna et al., 1986; Fernandez and Rustad, 1989), but few have considered the effects of decreased S inputs. Attention has recently begun to focus on the response of acidified ecosystems to decreases in SO_4^{2-} inputs that have resulted from decreases in emissions from the combustion of fossil fuels.

Sulfur emissions reached a maximum in the contiguous United States around 1970 and subsequently have been steadily decreasing (Husar et al., 1991). These decreases should continue with the passage of various environmental regulations such as the amendment to the Clean Air Act (Public Law 101-549) passed by the United States Congress in 1991. These decreases in S emission have been reflected in decreases in SO_4^{2-} concentrations of surface waters in the northeast United States (Driscoll and Van Dreason, 1992; Likens et al., 1990), eastern Canada (Dillon and LaZerte, 1992; Kelso and Jeffries, 1988), and Europe (Allot et al., 1992). Such decreases have also been confirmed by experimental studies, including a roof exclusion experiment of a small catchment (800 m^2) in southern Norway, which began in 1984. This exclusion caused a 50% decrease in SO_4^{2-} output after 3.5 years (Wright and Henriksen, 1990). A similar roof experiment was initiated in 1991 over a much larger (6500 m^2) catchment in Sweden (Hultberg and Grennfelt, 1992). A variety of experimental manipulations including addition and exclusion of S in small catchments and forested plots are in progress in Europe and North America (Hultberg et al., 1992).

With increasing attention being focused on decreasing S inputs, considering the potential responses of forest S cycles to lowered S inputs is especially timely. Prior to such a consideration, however, a brief overview of S cycling in forests and some important ecosystem S retention mechanisms is in order.

II. OVERVIEW OF SULFUR CYCLING IN FORESTS

The S cycle in forests can be represented as in Figure 1 (Johnson, 1984). As is the case for other nutrients, only a few of the fluxes shown in Figure 1 can be measured directly, and others must be calculated using various assumptions. Specifically, wet deposition, litterfall, and throughfall can be measured with relative ease, whereas measurement of dry deposition and leaching are made with considerably more difficulty and involve assumptions in order to scale up to an aerial (kg ha^{-1}) level. Other fluxes, such as foliar leaching, uptake, requirement, and translocation, must be calculated from direct measurements, including the estimates of dry deposition, the latter of which may be subject to considerable error.

The traditional method of calculating uptake, requirement, foliar leaching, and translocation is as follows (Cole and Rapp, 1981):

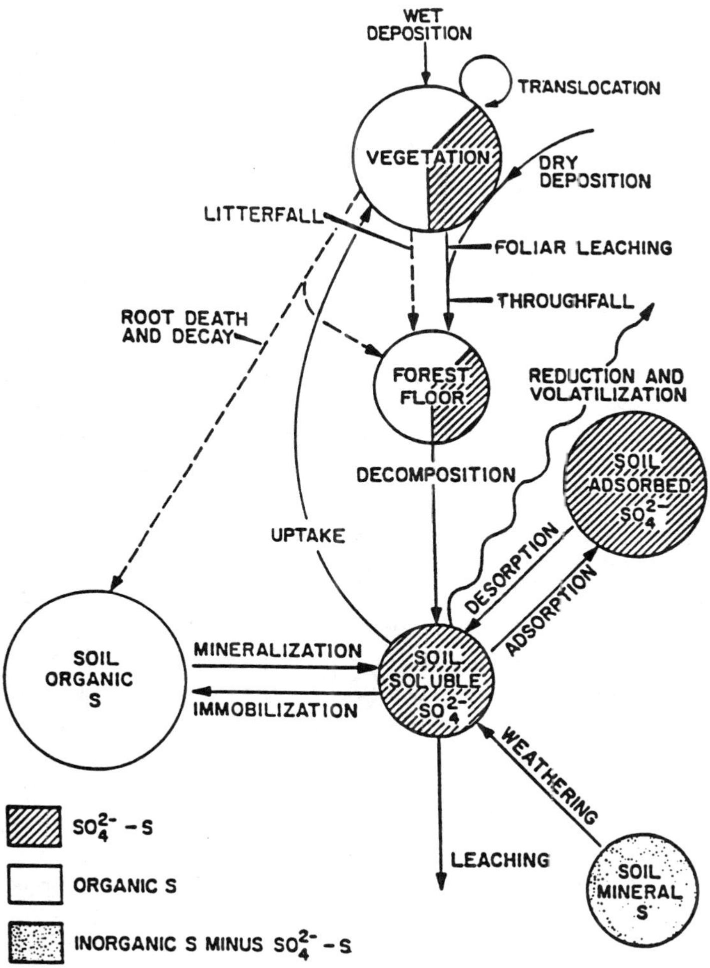

Figure 1 Schematic representation of S cycle in forests. (From Johnson, 1984. Reprinted with permission from Kluwer Academic Publishers, The Netherlands.)

$$\text{Requirement (R)} = \text{Nutrients used in new tissue} \tag{1}$$

$$= \text{Nutrients in new foliage (F)} + \text{wood increment (W)} + \text{root increment (RO)} \tag{2}$$

$$\text{Uptake (U)} = \text{Nutrient return} + \text{nutrient increment in perennial (woody) tissues (W)} \quad (3)$$

$$= \text{Litterfall (LF)} + \text{foliar leaching (FL)} + \text{root turnover (RT)} + \text{W} \quad (4)$$

$$\text{Folior leaching (FL)} = \text{Throughfall} + \text{stemflow} - \text{total translocation (T)}$$

$$= \text{R} - \text{U} \quad (5)$$

Atmospheric deposition is a very important component of the S cycle. This is true not only in polluted areas, where atmospheric S deposition is very high, but also in areas of low S input. Probert and Asmosir (1983) argue that any S in soil parent material will have leached away before the soil is weathered enough to accumulate S in either organic or adsorbed forms. Thus, they argue, atmospheric S inputs are the major source of S for terrestrial ecosystems.

Only in recent years has the great importance of dry deposition of S been recognized. In the past, failure to account for dry deposition of S to forests has sometimes resulted in serious error propagation making flux estimates within the S cycle. If dry deposition is not measured or underestimated, foliar leaching is overestimated (Eq. 5), uptake is overestimated (Eq. 4), and translocation is underestimated. For instance, Shriner and Henderson (1978) measured S cycling parameters without accurate estimates of dry deposition on Walker Branch Watershed (WBW), Tennessee, and found that foliar leaching (termed "net removal" by the authors) was very large relative to other fluxes (Table 1). This led to a large value for calculated uptake, and a very rapid rate of foliar S turnover by leaching (4–5 times per year), leading the authors to suspect that there was a large input of dry-deposited S not included in these calculations. This was later confirmed by Lindberg et al. (1986), who made detailed measurements of wet and dry deposition of S at WBW and obtained lower estimates of foliar leaching. A corrected version of the S cycle for the chestnut oak forest type was then published (Johnson et al., 1982). This showed foliar leaching (2 kg ha^{-1} yr^{-1}) to be more than an order of magnitude lower and, consequently, uptake to be 65% lower (22 kg ha^{-1} yr^{-1}) than the values obtained by Shriner and Henderson (1978) (44.2 kg ha^{-1} yr^{-1} for foliar leaching and 63.6 kg ha^{-1} yr^{-1} for uptake) (Table 1). Going back to the original data from Shriner and Henderson (1978) and inserting an assumed foliar leaching value of 2 kg ha^{-1} yr^{-1} gives an uptake value similar to that obtained later by Johnson et al. (1982), but a much higher total deposition value (63.3 versus 26 kg ha^{-1} yr^{-1}) (Table 1). The latter may have been

Table 1 Sulfur Fluxes on Walker Branch Watershed, Tennessee, Using Traditional Nutrient Cycling Methods and with Corrections by Dry Deposition

	Walker Branch Watershed[a]		
Flux	Traditional	Corrected[c]	Chestnut Oak[b]
Deposition	18.1	63.3[c]	26
Foliar leaching	44.2	2.0[c]	2
Litterfall	8.4	8.4	10
Root turnover	8.1	8.1	7
Wood increment	2.3	2.3	2.1
Uptake	63.6	18.5	22

[a]Data from Shriner and Henderson (1978).
[b]Data from Johnson et al. (1982).
[c]Assuming a 2 kg ha^{-1} yr^{-1} for foliar leaching.

due to a reduction in local S emissions due to the closure of a local power plant in the interim.

A considerable amount of effort was expended during the era of acidic deposition studies in order to obtain accurate estimates of dry deposition of S because of the great uncertainties created by the underestimation of dry deposition (Garten et al., 1988; Lindberg et al., 1990). Very intensive, detailed measurements of S deposition, along with the analysis of forest S cycles, eventually revealed that simple measurements of throughfall and stemflow yield highly accurate estimates of total S deposition, as was originally suggested by Mayer and Ulrich (1974). This method involves either ignoring foliar leaching or subtracting a nominal amount (<1.0 kg ha^{-1} yr^{-1}) for it (Lindberg and Lovett, 1992). Thus, an inexpensive and convenient way to monitor changes in total S deposition on a long-term basis is readily available to anyone with bottles and funnels and the capability of analyzing SO_4^{2-} accurately.

The atmospheric inputs of S to forest ecosystems vary widely, with less than 60 eq ha^{-1} yr^{-1} of SO_4^{2-} (1 kg S ha^{-1} yr^{-1}) in a Douglas fir site in Australia to over 5000 eq ha^{-1} yr^{-1} (80 kg S ha^{-1} yr^{-1}) polluted sites in central Europe such as in Poland (PO) and Czechoslovakia (DZ) (Fig. 2; Table 2). Examination of S inputs versus SO_4^{2-} leaching losses from forest ecosystems shows that for many forest ecosystems there is generally little net retention or loss of this element on an annual basis (Fig. 3). There is a net accumulation of SO_4^{2-} for some forest soils that has been attributed to sulfate adsorption processes that are a function of soil mineralogical

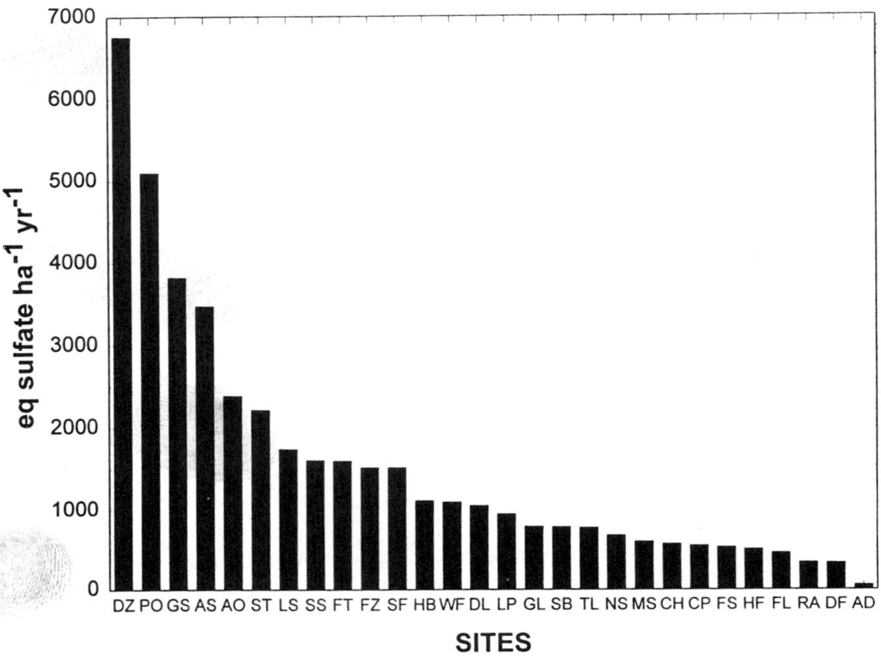

Figure 2 Atmospheric inputs for forested sites throughout the world. See Table 2 for explanation of abbreviations, locations, vegetation, and sources of information. (Reprinted with permission from Kluwer Academic Publishers, The Netherlands.)

characteristics (Johnson and Todd, 1983; Rochelle et al., 1987; Harrison et al., 1989) or possible immobilization by soil microbial processes (Swank et al., 1984). In contrast, net losses of SO_4^{2-} may be caused by desorption of SO_4^{2-} adsorbed during an early period of higher S deposition (Mitchell et al., 1993). Further details on the processes regulating S loss or retention are discussed below.

One interesting and unique feature of S cycles is that a significant proportion of total S can be cycled and accumulated as SO_4^{2-}, within both vegetation and soils. The relative amount of SO_4^{2-} as a proportion of total S increases with S availability. In vegetation, most carbon-bonded S is associated with the amino acids cysteine, cystine, and methionine, which are in turn used for protein synthesis (Kelly and Lambert, 1972). Other forms of organic S in the biota include biotin, thiamine, glutathione, sulfolipids and ester sulfates, with the latter form not being found in trees (Freney, 1967; Turner and Lambert, 1980). There is a very close relation-

Table 2a Sites of the Integrated Forest Study (IFS) Used in Analyzing Sulfur Budgets, Including Measurements of Atmospheric Deposition[a]. See Johnson and Lindberg (1991) for summary on these sites.

Abbreviation	Site location	Vegetation
DF	Thompson Forest, WA	*Pseudotsuga menziesii*
RA	Thompson Forest, WA	*Alnus rubra*
LP	Oak Ridge, TN	*Pinus taeda*
ST	Smoky Mountains, NC	*Picea rubens*
CP	Coweeta Hyd. Lab., NC	*Pinus strobus*
GL	BF Grant Forest, GA	*Pinus taeda*
FS	Gainesville, FL	*Pinus elliotii*
DL	Duke Forest, NC	*Pinus taeda*
TL	Turkey Lakes, Ontario	*Acer saccharum*, *Betula alleghaniensis*
HF	Huntington Forest, NY	*Acer saccharum*, *Fagus grandifolia*
WF	Whiteface MT., NY	*Picea rubens*
MS	Howland, ME	*Picea rubens*
NS	Nordmoen, Norway	*Picea abies*
FL[a]	Findley Lake, WA	*Abies amabilis*, *Tsuga heterophylla*
SS[a]	Smoky Mountains, NC	*Picea rubens*
SB[a]	Smoky Mountains, NC	*Fagus grandifolia*
CH[a]	Coweeta Hyd. Lab., NC	*Quercus* spp., *Carya* spp., *Acer* spp.

[a]Bulk deposition only sites.

ship between organic S and organic N in plant tissues due to their mutual involvement in protein synthesis. If N is limiting, even under conditions of relatively low S inputs, SO_4^{2-} may accumulate. Kelly and Lambert (1972) reported that the ratio of organic S to organic N in *Pinus radiata* foliage was a constant 0.03 on a molar basis (Kelly and Lambert, 1972; Turner et al., 1977; Turner and Lambert, 1980). Sulfur present in excess of that dictated by this ratio was present as SO_4^{2-}, extractable by dilute (0.01 M) HCl.

The potential importance of SO_4^{2-} in forest S cycles is illustrated in Figure 4, which depicts total S and SO_4^{2-} cycles for four forest ecosystems, two (chestnut oak [*Q. prinus*] and yellow poplar [*Liriodendron tulipifera*] mixed forest types) at Walker Branch Watershed, Tennessee, and two (red alder [*Alnus rubra*] and Douglas fir [*Pseudotsuga menziesii*]) at the Thompson site, Washington (Johnson et al., 1986b). At the more polluted

Table 2b Selected Non-IFS Forest Sites Used in Comparing Some Aspects of Sulfur Cycling

Name of site	Abbreviation	Location	Forest type	References
Vysoka Pec, Damaged Forest	DZ	Most, Czech.	*Quercus* spp. *Fagus*	Paces (1985)
Salacova Rural, Forest	FZ	Lhota, Czech.	*Picea*	Paces (1985)
Ardennes Oak	AO	Ardennes, France	*Quercus* spp.	Nys and Ranger (1988)
Ardennes Spruce	AS	Ardennes, France	*Picea abies*	Nys and Ranger (1988)
Ispina Mixed Oak	PO	Ispina, Poland	*Quercus robus*	Karkanis (1976)
Bago, NSW, Douglas-fir	AD	Australia	*Pseudotsuga menziesii*	Turner and Lambert (1980)
Solling, Beech	SF	West Germany	*Fagus silvatica*	Meiwes and Khanna (1981)
Soiling, Spruce	GS	West Germany	*Picea abies*	Meiwes and Khanna (1981)
Lake Gardsjon, Catchment	LS	Sweden	*Picea abies* *Pinus sylvestris*	Hultberg (1985)
Hubbard Brook White Mts.	HB	New Hampshire, U.S.A.	*Fagus grandifolia* *Acer* spp.	Likens et al. (1977, 1990), Mitchell et al. (1989)
Walker Branch (Fullerton soil)	FT	Tennessee, U.S.A.	*Quercus prinus* *Carya* spp.	Johnson et al. (1986), Lindberg et al. (1986)

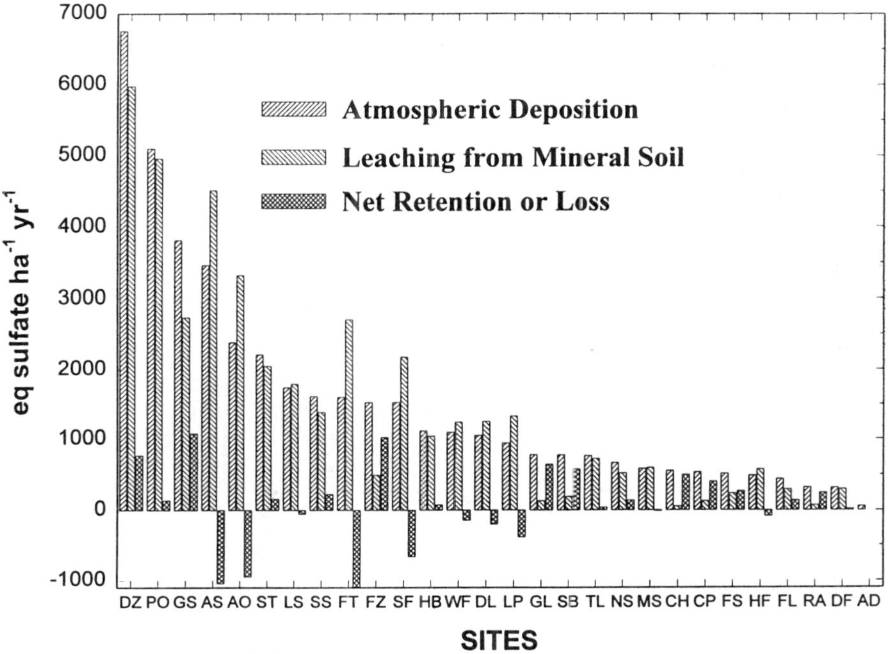

Figure 3 Net retention or loss of S in forest ecosystems. See Table 2 for explanation of abbreviations, locations, vegetation, and sources of information. (Reprinted with permission from Kluwer Academic Publishers, The Netherlands.)

Tennessee sites, SO_4^{2-} accounts for a somewhat greater proportion of total S in vegetation (25–30%) and soils (10–17%) than at the less polluted Washington sites (9–22% in vegetation and 6–11% in soils). Even more striking, however, is the much lower percentage of SO_4^{2-} in vegetation (9%) and soil (6%) in the N-fixing red alder stand than in the N-deficient Douglas fir stand (22% in vegetation and 11% in soil). This reflects incorporation of S into organic forms in the N-rich alder stand.

There has been some confusion in the literature as to the exact form of SO_4^{2-} in plant tissues. Some investigators found that the N:S ratio in tree tissues varied, especially following N fertilization (Turner et al., 1980; Kelly and Johnson, 1982; Mitchell et al., 1992a). Johnson et al. (1982) found that not all foliar SO_4^{2-} in some deciduous species could be extracted with water and suggested that some of this apparent SO_4^{2-} may be in organic forms (e.g., ester). However, Richter and Johnson (1983) found no difference between foliar SO_4^{2-} extracted by 0.01 *M* HCl and 0.01 *M*

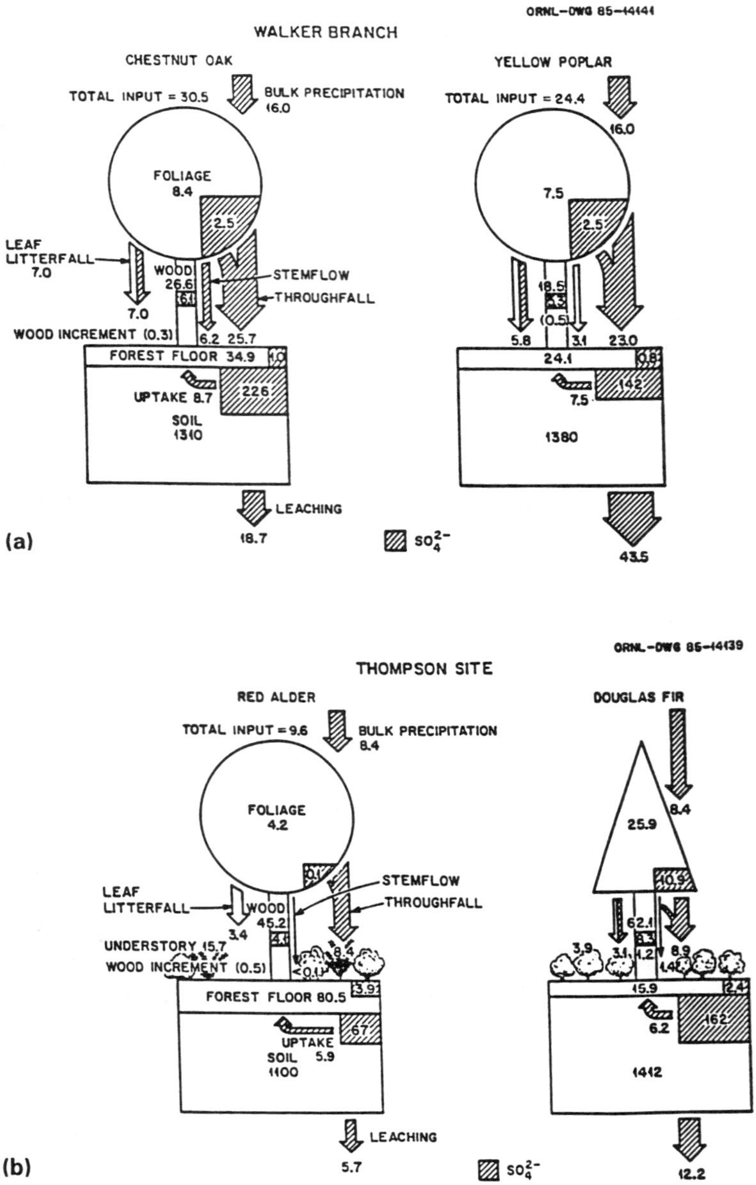

Figure 4 Sulfur cycles in a chestnut oak and yellow poplar forest type on Walker Branch Watershed, Tennessee (a) and a Douglas fir and red alder forest in Washington (b). (From Johnson et al., 1986b.)

KCl in deciduous species, indicating that foliar SO_4^{2-} was indeed inorganic SO_4^{2-} rather than an organic (ester) sulfate, again indicating the absence of this form of organic S in trees.

It is interesting that a certain level of foliar SO_4^{2-} seems to be necessary for adequate S nutrition. Sulfur deficiency is known to occur at foliar SO_4^{2-} levels of 80 μg/g (Turner and Lambert, 1980), suggesting that foliage SO_4^{2-} is not simply a reflection of excess S, but plays some important role in plant S nutrition. The most logical assumption for this role would be that foliar SO_4^{2-} acts as temporary storage; however, why this is not also the case for N is not known.

As is the case with many other nutrients, the soil is the major reservoir of S and site for S accumulation in forest ecosystems. Sulfur cycles in the four forest sites referred to above illustrate this (Fig. 4). In each case, the annual increment of S in vegetation is considerably less than deposition and leaching fluxes, indicating that sequestering by vegetation plays only a minor role in the retention (or loss) of S in these ecosystems. The same is true for a wide variety of forest ecosystems investigated as part of the Integrated Forest Study (Mitchell et al., 1992a) as well as for most other forest ecosystems throughout the world, except for a S-deficient Douglas fir stand in Australia (Mitchell et al., 1992b)

The distribution of S constituents among vegetation, forest floor, and mineral soil are given in Figure 5. The location of these sites and their vegetation composition is given in Table 2. Mineral soil S accounts for 80–97% of total ecosystem S in all sites but the Douglas fir stand in Australia, where vegetation contributes 37% of the total ecosystem content (Turner and Lambert, 1980). Regressing total ecosystem S content against total S input indicates that there is no direct relationship between these two parameters. However, the low input and low S content for the Australian site is noteworthy (Figs. 2 and 5). These data, however, may not be representative of forests on a global basis, because the vast majority of sites with information on S cycling are located in Europe and North America with no complete data from Asia, Africa, or South America (see Chapter 8, this volume) (Fig. 6).

Significant accumulations in the forest floor seem equally unlikely as a major factor in S retention or release, given the relative magnitudes of forest floor S content and annual deposition and leaching fluxes for most sites. However, for sites such as the Douglas fir in Australia (AD) and red alder (RA) and Douglas fir (DF) in the northwest United States with relatively low inputs (i.e., <15 kg S ha^{-1} yr^{-1}) the uptake of S by vegetation and its return via litter may play important roles in regulating S flux (Mitchell et al., 1992a). Analyses of S cycles in the Integrated Forest Study

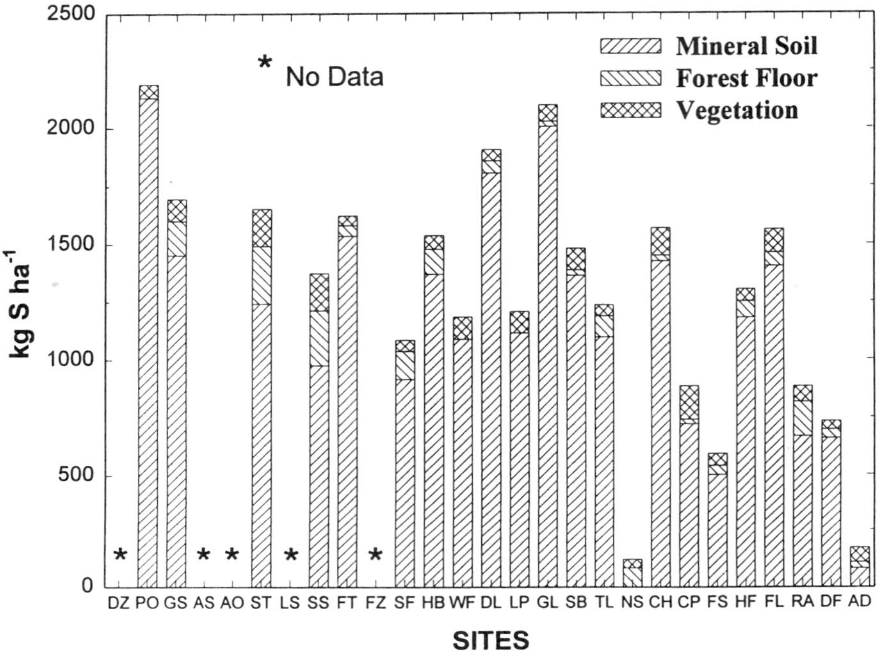

Figure 5 Distribution of S constituents among vegetation, forest floor, and mineral. See Table 2 for explanation of abbreviations, locations, vegetation, and sources of information.

indicated that the forest floor did not serve a major role for S accumulation for any of the sites (Mitchell et al., 1992b).

The limited studies of volatile S losses from forest soils suggest that these fluxes are very small relative to those via leaching. For example, Andrea and Jaeschke (1992) estimated that in the wet tropics, average emissions ranged from 0.16 to 0.52 kg S ha^{-1} yr^{-1} and for vegetated land surfaces there was an average emission of 0.4 kg S ha^{-1} yr^{-1}. In contrast, SO_4^{2-} leaching rates in forests are generally greater by up to two orders of magnitude, typically ranging from 5 to 60 kg S ha^{-1} yr^{-1} (Mitchell et al., 1992a). As mentioned previously, there is a paucity of data on tropical forests and it has been found for a few extreme cases that gaseous outputs may be as large as 10 to 20 kg S ha^{-1} yr^{-1}, mostly due to emissions from the vegetation (Mitchell et al., 1992a). In general, however, the soil is presumed to be the major site of S retention and release.

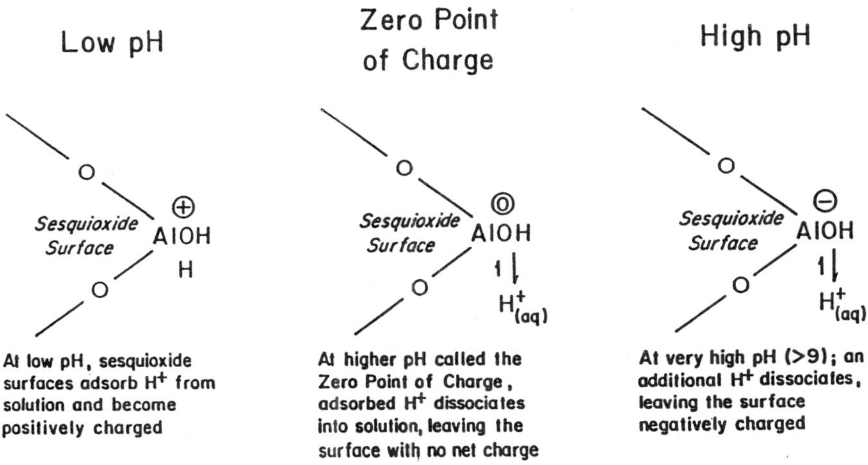

Figure 6 Schematic representation of the effect of pH on surface charge of Al oxides.

The relative importance of inorganic SO_4^{2-} adsorption or organic S incorporation formation as mechanisms for S retention has been the subject of considerable research and debate over the last decade (Johnson and Todd, 1983; Johnson, 1984; Fitzgerald et al., 1982; Swank et al., 1984; Mitchell et al., 1992a). In most forest soils, organic S (mostly in the form of carbon-bonded S) and ester sulfates are the dominant pools (>90%) except in those soils with high sulfate adsorption capacity and/or high levels of S deposition where SO_4^{2-} concentrations may approach that of organic S. In general, the relative contribution of the ester sulfate pools as a portion of the total S constituents of soils increases with soil depth (Mitchell et al., 1992a). The organic and inorganic mechanisms of S retention and release in soils are reviewed in the following sections.

III. SOIL S RETENTION AND RELEASE MECHANISMS

A. Inorganic Processes

Two major soil inorganic SO_4^{2-} retention mechanisms have been identified as being important in forest soils: (1) adsorption and (2) precipitation. Most studies of SO_4^{2-} mobility indicate that adsorption is more important than precipitation (Singh, 1984c; Rajan, 1978; David et al., 1987; Fasth et al.,

1991; Khanna et al., 1986), but precipitation of aluminum (Al) sulfates has been found to play an important role in some soil systems where both SO_4^{2-} and Al concentrations are high (e.g., Khanna et al., 1987).

1. Sulfate Precipitation Reactions

Sulfate can be retained in soils by precipitation in many forms. Calcium sulfate ($CaSO_4$) is the predominant form in arid soils, but in humid soils, the solubility product of $CaSO_4$ (log K = −4.62) is seldom reached, and aluminum sulfates are predominant. Reuss and Johnson (1986) describe precipitation reactions of the three minerals and the possibilities for their control over SO_4^{2-} and Al concentrations in soil solution:

Jurbanite (log K = −17.2):

$$Al(OH)SO_4.H_2O \longrightarrow Al^{3+} + SO_4^{2-} + OH^- + 5\ H_2O \tag{6}$$

Basalunite (log K = −117.5):

$$Al_4(OH)_{10}SO_4.H_2O \longrightarrow 4\ Al^{3+} + SO_4^{2-} + 10\ OH^- + 5\ H_2O \tag{7}$$

Alunite (log K = −85.4):

$$KAl_3(OH)_6SO_4 \longrightarrow K^+ + 3\ Al^{3+} + 2\ SO_4^{2-} + 6\ OH^- \tag{8}$$

Several investigators (Rajan, 1978; Singh, 1984c; Johnson et al., 1981, 1982; Fuller et al., 1987; Fasth et al., 1991) have concluded that sulfate adsorption rather than precipitation is the most important S retention mechanism in the soils that they studied. Others, however, have found evidence for a significant role for precipitation. Nilsson and Bergqvist (1983) investigated the possibilities for jurbanite and basaluminite control of SO_4^{2-} concentrations in soil solutions, groundwaters, and streamwaters from the Swedish coast. The authors concluded that Al^{3+} and SO_4^{2-} in lower B horizon soil solutions were near equilibrium with basaluminite, whereas Al^{3+} and SO_4^{2-} in upper B horizon and groundwater solutions were near equilibrium with jurbanite. Reuss and Johnson (1986), noting that Al^{3+} and SO_4^{2-} in both soil solutions and groundwaters often exceeded the solubility product for basaluminite, questioned whether basaluminite (or alunite, which is more soluble) could be controlling SO_4^{2-} concentrations. They did allow for the possibility that jurbanite controls Al^{3+} and SO_4^{2-}, however.

Khanna et al. (1987) presented convincing evidence that jurbanite controlled soil solution SO_4^{2-} dynamics over a 13-year period (1969–1982) in beech and spruce stands at the Solling sites in Germany. They report notably increased soil solution Al^{3+} and SO_4^{2-} concentrations following an acidification pulse in 1975–1976, which was consistent with the dissolution of jurbanite. Solubility diagrams also indicated jurbanite control of soil solution Al^{3+} and SO_4^{2-} in these sites.

The question of whether adsorption or precipitation is the major inorganic mechanism controlling SO_4^{2-} concentrations and S retention in soils has no single answer. In some cases, Al^{3+} and SO_4^{2-} concentrations are sufficiently high for precipitation, and such reactions may dominate SO_4^{2-} dynamics (e.g., Khanna et al., 1986). In other cases, soil solution Al^{3+} or SO_4^{2-} are not high enough for aluminum sulfate precipitation and SO_4^{2-} retention is consistent with adsorption models (e.g., Johnson et al., 1985).

2. Sulfate Adsorption

Anion adsorption has been associated with hydrous oxides of Fe and Al and, to a lesser extent, kaolinite (Gebhardt and Coleman, 1974; Hingston et al., 1967; Rajan, 1978; Parfitt and Smart, 1978). There are two basic types of adsorption: (1) nonspecific adsorption, in which anions are held as counter-ions in the diffuse double layer next to a positively charged surface, and (2) specific adsorption, in which the anion displaces $-OH$ or $-OH_2$ groups from hydrous oxide surfaces (Hingston et al., 1967). Iron and Al hydrous oxides can take on a net positive, neutral, or negative charge depending upon the concentration of the potential determining ion (usually H^+) (Hingston et al., 1967). At low pH (below the zero point of charge, or ZPC), surfaces become positively charged because of H^+ adsorption, and surfaces take on either a neutral (at ZPC) or negative (above ZPC) charge as pH increases and H^+ dissociates from the surface $-OH$ or $-OH_2$ ligands (Fig. 6). Thus, pH has a strong effect on anion adsorption, both because of the charge on adsorption surfaces and because of competition with OH^- for adsorption sites.

Rajan (1978) suggests two possible reactions for specific adsorption of sulfate: (1) a bi-dentate coordination when surfaces have a net positive charge (pH below ZPC) and (2) a mono-dentate coordination when surfaces are neutral (Fig. 7). No allowance is made for sulfate adsorption when surfaces have a net negative charge. Reaction one is favored when the surface has a positive charge (when pH is below ZPC) and reaction two is favored at neutral pH (when pH is at ZPC).

Specific adsorption of SO_4^{2-} (or any other anion) can result in the introduction of a net negative charge on the oxide surface, increasing cation exchange capacity and acidity (Hingston et al., 1967; Rajan, 1978; Parfitt and Smart, 1978; Coleman and Thomas, 1967).

Several soil chemical and physical factors affect sulfate adsorption capacity (see review by Harward and Reisenauer, 1966). As noted above,

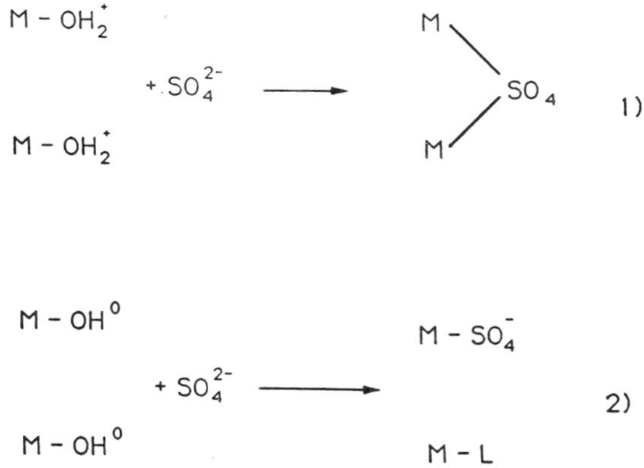

Figure 7 Suggested mechanism for sulfate adsorption on Fe or Al hydrous oxides. M = metal atom and L = ligand (usually OH^- or OH_2). (Adapted from Rajan, 1978.)

pH has a strong effect on sulfate adsorption, and liming is known to cause displacement of adsorbed sulfate. Because adsorption of SO_4^{2-} is often accompanied by the displacement of OH^- and the creation of cation exchange sites, it can contribute to soil acidification. This acidity is manifested when pH is raised (OH^- concentration is increased)—for instance, during liming. The added lime needed to displace adsorbed sulfate has been termed the "acidity due to anions" (Mehlich, 1964). Mehlich (1964) noted that soils containing adsorbed SO_4^{2-} had higher lime requirements and lower pHs than those containing adsorbed OH^-.

The importance of pH in regulating SO_4^{2-} flux has been demonstrated in experimental forested watersheds. At the Hubbard Brook Experimental Forest (HBEF) and other sites in the northeastern United States, both clear-cutting and whole-tree harvesting exerts a major influence on soil solution and streamwater chemistry, including SO_4^{2-} concentration and flux (Likens et al., 1969, 1978; Hornbeck et al., 1986). After vegetation removal, nitrification and resultant nitrate leaching are accelerated and the soil is acidified. This pH depression causes significant increases in sulfate adsorption and lowers streamwater SO_4^{2-} concentrations and fluxes (Nodvin et al., 1986b; Fuller et al., 1987; Mitchell et al., 1989). It has also been found that, in general, there is an inverse relationship between leaching rates of nitrate and SO_4^{2-} in forest ecosystems (Mitchell et al., 1992b).

Aside from the effects of pH, the cationic composition of the soil or soil solution also has a marked influence on sulfate adsorption. Chao et al.

(1963) found that sulfate adsorption increased in the order of $Ca^{2+} > K^+ > NH_4^+ > Na^+$ in solution and $Al^{3+} > Ca^{2+} > K^+$ on soil exchange sites. Chao et al. (1963) found that pH was a more important factor than cation composition, however.

Sulfate can be displaced by phosphate (Hingston et al., 1967; Mekaru and Uehara, 1972), and phosphate is an effective extractant for soil SO_4^{2-} (Tabatabai, 1982). However, SO_4^{2-} is ineffective in displacing phosphate (see review by Harward and Reisenauer, 1966). Previous treatment of soils with phosphate reduces sulfate adsorption capacity (Ensminger, 1954). Thus, specific adsorption of SO_4^{2-} and phosphate is not true anion exchange, as is the case with cation exchange and with nonspecific anion adsorption.

Because adsorption sites are primarily Fe and Al hydrous oxides, the presence and nature of these constituents in soils may be correlated with sulfate adsorption capacity, especially among similar soils (Johnson and Todd, 1983; Harrison et al., 1989; MacDonald and Hart, 1990). Johnson and Todd (1983) found that the single soil property most closely correlated with sulfate adsorption in a variety of forest soils was Fe_c (difference between Fe extracted by citrate-dithionite and oxalate). This was somewhat surprising, in that Fe_c is supposed to provide an index of crystalline Fe oxides, whereas several investigators found that sulfate adsorption was most closely associated with the more amorphous forms of both Fe (Parfitt and Smart, 1978) and Al (Rajan, 1978). Similarly, Fuller et al. (1985) found that sulfate adsorption in forested Spodosols in the northeastern United States was correlated with crystalline Fe. In contrast, Johnson et al. (1986a), Harrison et al. (1989), and MacDonald and Hart (1990) showed correlations between sulfate adsorption and amorphous forms of Al and/or Fe rather than crystalline forms. Such results are more consistent with those patterns predicted from mechanistic studies (Parfitt and Smart, 1978; Rajan, 1978).

The role of organic matter in sulfate adsorption has also been a subject of debate and conflicting results. Some studies have indicated that organic matter interferes with sulfate adsorption (Couto et al., 1979; Johnson and Todd, 1983; Singh, 1980; Gobran and Nilsson, 1988), others have indicated that organic matter enhances sulfate adsorption (Haque and Walmsley, 1973; Fuller et al., 1985); and some have found that it has little effect (Singh, 1984b). The confusion may lie in the juxtaposition of two conflicting processes: (1) the indirect enhancement of sulfate adsorption by organic matter by maintaining Fe and Al oxides in more amorphous forms with higher adsorption capacities and (2) the competition for and blockage of adsorption sites by organic ligands. In the latter context, the results of Gobran and Nilsson (1988) are of special interest; they demonstrated the decidedly negative effect of soil solution organic matter on sulfate adsorption in a Haplorthod, indicating strong competition of organic ligands for adsorption sites. They noted that laboratory-determined sulfate adsorption isopleths did not account for the effects of soil solution organic ligands and were of limited value for modeling or extrapolating to field conditions where concentrations of dissolved organic matter (DOM) may be quite high (e.g., surface horizons and Bh horizons of Spodosols).

The kinetics of sulfate adsorption and desorption have been explored by several investigators (Barrow and Shaw, 1977; Rajan, 1978; Johnson and Henderson, 1979; Singh, 1984d), but the most thorough studies have been done by Barrow and Shaw (1977) and Singh (1984d). In each case, the authors reported that sulfate adsorption occurred in two phases, an initial rapid one and a longer, slower one. The first phase occurs within 24 hours and can account for 60–85% of total adsorption (Singh, 1984d). Rajan (1978) reported that 95% of equilibrium sulfate adsorption was attained within 15 minutes in synthetic and natural allophane soils. The second phase of sulfate adsorption occurs over periods as long as 6 months, and although it typically accounts for less total sulfate adsorption than the first phase, it is accompanied by an increasing irreversibility of adsorbed sulfate (Sanders and Tinker, 1975; Barrow and Shaw, 1977; Singh, 1984d).

Of particular interest in terms of responses to lowered S emissions and hence lower SO_4^{2-} concentrations in soil solution is the reversibility of adsorbed sulfate. In some cases, a large fraction of adsorbed sulfate is retained against leaching with either water or dilute salt solutions containing nonspecifically adsorbing anions (Chao et al., 1962; Sanders and Tinker, 1975; Barrow and Shaw, 1977; Couto et al., 1979; Rajan, 1978; Johnson and Henderson, 1979; Singh, 1984c; Harrison et al., 1989). Several factors have been found to affect the reversibility of sulfate adsorption. As noted above, both pH and the contact time between soil solution and adsorbing soil prior to desorption has a major effect on the total amount desorbed. Rajan (1978) suggests that the often-observed hysteresis of sulfate adsorption–desorption isopleths is simply a reflection of pH changes (due to OH^- consumption or production) during the adsorption or desorption processes. This is an important point, as variations in pH have such a strong effect on sulfate adsorption.

We are aware of only one study that addressed the kinetics of sulfate desorption, that of Barrow and Shaw (1977). They found that most sulfate desorption was complete within 30 minutes at low solution to soil ratios; but at greater ratios, significant desorption continued to occur over the 60-hour measurement period. Unfortunately, it appears as if no research has been conducted on sulfate desorption over the longer periods that are more relevant to the issues of soil response to decreased S deposition. In this context, it is interesting to examine the S budget for the yellow poplar stand in Figure 4. The apparent net loss of S (as SO_4^{2-}) from this ecosystem—approximately 20 kg ha^{-1} yr^{-1}—is greater than could be expected from either an underestimate of S deposition (a factor of approximately two) or an overestimate of water flux (also a factor of approximately two). Thus, it appears as if SO_4^{2-} is being desorbed from this system, perhaps as a result of the previously higher inputs of S to the Walker Branch Watershed, as discussed above. Alternatively, the excess S leaching could have originated from organic S pools; S mineralization rates on WBW are known to be very high (Johnson et al., 1982).

Similar conclusions were drawn by Mitchell et al. (1993) in analyzing the processes regulating the spatial and temporal patterns of SO_4^{2-} flux in a northern hardwood stand at the Huntington Forest in the Adirondack Mountains of New York. An increase in SO_4^{2-} flux with depth in the mineral soil and the lack of variability in solute concentration below the rooting zone all suggest that SO_4^{2-} stored during an early period of greater atmospheric inputs is now being desorbed. For this site the potential importance of organic S in effecting SO_4^{2-} leaching is shown by the important contribution of organic S mineralization in contributing to SO_4^{2-} leaching within the forest floor (David et al., 1987).

3. Modeling sulfate adsorption and desorption

All models of sulfate adsorption and desorption assume that adsorption is concentration-dependent. Thus, the capacity of soils to adsorb sulfate at any given time is a function of the slope of the sulfate adsorption isopleth and the position of the soil system on that isopleth at the time in question. The slope of the isopleth is a function of certain soil chemical properties (discussed above) and the position on the curve is a function of the history of S inputs.

Several equations have been used to model sulfate adsorption (see review by Singh, 1984d). The most commonly used are the Freundlich and the Langmiur equations:

$$\text{Freundlich:} \quad X = A(SO_4^{2-})^b \tag{9}$$

where X = adsorbed sulfate on soil (mol g^{-1})
(SO_4^{2-}) = solution sulfate activity (mol l^{-1}) and
A,b = constants.

$$\text{Langmiur:} \quad X = \frac{(X_{max})(b)(SO_4^{2-})}{1 + (b)(SO_4^{2-})} \tag{10}$$

where X_{max} = maximum sulfate that can be adsorbed on soil (a constant) and

b = a constant.

Some studies have shown the Langmuir equation to provide the best depiction of laboratory-determined sulfate adsorption isopleths (e.g., Johnson and Cole, 1977), and some have shown the Freundlich to provide the best depiction (e.g., Khanna et al., 1986; Singh, 1984a).

Nodvin et al. (1986a) described a very different but very simple equation involving a partition coefficient, which best depicted the sulfate adsorption isopleth for a northern hardwood soil. They named their isopleth the "initial mass isotherm":

$$\text{IM:} \quad RE = mX_i - b \tag{11}$$

where
RE = amount of sulfate removed from or released to solution (mmol/kg)
m = partition coefficient (unitless)
X_i = initial concentration of SO_4^{2-} in solution, expressed per unit soil (mmol/kg) and
b = intercept (mmol/kg)

This formulation assumes that there is a total pool of reactive SO_4^{2-} that consists of readily soluble SO_4^{2-} in the soil (RSP) and that added in solution (X_i):

$$\text{Total reactive sulfate} = RSP + Xi \tag{12}$$

The readily soluble sulfate pool in soil can be calculated from isopleth data as follows:

$$RSP = b/(1 - m) \tag{13}$$

In the case where RSP = 0, this reduces to a simple linear isopleth

$$RE = Kd(Cf)$$

where
Kd = distribution coefficient ($L\ kg^{-1}$)
Cf = equilibrium solution concentration ($mmol\ L^{-1}$) and

$$Kd = \frac{(m)(\text{volume of solution})}{(1-m)(\text{mass of soil})}$$

Figure 8 depicts three generic curves for the three types of isopleths. The Langmuir and Freundlich isopleths both show decreasing adsorption with increasing solution concentrations, whereas the linear isopleth does not. Only the Langmuir isopleth shows a maximum adsorption, and thus it is

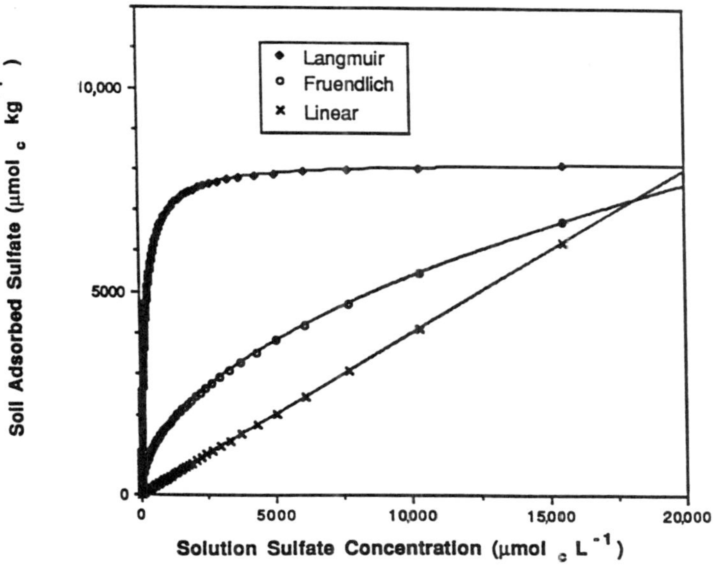

Figure 8 General shapes of Langmuir, Freundlich, and linear sulfate adsorption isopleths.

the most realistic at very high soil solution concentrations. However, typical soil solution concentrations seldom reach values that would produce maximum soil adsorbed sulfate, and thus both the Freundlich (Singh, 1984c) and linear (Nodvin et al., 1986b) isopleths have been found to best describe sulfate adsorption at lower concentrations.

Reuss and Johnson (1986) considered the potential effects of differing degrees of reversibility of sulfate adsorption on recovery of soils and soil solutions using the Langmuir model. For the purposes of illustration, a square wave input pulse was considered with three reversibility scenarios: (1) complete reversibility with no hysteresis, (2) partial reversibility, and (3) complete reversibility (Figure 9). If sulfate adsorption is completely reversible with no hysteresis (scenario *a* in Fig. 9), there is theoretically no difference in the response slope of soil solution to increases or decreases in SO_4^{2-} input (scenario *a* in Fig. 10), and no residual CEC or acidity due to anions remaining in the soil. The soil has, theoretically, been acidified to the amount equivalent to the SO_4^{2-} lost on desorption. With partial reversibility (scenario *b* in Fig. 9), soil solution response is more rapid (Fig. 10), and some residual CEC and acidity due to anions remains within the soil. With no reversibility (all adsorbed SO_4^{2-} is permanently retained;

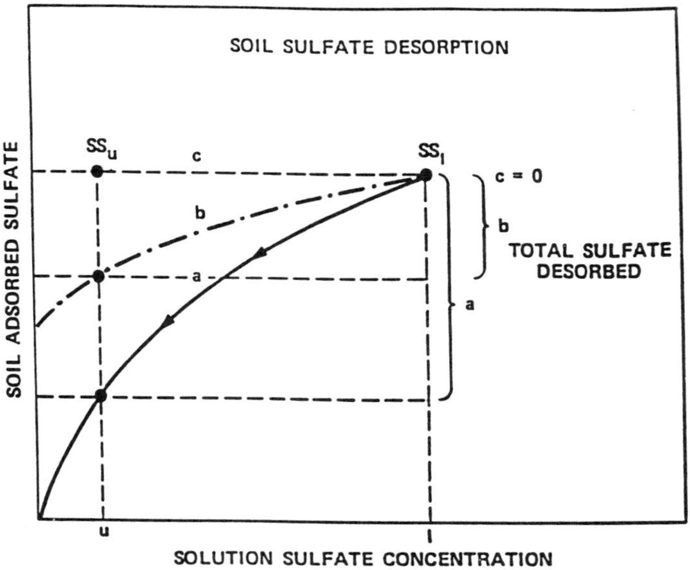

Figure 9 Schematic representation of different degrees of reversibility of sulfate adsorption: (a) completely reversible (no hysteresis); (b) partially reversible; (c) irreversible. (From Reuss and Johnson, 1986. Reprinted with permission from Springer-Verlag, New York, New York.)

scenario *c* in Fig. 9), soil solution recovery is instantaneous, and all CEC and acidity due to adsorbed sulfate remains within the soil.

IV. ORGANIC S RETENTION MECHANISMS

Organic S accumulates in soil if immobilization plus organic S inputs is greater than mineralization and/or dissolved organic S constituents show a net retention in soil as shown conceptually in Figure 11. It is obvious that over extended periods, there is a net accumulation of organic S in soil, resulting in the formation in this large pool of organic S. For example, Mitchell et al. (1989) estimated within a northern hardwood forest at the Hubbard Brook Experimental Forest that the net retention of dissolved organic S since the period of most recent glaciation, which occurred about 10,000 years ago, could account for the present organic S concentration. The importance of dissolved organic S being retained in soils has also been proposed for other soils (Schoenau and Bettany, 1987)

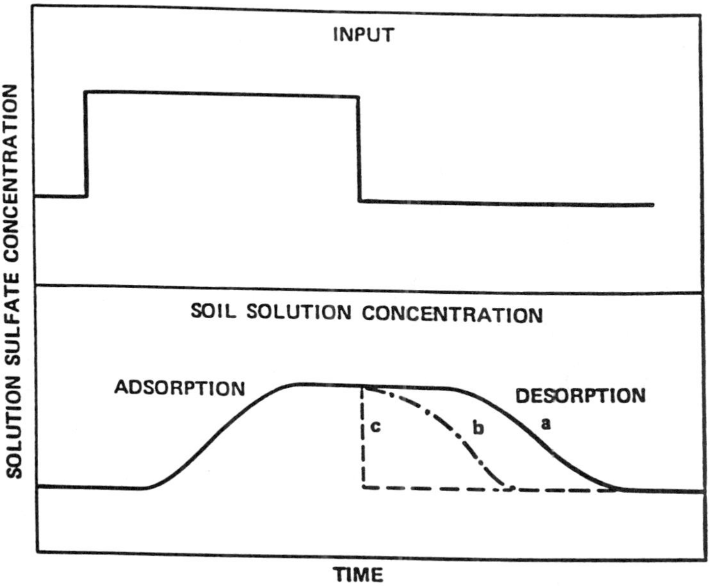

Figure 10 Theoretical response of soil solution sulfate concentration to a square wave pulse input and different degrees of reversibility as in Figure 7. (From Reuss and Johnson, 1986. Reprinted with permission from Springer-Verlag, New York, New York.)

The use of the radioisotope of sulfur ^{35}S has been used extensively to study the dynamics of organic S. Laboratory (Strickland et al., 1986b; Schindler and Mitchell, 1987; Dhamala et al., 1990) and field studies (Strickland et al., 1986a; David et al., 1987) have both shown that there is a rapid conversion of ^{35}S- SO_4 to both ester sulfates and carbon-bonded S. However, these pools may be highly labile and there may also be a rapid mineralization of this newly formed organic S back to inorganic sulfate. The ester sulfate pools appear to be particularly labile, but depending on the actual chemical species, carbon-bonded S may also be rapidly mineralized (Strickland et al., 1984).

When interpreting rates of sulfur incorporation using ^{35}S techniques, it is important to clearly define whether gross or net rates are being estimated. If gross rates are used to estimate the potential for SO_4^{2-} incorporation and retention into organic pools and the reverse mineralization reaction is not evaluated concomitantly, then these estimates would overestimate the actual net incorporation of inorganic SO_4^{2-} into organic pools. Such estimates may be useful for estimating the potential for organic S

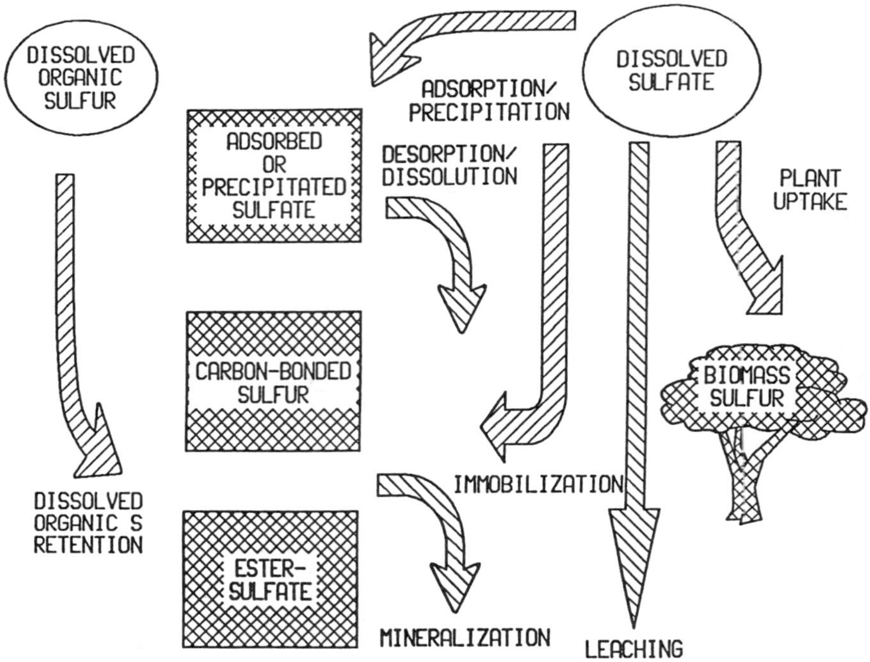

Figure 11 Schematic diagram of mechanisms for S retention in a forest ecosystem.

formation over short periods (i.e., days or weeks), but they do not give the actual rate of organic S accumulation over longer periods (i.e., greater than one year). The more extended periods are most important in considering the long-term reversibility of sulfur retention in soils.

Fuller et al. (1986a), incorporating laboratory data into a simulation model of S fluxes among inorganic and organic pools including adsorption-desorption as well as immobilization-mineralization reactions, found that residence time was much shorter in adsorbed pools than in either carbon-bonded or ester sulfate pools. In addition, within the mineral soil, ester sulfates had significantly greater turnover rates than the total pool of carbon-bonded S.

Stable isotopes of sulfur have begun to be used more commonly to examine S transformations and fluxes directly in the field. Sulfur isotope ratios are expressed in the standard ^{34}S notation:

$$^{34}S(‰) = [(R_{sample}/R_{standard}) - 1] \times 1000 \quad (14)$$

where R_{sample} and $R_{standard}$ refer to the $^{34}S/^{32}S$ abundance ratio of the sample and the Canyon Diablo meteorite standard, respectively. Natural abundances of ^{34}S range from −40 to +40‰ (Ehleringer and Rundel, 1988). Stable isotope abundances have been especially useful in ascertaining the contribution of anthropogenic sources (i.e., combustion of fossil fuels) to atmospheric inputs of S if the source of fossil fuel has a ^{34}S signature that is distinctive from natural levels (Krouse et al., 1991). Analysis of ^{34}S in forested watersheds has also been used to evaluate which processes are regulating SO_4^{2-} flux in solutions (Stam et al., 1992), although in some cases it appears that isotopic discrimination processes may be affecting the distribution of S isotopes among different ecosystem components (Fuller et al., 1986b; Heslein et al., 1988). An analysis of ^{34}S values of precipitation and streamwaters at Bear Brook Watershed in Maine (BBWM) suggested that inorganic SO_4^{2-} dynamics was primarily responsible in regulating seasonal changes of SO_4^{2-} in surface waters at this site (Stam et al., 1992).

A major problem in estimating S turnover in soil organic pools, especially carbon-bonded S, is the assumption of pool homogeneity. It has been established that only a small fraction of the organic S pool is biologically active, with less than 5% of the S represented by microbial biomass (Nakas, 1986). Similarly, it is generally assumed for agronomic soils that only 1–3% of the organic S is converted to inorganic SO_4^{2-} annually (Freney and Williams, 1983).

More studies have been done on the transformations of organic S in grassland soils than in forest soils. McGill and Cole (1981) proposed a model that separates biochemical and biological mineralization processes for S in soil systems, and portions of this model have been incorporated into more complex conceptual (Maynard et al., 1984) and simulation (Hunt et al., 1986) models. In general, these models (which incorporate organic S) have been utilized only for grasslands whereas models of SO_4^{2-} dynamics in forest soils have focused on adsorption-desorption processes (Mitchell and Fuller, 1988).

Most measurements of SO_4^{2-} incorporation under field conditions have focused on evaluating changes in soil water or surface water chemistry and using mass balance calculations to quantify the contribution of soil sulfate adsorption. A few studies have made direct measurements of changes in soil pools. At Hubbard Brook Experimental Forest (HBEF), Mitchell et al. (1989) combined measurements of soil solution chemistry, stream chemistry, and changes in soil S constituents to establish that the reduction of SO_4^{2-} in solution could be attributed to adsorption of SO_4^{2-}, especially in the upper B horizon. David et al. (1990), using buried mineral soil bags installed at Bear Brook (BBMW) in experimental plots treated with sulfuric and/or nitric acid, was able to establish that adsorbed sulfate significantly increase as a function of SO_4^{2-} loading. Moreover, this approach showed promise in detecting smaller changes in organic S pools than could be detected using more traditional soil sampling procedures. Small changes in these large organic S pools may be significant in evaluating S storage over extended periods (i.e., >5 years). At BBWM, HBEF, and the Adirondack Mountains, a series of studies have been recently initiated using buried mineral soil bags that are evaluating changes in S pools under various additions of N and S.

V. SIMULATION MODELING OF SULFATE RETENTION AND RELEASE IN TWO FOREST ECOSYSTEMS

The effects of increased and decreased S inputs on N, P, K, Ca, Mg, and S cycling; soil S retention; and soil acidification/alkalization were simulated for two forest ecosystems—a red spruce system in the Great Smoky Mountains, North Carolina, and a mixed hardwood forest at Coweeta Hydrologic Laboratory in the southern Appalachians of North Carolina—using the Nutrient Cycling Model (NuCM). Following is a brief description of the model; see Liu et al. (1991) for more detail.

A. The NuCM Model

NuCM was developed as part of the Electric Power Research Institute's Integrated Forest Study (Liu et al., 1991; Johnson and Lindberg, 1991). The NuCM model incorporates state-of-the-art understanding of the biogeochemical and transport processes controlling nutrient cycles. The available nutrients in soil strata and vegetation pools and the fluxes between them are explicitly tracked and provided as model output. The model can be used to simulate the response of forests to atmospheric deposition and to various management practices (e.g., application of fertilizers). Factors are included in the model that allow the user to easily increase or decrease atmospheric deposition loads. The forested ecosystem is represented as a series of vegetation and soil components. The model provides for both an overstory and understory, each of which can be divided into canopy, bole, and roots.

Tree growth in the model is a function of user-defined developmental stage and the availability of nutrients and moisture. Translocation of nutrients prior to senescence is included. The understory is simulated in a similar manner to the overstory, except that its "incident precipitation" is the overstory's throughfall and its biomass nutrient concentrations are allowed to be different. For both the overstory and the understory, variable tissue nutrient concentrations are allowed to accommodate multiple species and availability of nutrients.

Using mass balance and transport formulations, the model tracks 16 solution-phase components, including the major cations and anions (analytical totals), ANC (acid-neutralizing capacity), an organic acid analogue, and total Al (Liu et al., 1991). The model routes precipitation through the canopy and soil layers and simulates evapotranspiration, deep seepage, and lateral flow. The soil includes multiple layers (up to ten), and each layer can have different physical and chemical characteristics. The nutrient pools associated with soil solution, the ion-exchange complex, minerals, and soil organic matter are all tracked explicitly. The processes that govern interactions among these pools include decay, nitrification, anion adsorption, cation exchange, and mineral weathering. Litter decay is represented as a series reaction with first-order dependencies on the reactant concentrations and C/N ratios. The decay products include nutrients and organic matter—both solid (e.g., humus) and solution phase (e.g., organic acids). The nutrients produced enter the solution phase, where they are available for uptake by vegetation or the exchange complex, and for transport from the forest floor by percolation and/or lateral flow.

The model simulates the noncompetitive adsorption of sulfate, phosphate, and organic acid. Sulfate adsorption can be simulated in NuCM using either linear or Langmuir (saturation) adsorption isopleths; Langmuir equations were used for these simulations. Figure 12 shows the isopleths input for the Smokies and Coweeta sites.

Cation exchange is represented by the Gapon equation. Mineral weathering reactions are normally slow and are described in the model using rate expressions with dependencies on the mass of mineral present and solution-phase hydrogen-ion concentration taken to a fractional power.

Model input is based on measurable parameters. The model uses such data to compute selectivity coefficients for each soil layer simulated. Model output options include nutrient pool sizes, fluxes between components, the relative contribution or loss by process, and soil solution and adsorbed concentrations versus time. Long-term nutrient loss or accumulation can be tracked by following annual pool and flux charts.

B. Simulation Results

Thirty-year simulations were run that included current (no change) S inputs, 100% increase, 50% reduction, and 90% reduction for both the Smokies and Coweeta sites. Simulated soil SO_4^{2-} leaching and soil-adsorbed SO_4^{2-} for each site are summarized in Figures 13 and 14.

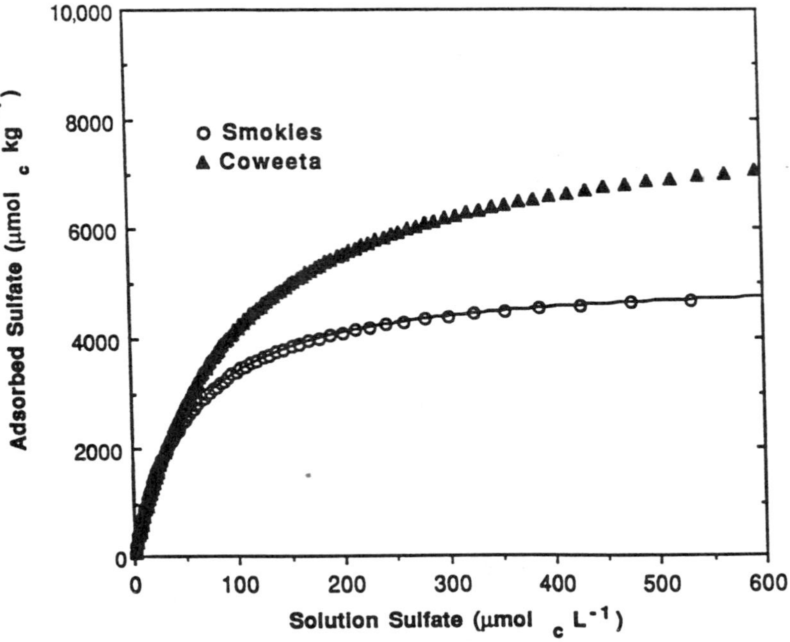

Figure 12 Sulfate adsorption isopleths (Langmuir) used for simulations for the Coweeta and Smokies red spruce sites.

The effects of soil SO_4^{2-} adsorption characteristics (Fig. 13) on response times to changes in atmospheric S loading are very clear. Sulfate leaching at the Smokies site increased or decreased with changes in S deposition almost immediately, and there was little or no net retention or loss of S from the system during most of the simulations (Fig. 13a). Soil SO_4^{2-} reached a steady-state condition rapidly under all but the 90% decrease scenario, where it declined slowly (Fig. 14a). On the other hand, the Coweeta site accumulated S throughout all scenarios except for the 90% decrease, where it was approximately at steady state (Figs. 13b and 14b). At Coweeta, the rates of increase in SO_4^{2-} leaching and soil SO_4^{2-} increased with increasing deposition but showed no signs of leveling off except in the 90% decrease scenario, which was approximately in steady state to begin with (Figs. 13b and 14b). Sulfur continued to accumulate in the soil throughout the 100% increase, no change, and 50% decrease scenarios (Fig. 13b).

Johnson and Reuss (1984) note that "because sulfate adsorption is concentration-dependent, the time required to achieve equilibrium may not

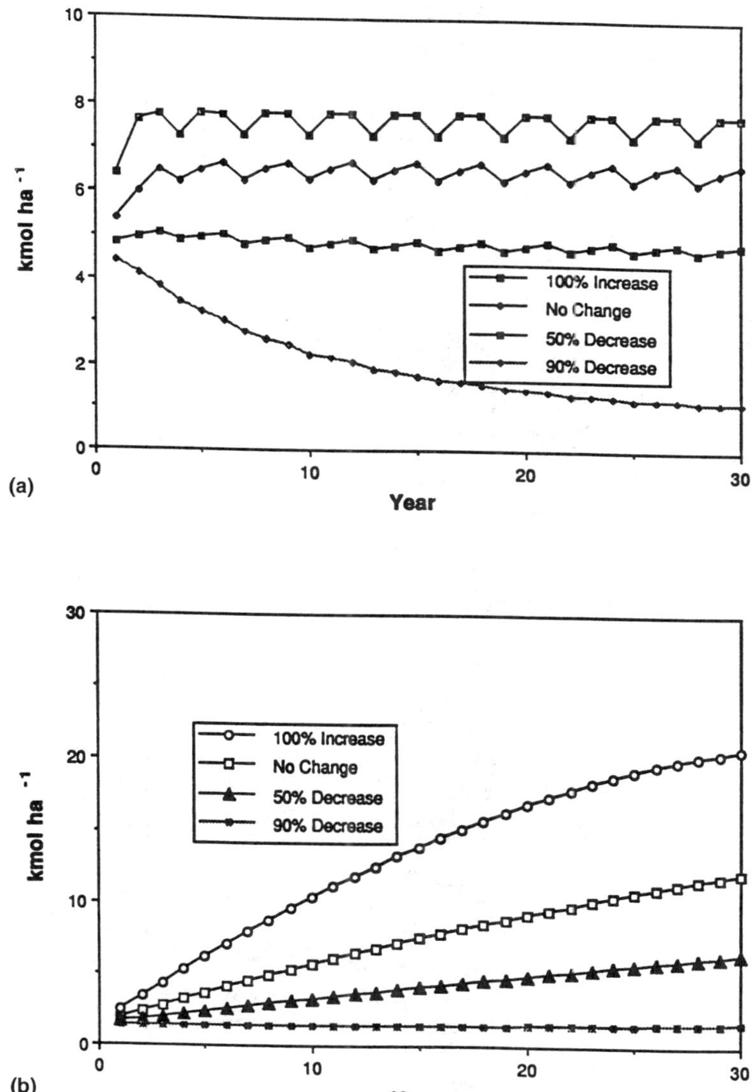

Figure 13 Simulated adsorbed sulfate in the Smokies (a) and Coweeta (b) sites.

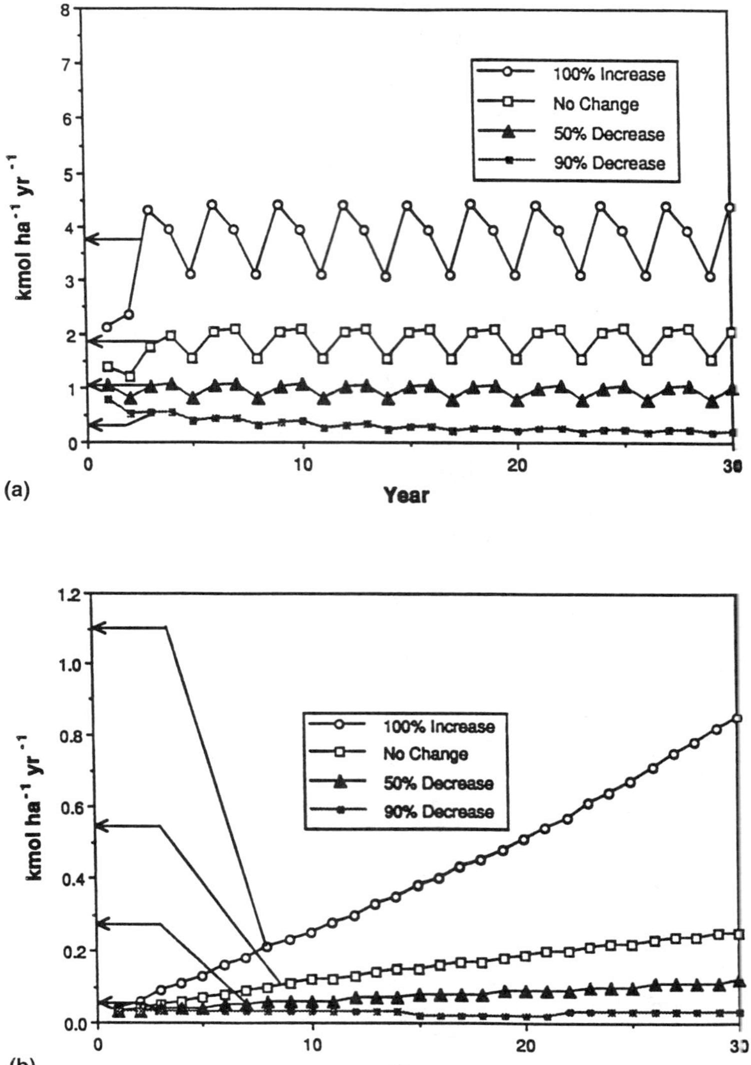

Figure 14 Simulated soil solution sulfate concentrations in the Smokies (a) and Coweeta (b) sites.

be materially shortened by an increase in the impact dosage." The results of these simulations would tend to support this contention. However, the time required to achieve steady state (equilibrium) is not entirely independent of the impact dosage, in that the latter may affect the amount of soil-adsorbed sulfate. Soil-solution SO_4^{2-} concentration becomes increasingly responsive to changes in soil adsorbed sulfate as the latter increases, as can be seen from the slope (first derivative) of the adsorption isopleth. Solving the Langmuir equation (Eq. 10) for (SO_4^{2-}) and taking the first derivative yields:

$$d(SO_4^{2-})/dX = X_{max}/[b(X_{max} - X)^2], \qquad \text{for} \quad X < X_{max} \tag{14}$$

Thus, the responsiveness of soil-solution SO_4^{2-} to increases in soil-adsorbed sulfate (and, therefore, to increased sulfate inputs to the system) increases as the soil becomes "saturated" with SO_4^{2-}; i.e., as X increases and approaches X_{max}, $d(SO_4^{2-})/dX$ increases. This can be seen in Figure 8 where as the curve of soil-adsorbed sulfate versus solution-SO_4^{2-} concentration flattens out. In practice, however, soils subjected to atmospheric S inputs are far below saturation and typically within the region where the adsorption isopleth is still increasing sharply, and the statement of Johnson and Reuss (1984) is valid under those conditions.

These simulations indicate the great importance of the nature of sulfate-adsorption isopleths in determining the response of soils to changes in S inputs. Other S retention factors were either not built in to the model (precipitation) or were unimportant in the simulations (net changes in storage in soil organic matter and vegetation were negligible in all cases). Thus, these simulations in effect show the theoretical responses of these systems to changing S inputs due to soil adsorption–desorption alone. Whether the predicted responses are accurate or not remains to be tested with long-term monitoring, manipulations, or both (bearing in mind that short-term manipulative studies may produce little change at Coweeta, even with large increases in input).

VI. OTHER MODELS

Mitchell and Fuller (1988) reviewed a wide range of models of both forest and grassland systems and outlined how sulfur processes were formulated in various applications (Table 3). For those models that have been applied to forest ecosystems, the major focus with respect to sulfur retention has been on sulfate adsorption-desorption processes with no attention being paid to organic sulfur dynamics. It has also been assumed that the sorption processes are completely reversible. There has been some attempt to do

Table 3 Comparison of the Treatments of Sulfur Dynamics in Forest and Grassland Ecosystem Models

Model (Reference)	General model type	Inorganic S dynamics	Organic S dynamics	Hydrology
McGill and Cole, 1981	Conceptual (general soil)	Two-phase (solid and soluble)	Linked to C and N dynamics	Not considered
Bettany and Stewart, 1983; Maynard et al., 1984	Conceptual (grasslands)	Adsorbed and occluded sulfate. Precipitated sulfide, elemental S, labile organic S	Extension of McGill and Cole (1981) and inclusion of microbial S, labile organic S, stable organic S, plant uptake	Not considered
David et al. 1984	Conceptual (forest soils)	Adsorbed sulfate, sulfate in solution	Carbon-bonded S, ester sulfate in solution and solid phases, mineralization and immobilization	Effects of solution transport
Johnson, 1984	Mass balance (forest ecosystems)	Adsorbed sulfate, soluble sulfate, weathering	Organic constituents combined, plant uptake, decomposition	Leaching effects
MINEQL Westhall et al., 1985	Chemical equilibrium	Complexation and precipitation of some inorganic sulfate constituents	Not considered	Not explicit part of model
ALCHEMI Driscoll and Schecher, 1985	Chemical equilibrium	Complexation and precipitation of some inorganic sulfate constituents; pH dependent-anion adsorption	Not considered	Not explicit part of model

ILWAS Chen et al., 1983	Simulation (forest watershed)	Sulfate adsorption fitted to linear isotherm; weathering based on stoichiometric release from minerals	Only mineralization from organic matter and plant uptake based on decomposition rates and plant uptake nutrient needs	Linked to hydrology sub-model
Birkenes Christophersen and Wright, 1981; Wright, 1984	Simulation (forest watershed)	Soluble sulfate, combined effect of both organic and inorganic S in solid phase; linear adsorption assumed	See inorganic S	Linked to hydrology sub-model
Storgama Model Christophersen et al., 1984	Simulation (forest watershed); modification of Birkenes	Formulation for removal of water-soluble sulfate during snowmelt, which could include both organic and inorganic S transformation	See inorganic S	Linked to hydrology sub-model
Trickle-Down Schnoor et al., 1984 Schnoor, pers. comm.	Simulation (watershed)	Sulfate adsorption treated as linear isotherm	Not considered	Linked to hydrology sub-model
MAGIC Cosby et al., 1985	Simulation (forest watershed)	Chemical equilibria reactions used to describe relationships among dissolved and adsorbed ions; some coefficients must be ascertained for each watershed	Not considered	Linked to hydrology sub-mold

Table 3 Continued

Model (Reference)	General model type	Inorganic S dynamics	Organic S dynamics	Hydrology
Coughenour et al., 1980	Simulation (grassland)	Sulfate adsorption by Freudlich isotherm	Uptake of sulfate by Michaelis-Menton; elemental ratios in microbes and plants used for determining demand; mineralization function of uptake and microbial S content; ester sulfate formation not integrated directly into model	Linked to soil water sub-model
Fuller et al., 1986a	Simulation (Spodosols)	Sulfate adsorption by kinetic formulation	Transformation rates of sulfate between ester sulfates and carbon-bonded S fitted to first-order reactions	Not considered
Hunt et al., 1986	Simulation (grassland soil)	Not considered	Relationships between microbial activity, including esterases, in effecting formation of carbon-bonded S and ester sulfates	Not considered

Source: Mitchell and Fuller, 1988.

historical reconstructions using models and it has been shown in some simulations using Model of Acidification of Groundwater in Catchments (MAGIC) (e.g., Jenkins et al., 1990; Wright et al., 1986) that acidification processes, including sulfate adsorption-desorption, appear to be reversible.

As part of the National Acid Precipitation Assessment Program, there was an extensive evaluation of a set of models that predict surface water chemistry as part of the Direct/Delayed Response Project for three regions (Northeast, Mid-Appalachian, and Southern Blue Ridge) in the United States (Thornton et al., 1990). This research focused on the utilization of three models: Enhanced Trickle Down (ETD) (Lee and Schnoor, 1988), Integrated Lake-Watershed Acidification Study (ILWAS) (Gherini et al., 1985) and MAGIC (Cosby et al., 1985). In the review of the application of these models for predicting future changes in surface water chemistry, it was concluded by Thorton et al. (1990) that any irreversible adsorption of SO_4^{2-} could alter predictions on the rate of recovery of forested watersheds from acidification.

VII. SUMMARY AND CONCLUSIONS

There was little information on S cycling in forests compared to other nutrients (especially N) until the past two decades, when interest in S nutrition and cycling in forests was heightened with the discovery of deficiencies in some unpolluted regions and excesses associated with air pollution in other regions. Several studies have addressed the effects of increased S inputs in forest soils and forest ecosystems, but few have considered the effects of decreased S inputs, which are now occurring as S emissions decline in many polluted regions. These decreases in S emission have been reflected in decreases in SO_4^{2-} concentrations of surfaces waters in the northeast United States, eastern Canada, and Europe; however, there are no data as to effects on soils or forest ecosystems.

One interesting and unique feature of S cycles is that a significant proportion of total S can be cycled and accumulated as SO_4^{2-}, within both vegetation and soils. The relative amount of SO_4^{2-} as a proportion of total S increases with S availability. In vegetation, most carbon-bonded S is associated with the amino acids cysteine, cystine, and methionine, which are in turn used for protein synthesis, and thus there is a very close relationship between organic S and organic N in plant tissues. As is the case with many other nutrients, the soil is the major reservoir of S and site for S accumulation in forest ecosystems. Mineral soil S accounts for >80% of total ecosystem S in all but the most S-deficient systems. In most forest soils, organic S—mostly in the form of carbon-bonded S and ester sulfates—makes up the dominant pools (>90%), except in those soils with high sulfate adsorption capacity and/or high levels or S deposition where SO_4^{2-} concentrations may approach that of organic S. In soils, S can accumulate by incorporation into organic matter, precipitation, and adsorption, mostly as adsorbed SO_4^{2-}. In cases in which S deposition is high, the capacity for organic S incorporation is often exceeded, and precipitation and adsorption reactions predominate. Precipitation reactions tend to be important when soil solution SO_4^{2-} concentrations are very high, whereas adsorption predominates in most moderately polluted systems. The degree to which soils can accumulate SO_4^{2-} is a function of Fe and Al hydrous oxide content and the concentration of SO_4^{2-} in soil solution. Soil SO_4^{2-} adsorption can be modeled by adsorption isotherms of the Langmuir, Freundlich, or linear type.

A major remaining gap in our understanding of S cycling in forests is the response to the reduced S deposition that is currently under way. The degree and duration of SO_4^{2-} release from organic matter, precipitates, and adsorption sites under lower deposition will have profound impacts upon the effectiveness of S emission controls on both terrestrial and aquatic ecosystems. We know little about the potential release of S from organic matter under reduced S availability. Models are available that can predict the release of SO_4^{2-} from precipitates and adsorption sites; however, none of the current models incorporate the significant hysteresis always noted in the adsorption–desorption of SO_4^{2-} from soil adsorption sites.

REFERENCES

Allot, T. E. H., R. Harriman, and R. W. Battarbee. 1992. Reversibility of lake acidification at the Round Loch of Glenhead, Galloway, Scotland. Environmental Pollution 77:219–225.

Andrea, M. O., and W. A. Jaeschke. 1992. Exchange of sulphur between biosphere and atmosphere over temperate and tropical regions. pp. 27–66, Chapter 3. In: Howarth, R. W., J. W. B. Stewart, and M. V. Ivanov (eds.), Sulphur Cycling on the Continents: Wetlands, Terrestrial Ecosystems and Associated Water Bodies. SCOPE 48. John Wiley and Sons, Chichester.

Barrow, N. J., and T. C. Shaw. 1977. The slow reactions between soil and anions: 7. Effect of time and temperature of contact between an adsorbing soil and sulfate. Soil Sci. 124:347–354.

Bettany, J. R., and J. W. B. Stewart. 1983. Dynamics of organic sulphur in semi-arid agroecosystems. pp. 767–785. In: A. I. More (ed.), Sulphur-82: Proceedings of the International Conference, Vol. II. British Sulphur Corporation, London.

Chao, T. T., M. E. Harward, and S. C. Fang. 1962. Adsorption and desorption phenomena of sulfate in soils. Soil Sci. Soc. Amer. Proc. 26:234–237.

Chao, T. T., M. E. Harward, and S. C. Fang. 1963. Cationic effects of sulfate adsorption by soils. Soil Sci. Soc. Amer. Proc. 27:35–41.

Chen, C. W., S. A. Gherini, R. J. M. Hudson, and J. D. Dean. 1983. The integrated lake-watershed acidification study. Vol. 1. Model principles and application procedures. EPRI Report, EA-32221, Vol. I. Res. Proj. 1109–5. Palo Alto, CA.

Christopherson, N., and R. F. Wright. 1981. Sulfate budget and model for sulfate in streamwater at Birkenes, a small forested catchment in Southern Norway. Water Resources Research 17:337–389.

Christophersen, N., L. H. Dymbe, M. Johannssen, and H. M. Seip. 1984. A model for sulphate in streamwater at Storgama, Southern Norway. Ecological Modeling 21:35–61.

Cole, D. W., and M. Rapp. 1981. Elemental cycling in forest ecosystems. pp. 341–409. In D. E. Reichle (ed.), Dynamic Properties of Forest Ecosystems. Cambridge University Press, London

Coleman, N. T., and G. W. Thomas. 1967. The basic chemistry of soil acidity. pp. 1–41. In: R. W. Pearson and F. Adams (eds.), Soil Acidity and Liming. American Society of Agronomy, Madison, WI.

Cosby, B. J., R. F. Wright, G. M. Hornberger, and J. N. Galloway. 1985. Modeling effects of acid deposition: estimation of long-term quality responses in a small forested catchment. Water Resources Research 21:1591–1601.

Coughenour, M. B., W. J. Parton, W. K. Lauenroth, J. L. Dodd, and R. G. Woodmansee. 1980. Simulation of a grassland sulfur-cycle. Ecological Modeling 9:179–213.

Couto, W., D. J. Lathwell, and D. R. Bouldin. 1979. Sulphate sorption by two Oxisols and an Alfisol of the tropics. Soil Sci. 127:108–116.

David, M. B., and M. J. Mitchell. 1987. Transformations of organic and inorganic sulfur: Importance to sulfate flux in an Adirondack forest soil. JAPCA 37: 39–44.

David, M. B., M. J. Mitchell, and T. J. Scott. 1987. Importance of biological processes in the sulfur budget of a northern hardwood ecosystem. Biol Fert. Soils. 5:258–264.

David, M. B., M. J. Mitchell, and S. C. Schindler. 1984. Dynamics of organic and inorganic sulfur constituents in hardwood forest soils. pp. 221–345. In: E. L. Stone (ed.), Forest Soils and Treatment Impacts. Proceedings of the Sixth North American Soils Conference. The University of Tennessee, Knoxville, Tenn.

David, M. B., R. D. Fuller, I. J. Fernandez, M. J. Mitchell, L. E. Rustad, G. F. Vance, A. C. Stam, and S. C. Nodvin. 1990. Spodosol variability and assessment of response to acidification. Soil Science Society of America Journal 54:541–548.

Dhamala, B. R., M. J. Mitchell, and A. C. Stam. 1990. Sulfur dynamics of two northern hardwood soils: a column study with ^{35}S. Biogeochemistry 10: 143–160.

Dillon, P. J., and B. D. LaZerte. 1992. Response of Plastic Lake catchment, Ontario, to reduced sulphur deposition. Environmental Pollution 77:211–217.

Driscoll, C. T., and W. D. Schecher. 1985. Nature and rates of aluminum removal in acidic surface waters. pp. 107–128. In: Comparative Interregional Analysis of Aluminum Biogeochemistry in Forest Watersheds Exposed to Acidic Deposition. EPRI Progress Report 2365-01, Palo Alto, California.

Driscoll, C. T., and R. Van Dreason. 1993. Seasonal and long-term temporal patterns in the chemistry of Adirondack Lakes. Water, Air and Soil Pollution 67: 319–344.

Ehleringer, J. R., and P. W. Rundel. 1988. Stable isotopes: history, units and instrumentation. pp. 1–19. In: P. W. Rundel, J. R. Ehleringer, and K. A. Nagy (eds.), Stable Isotopes and Ecological Research. Ecological Studies 68. Springer-Verlag, New York.

Ensminger, L. E. 1954. Some factors affecting sulfate adsorption by Alabama soils. Soil Sci. Soc. Amer. Proc. 18:259–264.

Fasth, W. J., M. B. David, and G. F. Vance. 1991. Sulfate retention and cation leaching of forest soils in response to acid additions. Can. J. For Res. 21: 32–41.

Fernandez, I. J., and L. E. Rustad. 1989. Soil response to S and N treatments in a northern New England and low elevation coniferous forest. Water, Air, and Soil Pollut. 52:23–39.

Fitzgerald, J. W., T. C. Strickland, and W. T. Swank. 1982. Metabolic fate of inorganic sulphate in undisturbed and managed forest ecosystems. Soil Biol. Biochem. 14:529–536.

Freney, J. R. 1967. Sulfur-containing organics. pp. 229–259. In: A. D. McLaren and G. H. Peterson (eds.), Soil Biochemistry. Marcel Dekker, New York.

Freney, J. R., and C. H. Williams. 1983. The sulfur cycle in soil. pp. 129–201. In: M. V. Ivanov and J. R. Freney (eds.), The Global Biogeochemical Sulphur Cycle. SCOPE 19. John Wiley & Sons, Chichester.

Fuller, R. D., M. B. David, and C. T. Driscoll. 1985. Sulfate adsorption relationships in forested Spodosols of the northeastern USA. Soil Sci. Soc. Amer. J. 49: 1034–1040.

Fuller, R. D., C. T. Driscoll, S. C. Schindler, and M. J. Mitchell. 1986a. A simulation model of sulfur transformations in forested Spodosols. Biogeochemistry 2:313–328.

Fuller, R. D., M. J. Mitchell, H. R. Krouse, B. J. Wyskowski, and C. T. Driscoll. 1986b. Stable sulfur isotope ratios as a tool for interpreting ecosystem sulfur dynamics. Water, Air and Soil Pollution 28:163–171.

Fuller, R. D., C. T. Driscoll, G. B. Lawrence, and S. C. Nodvin. 1987. Processes regulating sulphate flux after whole-tree harvesting. Nature 325:707–709.

Garten, C. T., E. A. Bondietti, and R. D. Lomax. 1988. Contribution of foliar leaching and dry deposition to sulfate in net throughfall below deciduous trees. Atmospheric Environment 22:1425–1432.

Gebhardt, H., and N. T. Coleman. 1974. Anions adsorption by allophanic tropical soils: II. Sulfate adsorption. Soil Sci. Soc. Amer. Proc. 38:259–262.

Gherini, S. A., L. Mok, R. J. M. Hudson, G. F. Davis, C. W. Chen, and R. A. Goldstein. 1985. The ILWAS model: formulation and application. Water, Air and Soil Pollution 26:425–459.

Gobran, G. R., and S. I. Nilsson. 1988. Effects of forest floor leachate on sulfate retention in a Spodosol soil J. Environ. Qual. 17:235–239.

Haque, I., and D. Walmsley. 1973. Adsorption and desorption of sulfate in some soils of the West Indies. Geoderma 9:326–335.

Harrison, R. B., D. W. Johnson, and D. E. Todd. 1989. Sulfate adsorption and desorption in a variety of forest soils. J. Environ. Qual. 18:419–426.

Harward, M. E., and H. M. Reisenauer. 1966. Reactors and movement of inorganic soil sulfur. Soil Sci. 101:326–335.

Heslein, R. H., M. J. Capel, and D. E. Fox. 1988. Stable isotopes in sulfate in the inputs and outputs of a Canadian Shield watershed. Biogeochemistry 5: 263–273.

Hingston, F. J., R. J. Atkinson, A. M. Posner, and J. P. Quirk. 1967. Specific adsorption of anions. Nature 215:1459–1461.

Hornbeck, J. W., C. W. Martin, R. S. Pierce, F. H. Bormann, G. E. Likens, and J. S. Eaton. 1986. Clearcutting northern hardwoods: effects on hydrologic and nutrient ion budgets. Forest Science 32:667–686.

Hultberg, H. 1985. Budgets of base cations, chloride, nitrogen and sulphur in the acid Lake G rds n catchment, SW Sweden. pp. 133–157. In: F. Anderson and B. Olson (eds.), Lake G rds n—an acid lake and its catchment. Ecological Bulletin (Stockholm) 37.

Hultberg, H., and P. Grennfelt 1992. Sulphur and sea salt deposition as reflected by throughfall and runoff chemistry in forested catchments. Environmental Pollution 75:215–222.

Hultberg, H., H. ApSimon, R. M. Church, P. Greenfelt. M. J. Mitchell, F. Moldan, and H. B. Ross. 1992. Chapter 10. Sulphur. pp. 229–254. In: B. Moldan and J. Cerny (eds.), Biogeochemistry of Small Catchments: a Tool for Environmental Research. SCOPE. John Wiley & Sons, Chichester.

Humphreys, F. R., M. J. Lambert, and J. Kelly. 1975. The occurrence of sulphur deficiency in forests. In: (K. D. McLachlan (ed.), Sulphur in Australasian Agriculture. pp. 154–162. Sydney Univ Press, Sydney, Australia.

Hunt, W. B., J. W. B. Stewart, and C. V. Cole. 1986. Concepts of sulfur, carbon, and nitrogen transformations in soil: evaluation by simulation modeling. Biogeochemistry 2:163–177.

Husar, R. B., T. J. Sullivan, and D. F. Charles. 1991. Historical trends in atmospheric sulfur deposition and methods for assessing long-term trends in surface water chemistry. pp. 65–82. In: D. F. Charles (ed.), Acidic Deposition and Aquatic Ecosystems: Regional Case Studies. Springer-Verlag, New York.

Jenkins, A., P. G. Whitehead, B. J. Cosby, and H. J. B. Birks. 1990. Modelling long-term acidification: a comparison with diatom reconstructions and the implications for reversibility. Phil. Trans. R. Soc. Lond. B 327:435–440.

Johnson, D. W. 1984. Sulfur cycling in forests. Biogeochemistry 1:29–43.

Johnson, D. W., and D. W. Cole. 1977. Sulfate mobility in an outwash soil in western Washington. Water, Air, and Soil Pollution 7:489–495.

Johnson, D. W., and G. S. Henderson. 1979. Sulfate adsorption and sulfur fractions in a highly weathered soil under a mixed deciduous forest Soil Sci. 128: 34–40.

Johnson, D. W., G. S. Henderson, D. D. Huff, S. E. Lindberg, D. D. Richter, D. S. Shriner, D. E. Todd, and J. Turner. 1982. Cycling of organic and inorganic sulphur in a chestnut oak forest. Oecologia 54:141–148.

Johnson, D. W., G. S. Henderson, and D. E. Todd. 1981. Evidence of modern accumulation of sulfate in an east Tennessee forested Ultisol. Soil Science 132:422–426.

Johnson, D. W., and S. E. Lindberg (eds.). 1991. Atmospheric Deposition and Forest Nutrient Cycling. Ecological Series 91, Springer-Verlag, New York. 707 pages.

Johnson, D. W., D. W. Cole, H. Van Miegroet, and F. W. Horng. 1986a. Factors affecting anion movement and retention in four forest soils. Soil Sci. Soc. Am. J. 50:776–783.

Johnson, D. W., D. D. Richter, H. Van Miegroet, D. W. Cole, and J. M. Kelly. 1986b. Sulfur cycling in five forest ecosystems. Water, Air Soil Pollut. 30: 965–979.

Johnson, D. W., and J. O. Reuss. 1984. Soil-mediated effects of atmospherically-deposited sulphur and nitrogen. Phil. Trans. R. Soc. Lond. B. 305:383–392.

Johnson, D. W., D. D. Richter, G. M. Lovett, and S. E. Lindberg. 1985. The effects of atmospheric deposition on potassium, calcium, and magnesium cycling in two deciduous forests. Can. J. For. Res. 15:773–782.

Johnson, D. W., H. Van Miegroet, S. E. Lindberg, R. B. Harrison, and D. E. Todd. 1991. Nutrient cycling in red spruce forests of the Great Smoky Mountains. Can. J. For. Res. 21:769–787.

Johnson, D. W., and D. E. Todd. 1983. Relationships among iron, aluminum, carbon, and sulfate in a variety of forest soils. Soil Sci. Soc. Amer. J. 47: 792–800.

Karkanis, M. 1976. The circulation of sulphur in the forest ecosystem Tilo-Carpinetum in the northern part of Puszcza Niepolomicka near Ispina. Fragmenta Floristica et Geobotanica 22:351–363.

Kelly, J., and M. J. Lambert. 1972. The relationship between sulphur and nitrogen in the foliage of Pinus radiata. Plant Soil 37:395–408.

Kelly, J. M., and D. W. Johnson. 1982. Sulfur and nitrogen uptake by loblolly pine seedlings as influenced by soil nitrogen and sulfur additions. For. Science 28: 725–731.

Kelso, J. R. M. and D. S. Jeffries. 1988. Response of headwater lakes to varying atmospheric deposition in north-central Ontario, 1979–85. Canadian Journal of Fisheries and Aquatic Sciences. 45:1905–1911.

Khanna, P. K., and F. Beese. 1978. The behavior of salt input in podzolic brown earth. Soil Sci. 125:16–22.

Khanna, P. K., G. T. Weaver, and F. Beese. 1986. Effect of sulfate on ionic transport and balance in a slightly acidic forest soil. Soil Sci. Soc. Amer. J. 50: 770–776.

Khanna, P. K., J. Prenzel, K. J. Meiwes, B. Ulrich, and E. Matzner. 1987. Dynamics of sulfate retention by acid forest soils in an acidic deposition environment. Soil Sci. Soc. Amer. J. 51:446–452.

Krouse, H. R., L. N. Grinenko, V. A. Greinenko, L. Newman, J. Forest, N. Nakai, Y. Tsuji, et al. 1991. Case Studies and Potential Applications. pp. 307–422. In: H. R. Krouse, and V. A. Grinenko (eds.), Stable Isotopes: Natural and Anthropogenic Sulphur in the Environment. SCOPE 43. John Wiley and Sons, Chichester.

Lee, J. J., and D. E. Weber. 1982. Effects of simulated acid rain on major cation and sulfate concentrations of water percolating through two model hardwood forests. Journal of Environmental Quality 11:57–65.

Lee, S. K., and J. L. Schnoor. 1988. Reactions that modify chemistry in lakes of the National Surface Water Survey. Environmental Science and Technology 22:190–195.

Likens, G. E., F. H. Bormann, R. S. Pierce, J. S. Eaton, and N. M. Johnson. 1977. Biogeochemistry of a Forested Ecosystem. Springer-Verlag, New York.

Likens, G. E., F. H. Bormann, and N. M. Johnson. 1969. Nitrification: Importance to nutrient losses from a cut-over forest ecosystem. Science 163:1205–1206.

Likens, G. E., F. H. Bormann, N. M. Johnson, and R. S. Pierce. 1978. Recovery of a deforested ecosystem. Science 199:492–496.

Likens, G. E., F. H. Bormann, L. O. Hedin, C. T. Driscoll, and J. S. Eaton. 1990. Dry deposition of sulfur: a 23 year record for the Hubbard Brook Forest Ecosystem. Tellus 42B:319–329.

Lindberg, S. E., M. Bredemeir, D. A. Schaefer, and L. Qi. 1990. Atmospheric concentrations and deposition during the growing season in conifer forests in the United States and West Germany. Atmospheric Environment 24A: 2207–2220.

Lindberg, S. E., and R. C. Harriss. 1981. The role of atmospheric deposition in an eastern U.S. deciduous forest. Water Air Soil Pollut. 16:13–31.

Lindberg, S. E., and G. M. Lovett. 1992. Deposition and forest canopy interactions of airborne sulfur: Results of the Integrated Forest Study. Atmos. Environ. 26A:1477–1492.

Lindberg, S. E., G. M. Lovett, D. D. Richter, and D. W. Johnson. 1986. Atmospheric deposition and canopy interactions of major ions in a forest. Science 231:141–145.

Liu, S., R. Munson, D. Johnson, S. Gherini, K. Summers, R. Hudson, K Wilkinson, and L. Pitelka. 1991. Application of a nutrient cycling model (NuCM) to northern mixed hardwood and southern coniferous forest. Tree Physiology 9: 173–182.

MacDonald, N. W., and J. B. Hart. 1990. Relating sulfate adsorption to soil properties in Michigan forest soils. Soil Sci. Soc. Amer. J. 54:238–245.

Mayer, R., and B. Ulrich. 1974. Conclusions on the filtering action of forests from ecosystem analysis. Oecologia Plantarum 9:157–168.

Maynard, D. G., J. W. B. Stewart, and J. R. Bettany. 1984. Sulfur cycling in grassland and parkland soils. Biogeochemistry 1:97–111.

McGill, W. B., and C. V. Cole. 1981. Comparative aspects of cycling of organic C, N, S and P through soil organic matter. Geoderma 26:267–286.

Mehlich, A. 1964. Influence of sorbed hydroxyl and sulfate on liming efficiency, pH, and conductivity. Soil Sci. Soc. Amer. Proc. 27:496–499.

Meiwes, K. J., and P. K. Khanna. 1981. Distribution and cycling of sulphur in the vegetation of two forest ecosystems in an acid rain environment. Plant Soil 60:369–375.

Mekaru, T., and G. Uehara. 1972. Anion adsorption in ferruginous tropical soils Soil Sci. Soc. Am. Proc. 36:296–300.

Mitchell, M. J., C. T. Driscoll, R. D. Fuller, M. B. David, and G. E. Likens. 1989. Effect of whole-tree harvesting on the sulfur dynamics of a forest soil. Soil Sci. Soc. Amer. J. 53:933–940.

Mitchell, M. J., M. B. David, and R. B. Harrison. 1992a. Sulphur Dynamics of Forest Ecosystems. pp. 215–254, Chapter 9. In: R. W. Howarth, J. W. B. Stewart, and M. V. Ivanov (eds.), Sulphur Cycling on the Continents: Wetlands, Terrestrial Ecosystems and Associated Water Bodies. SCOPE 48. John Wiley and Sons, Chichester.

Mitchell, M. J., R. B. Harrison, J. W. Fitzgerald, D. W. Johnson, S. E. Lindberg, Y. Zhang, and A. Autry. 1992b. Sulfur distribution and cycling in forest ecosystems. pp. 90–128, Chapter 5. In: D. W. Johnson, and S. E. Lindberg (eds.), Atmospheric Deposition and Forest Nutrient Ecological Series 91. Springer-Verlag, New York. 707 pages.

Mitchell, M. J., M. K. Burke, and J. P. Shepard. 1993. Seasonal and spatial patterns of S, Ca, and N dynamics of a northern hardwood forest ecosystem. Biogeochemistry 17:165–189.

Mitchell, M. J., and R. D. Fuller. 1988. Sulfur models of forest and grassland ecosystems with emphasis on soil processes. Biogeochemistry 5:133–163.

Nakas, J. P. 1986. Role of microflora in terrestrial sulfur cycling. pp. 285–316. In: M. J. Mitchell, and J. P. Nakas (eds.), Microfloral and Faunal Interactions in Natural and Agro-ecosystems. Martinus Nijhoff Publ., The Hague.

Nilsson, S. I., and B. Bergkvist. 1983. Aluminum chemistry and acidification processes in a shallow podzol on the Swedish west coast. Water Air Soil Pollut. 20:311–329.

Nodvin, S. C., C. T. Driscoll, and G. E. Likens. 1986a. Simple partitioning of anions and dissolved organic carbon in a forest soil. Soil Sci. 142:27–35.

Nodvin, S. C., C. T. Driscoll, and G. E. Likens. 1986b. The effect of pH on sulfate adsorption by a forest soil. Soil Sci. 142:69–75.

Nys, C., and J. Ranger. 1988. Influence d'une substitution d'esp sur le fontionnement biogeochimique de l'ecosysteme foret l'exemple du cycle du souffre. Ann. Sci. For. (Paris): 169–188.

Paces, T. 1985. Sources of acidification in Central Europe estimated from elemental budgets in small basins. Nature 315:31–36.

Parfitt, R. L., and R. S. C. Smart. 1978. The mechanism of sulfate adsorption on iron oxides. Soil Sci. Soc. Amer. J. 42:48–50.

Probert, M. E., and S. Asmosir. 1983. Sulfur in non-flooded tropical soils. pp. 15–27. In: G. J. Blair, and A. R. Till (eds.), Sulfur in South-east Asian and South Pacific Agriculture. The University of New England, Armidale, N.S.W. Australia.

Rajan, S. S. S. 1978. Sulfate adsorbed on hydrous alumina, ligands displaced, and changes in surface charge. Soil Sci. Soc. Amer. J. 42:39–44.

Reuss, J. O., and D. W. Johnson. 1986. Acid deposition and the acidification of soil and water. Springer-Verlag, New York.

Richter, D. D, and D. W. Johnson. 1983. Determination of inorganic sulfate in foliage with barium chloranilate. Soil Sci. Soc. Amer. J. 47:522–524.

Rochelle B. P., M. R. Church, and M. B. David. 1987. Sulfur retention at intensively studied sites in the U.S. and Canada. Water, Air and Soil Pollution 33:73–83.

Sanders, F. E., and P. H. B. Tinker. 1975. Adsorption of sulphate by a sandy loam (Calcic cambisol). Geoderma 13:317–324.

Schindler, S. C., and M. J. Mitchell. 1987. Dynamics of ^{35}S in horizons and leachates from a hardwood forest spodosol. Soil Biology and Biochemistry 19: 531–538.

Schnoor, J. L., W. D. Palmer, and G. E. Glass. 1984. Modeling impacts of acid precipitation for northeastern Minnesota. pp. 155–173. In: J. L. Schnoor (ed.), Modeling of Total Acid Precipitation Impacts, Butterworth Publ, Boston, MA.

Schoenau, J. J., and J. R. Bettany. 1987. Organic matter leaching as a component of carbon, nitrogen, phosphorus, and sulfur cycles in a forest, grassland, and gleyed soil. Soil Science Society of America Journal 51:646–651.

Shriner, D. S., and G. S. Henderson. 1978. Sulfur distribution and cycling in a deciduous forest watershed J. Environ. Qual. 7:392–397.

Singh, B. R. 1980. Distribution of total and extractable S and adsorbed SO_4^{2-} in some acid forest soil profiles of southern Norway. Acta Agric. Scand. 28: 313–322.

Singh, B. R. 1984a. Sulfate adsorption by acid forest soils. 1. Sulfate adsorption isotherms and comparison of different adsorption equations in describing sulfate adsorption. Soil. Sci. 138:189–197.

Singh, B. R. 1984b. Sulfate adsorption by acid forest soils. 2. Sulfate adsorption isotherms with and without organic matter and oxides of aluminum and iron. Soil. Sci. 138:294–297.

Singh, B. R. 1984c. Sulfate adsorption by acid forest soils. 3. Desorption of sulfate from adsorbed surfaces as a function of time, desorbing ion, pH, and amount of adsorption. Soil. Sci. 138:294–297.

Singh, B. R. 1984d. Sulfate adsorption by acid forest soils. 4. Kinetics and effects of temperature and moisture. Soil. Sci. 138:346–353.

Singh, B. R., G. Abrahamsen, and A. O. Stuanes. 1980. Effect of simulated acid rain on sulfate movement in acid forest soil. Soil Science Society of America Journal 44:75–80.

Stam, A. C., M. J. Mitchell, H. R. Krouse, and J. S. Kahl. 1992. Stable sulfur isotopes of sulfate in precipitation and stream solutions in a northern hardwood watershed. Water Resources Research 28:231–236.

Strickland, T. C., J. W. Fitzgerald, and W. T. Swank. 1984. Mobilization of recently formed forest soil organic sulfur. Canadian Journal of Forest Research. 14: 63–67.

Strickland, T. C., J. W. Fitzgerald, and W. T. Swank. 1986a. In situ mobilization of 35S-labelled organic sulphur in litter and soil from a hardwood forest. Soil Biol. Biochem. 18:463–468.

Strickland, T. C., J. W. Fitzgerald, and W. T. Swank. 1986b. In situ measurements of sulfate incorporation into forest floor and soil organic matter. Canadian Journal of Forest Research 16:549–553.

Swank, W. T., J. W. Fitzgerald, and J. T. Ash. 1984. Microbial transformation of sulfate in forest soils. Science 223:182–184.

Tabatabai. M. A. 1982. Sulfur. pp. 501–538. In: A. L. Page, R. H. Miller, and D. R. Keeney (eds.), Methods of Soil Analysis. Part 2: Chemical and microbiological properties. Second Edition. American Society of Agronomy, Madison, WI.

Thornton, K. W., D. Marmorek, P. F. Ryan, K. Heltcher, and D. Robinson. 1990. Methods for projecting future changes in surface water acid-base chemistry. NAPAP Report 14. Government Printing Office, Washington, D.C.

Turner, J., D. W. Johnson, and M. J. Lambert. 1980. Sulphur cycling in a Douglas-fir forest and its modification by nitrogen applications Oecologia Plantarum 15:27–35.

Turner, J., and M. J. Lambert. 1980. Sulfur nutrition of forests. pp. 321–334. In: D. S. Shriner, C. R. Richmond, and S. E. Lindberg (eds.), Atmospheric Sulfur Deposition: Environmental Impact and Health Effects. Ann Arbor Science, Ann Arbor, MI.

Turner, J., M. J. Lambert, and S. P. Gessel. 1977. Use of foliage sulphate concentrations to predict response to area application by Douglas-fir. Can. J. For. Res. 7:476–480.

Westhall, J. C., J. L. Zachary, and F. M. M. Morel. 1985. MINEQL: A computer program for the calculation of chemical equilibrium composition of aqueous systems. Water Quality Laboratory, Department of Civil Engineering. Massachusetts Institute of Technology. Technical Note No. 18.

Wright, R. F. 1984. Norwegian models for surface water chemistry: an overview. pp. 73–87. In: J. L. Schnoor (ed.), Modeling of Total Acid Precipitation Impacts. Butterworth Publ., Boston, MA.

Wright, R. F., B. J. Cosby, G. M. Hornberger, and J. N. Galloway. 1986. Comparison of paleolimnological with magic model reconstructions of water acidification. Water, Air and Soil Pollution 30:367–380.

Wright, R. F., and A. Henriksen. 1990. The RAIN project—an overview, pp. 161–161. In: B. J. Mason (ed.), The Surface Water Acidification Program. University of Cambridge Press, Cambridge.

8

Sulfur Nutrition and Cycling in Southern Hemisphere Temperate and Subtropical Forest Ecosystems

Marcia J. Lambert and John Turner
FORSCI Pty. Ltd., Eastwood, New South Wales, Australia

I. INTRODUCTION

Southern Hemisphere temperate and subtropical forests are highly variable in floristic composition, structure, productivity and nutritional status. The main forest areas are in South America, southern Africa, and Australasia. South America has more than 860 million hectares of forested land. The environment is extremely diverse, and broad forest types include tropical broadleaf evergreen (rainforest), tropical broadleaf deciduous, mountain broadleaf evergreen, temperate coniferous and broadleaf, and mangrove (Barkow, 1956). In southern Africa, the extent of natural high tree forest is limited, but this has now been supplemented by establishment of plantations (Adamson, 1956).

Floristically, native forests in Australia are dominated by the sclerophyllous broad-leaved *Eucalyptus* and *Acacia*, and include some coniferous genera (such as *Araucaria*, *Callitris*, *Agathis*) and a range of rainforest and woodland communities (Boland et al., 1992). Broad-leaved trees predominate in about 97% of the total wooded area of Australia, and about 94% of the broad-leaved woodland is dominated by eucalypts. Structurally, these communities range from open woodland, through tall forest to multilayered rainforests. Productivity of native forests varies according to environment and forest type but ranges from very low to very high (for example, Attiwill (1991) reported net production of 36 t ha^{-1} yr^{-1} in fire regeneration of *E. regnans*).

The indigenous fauna and flora of Australia are unique and result from a long period of ecological isolation (Duffy and Shepherd, 1969). Most of the land is characterized by soils with low nutrient status and poor physical properties. Evaporation exceeds precipitation in most places, hence commercial forests are found only in Tasmania and along the south-eastern and eastern fringe and in the southwest corner of the continent. Approximately 4% of the land surface in Australia is capable of supporting these forests.

In addition to native forests in the Southern Hemisphere, there are very significant areas of fast growing plantations, both coniferous and hardwood. Coniferous plantations are dominated by *Pinus radiata* (planted in Chile, New Zealand, Australia, and South Africa) and there are significant areas of other *Pinus* species (for example, *P. caribaea*, *P. taeda*, *P. elliottii*, *P. pinaster*, and *P. patula*), generally all being grown on 25- to 35-year rotations. The hardwood plantations are primarily of the genus *Eucalyptus* with species such as *E. grandis*, *E. globulus*, *E. nitens*, *E. pilularis*, and *E. saligna* in most situations on relatively short rotations, that is, less than 20 years. Plantations generally produce in the range of 15–25 m^3 ha^{-1} yr^{-1} mean annual increment; however, with shorter rotations, they produce much higher levels. There are now in excess of 15 million hectares of plantations, about 9 million ha of which are coniferous and the remainder hardwoods (Table 1).

When considering sulfur (S) nutrition, these forests are primarily affected by natural precipitation inputs or S available from soils. Except in defined relatively small areas, the excessive anthropogenic S inputs of the Northern Hemisphere are not a significant feature in the Southern Hemisphere. There are relatively low atmospheric S inputs in conjunction with soils with low S availability derived from volcanic sources and highly weathered soils, such as those in Australia. These lead to areas of native forest adapted to low S conditions or, as is the case for some plantations, occurrence of S deficiency. Although S is an essential plant nutrient, only limited information is available on S requirements and patterns of S cycling within Southern Hemisphere forest ecosystems. This chapter attempts to integrate that information.

Table 1 Area (10^3) of Coniferous and Hardwood Plantations in the Southern Hemisphere; Countries with the Larger Areas of Plantings Have Been Listed Individually

	Plantations		
Country	Coniferous	Hardwood	Total
Australia	955	123	1,083
New Zealand	1,375	22	1,400
Total Australasia	2,362	163	2,525
Brazil	3,383	3,617	7,000
Chile	1,385	215	1,600
Total South America	5,419	4,509	9,928
South Africa	842	658	1,500
Total Africa	1,585	1,150	2,735
TOTAL	9,366	5,822	15,188

II. S NUTRITION OF FORESTS

Sulfur is an essential major plant nutrient. However, within forest ecosystems limited information is available on S requirements and patterns of cycling, and this element has received much less attention in laboratory, glasshouse, and field research in comparison to that received by nitrogen (N) and phosphorus (P).

Sulfur is principally required by plants for the synthesis of three amino acids—cystine, cysteine, and methionine—which are essential components of protein and contain approximately 90% of the organic sulfur found in plants (Anderson, 1975; Turner and Lambert, 1979). Sulfur is of the same importance to trees as it is to most other plants, and the level of S in tree foliage is similar to that of P. However, for the conifer species examined in Australia, total P and total S are not significantly correlated, except where superphosphate has been applied (Humphreys et al., 1975).

In the foliage and seeds of *P. radiata*, Kelly and Lambert (1972) found that S and N occurred in organic forms in a constant ratio of 0.030 on a gram atom basis. This ratio was the same as the S:N ratio in the proteins in the organic material. A constant ratio of 0.030 on a gram atom basis was also found between organic S and total N in the foliage of *P. radiata*, other *Pinus* species (e.g., *P. taeda*, *P. elliottii*, *P. pinaster*, and *P. ponderosa*), other coniferous species such as Douglas fir (*Pseudotsuga menziessii*) (Kelly and Lambert, 1972) and *Eucalyptus* spp. (Lambert, 1979) and this is the same ratio as that in their foliar protein. The ratio was found to be constant for a very large number of samples collected over a wide geographic area of southeastern Australia. These samples include those collected from the same trees over a number of years; at different times of the year; from various parts of the tree crown; and for whole seeds, seed coats, and the fleshy parts of seeds of *P. radiata*.

In coniferous and broad-leaved forest trees, unlike many agricultural crops, foliar total N in forest trees has been found to be equivalent to foliar organic N and there is no evidence of nitrate-N accumulation even when S is limiting. Nitrogen appears to be taken up by most forest tree species only at the rate at which S is available and therefore protein formation is limited by the amount of S available. When the S supply is adequate, the trees accumulate S as sulfate, beyond that which is required to balance the available N, and protein formation proceeds at the rate at which N becomes available. The sulfate-S status of the foliage provides an indicator of the S status of both the tree and the site and is a much more sensitive indicative measure than foliar total S (Lambert and Turner, 1977). The situation of N supply is modified in species in which symbiotic N fixation occurs (for example, *Acacias* and *Casuarinas*).

For nutritional assessments, the relationship between S and N has indicated that neither element can be adequately assessed without reference to the other and that for diagnostic purposes, estimates of either S or N alone are poor parameters. General S ratings based on foliage sulfate-S concentrations have been derived for various species (Table 2).

In contrast with P and N, there do not appear to be any unequivocal demonstrations of growth responses to S by trees in the forest situation, although such have been demonstrated in culture solution and pot experiments for many commercial species. Ingestad (1962) showed that there was a clear relationship between growth and S for *P. sylvestris* in culture solution. Beaton (1966) reported an S response in the greenhouse on pumice soils for *P. ponderosa* and found some evidence of a response in the field. Lyle (1969), working with culture solutions, showed that in solutions lacking S, *P. taeda* developed yellowing that started with the terminal needles and spread over the seedlings. Bolton and Benzian (1970) reported that S fertilizers improved the color but not the height of Sitka spruce *(Picea sitchensis)* seedlings in the field and increased S concentrations in the needles above those considered by Ingestad (1962) to indicate S deficiency. However, S deficiency symptoms and some growth reductions were achieved in *P. radiata* seedlings by Afzal (1972) working with a vermiculite-perlite culture solution system. The symptoms he reported were needles light yellow in color, length of needles decreasing toward the apex, terminal foliage chlorotic, and apical buds of all plants healthy but generally resinous. Attempts to produce S deficiency symptoms in *P. radiata*

Table 2 Reported Total-S and Sulfate-S Levels in Foliage Associated with S Deficiency and Sufficiency in Plantations

Sulfur status	*P. radiata* Sulfate-S (ppm)	*P. radiata* Total-S (ppm)	*E. grandis* Sulfate-S (ppm)	*E. maculata* Total-S (ppm)
Deficiency to incipient deficiency	0–80	1000[a]	1000	1200–1300
Marginal to adequate	80–200	1100	—	1400–1600
Adequate to high (optimum)	200–400	1250	2000[b]	1800–4200
High (possible N deficiency)	400+	1520[a]	—	—
Reference	Turner et al. (1977)		Herbert (1992)	Dell and Robinson (1993)

[a]Based on 1.5% N.
[b]Optimum nitrogen of 2.6% N.
Deficient = Plants with severe symptoms.
Marginal = Plants where symptoms appear or dry matter yield is depressed.
Adequate (optimum) = Plants without symptoms.

in sand culture in a glasshouse in the Sydney city area in Australia failed, presumably because of the high sulfur dioxide (SO_2) concentration in the air, although a number of eucalypts readily developed symptoms in the same location using the same cultural solutions and techniques (Humphreys et al., 1975).

Symptoms of S deficiency are generally similar to those for N deficiency. In *P. radiata*, moderate S deficiency is associated with overall yellowing, most pronounced at base of needles, especially on cold exposed sites. In severe cases, there are lesions and resin bleeding on the stem. When severe deficiency occurs, one or two meters of the leading shoot, or sometimes merely the terminal buds, die back, often with spectacular brownish red or red-orange coloration. Needle bases are frequently much yellower than the remainder of the needle (Lambert and Turner, 1977). Symptoms of S toxicity are somewhat harder to define. Relatively high concentrations of foliage total-S or sulfate-S do not necessarily cause apparent damage symptoms unless the S source is as SO_2. Damage to trees from high SO_2 levels has been described as discoloration of foliage including bleaching, yellowing, and browning of needles or leaves (Linzon, 1965, 1966; Sidhu and Singh, 1977). This can be followed by irregular necrotic patches on leaf apices and margins in broad-leaved species; in conifers, necrosis appears at the needle tip and extends to the needle base. There is also often premature leaf shedding leading to partial or complete defoliation. Although tree damage has been attributed to excessive SO_2 levels, foliar sulfate-S levels can be accumulated to very high concentrations from soil sources before tip scorch, similar to salt damage, occurs.

In the field situation, the incidence of and variation in dieback in high site quality *P. radiata* plantations has been associated with S and/or boron (B) deficiency in the trees (Lambert and Turner, 1977). Deaths occur in patches (0.5 ha+), and because of the necrosis of the buds and growing shoots, extensive areas of surviving trees sustain deformities. Typically in these areas, more than 40% of the seedlings have sustained permanent deformities. Lambert and Turner (1977) suggested that dieback can occur directly as a result of S deficiency or because the deficiency has predisposed the trees to other agent that can cause dieback. The dieback occurs because the vascular system becomes ineffective either directly because insufficient S or boron or both are supplied from the soil (leading to structural weakness in the plant) or indirectly because due to these nutrient deficiencies, the fungus *Diplodia pinea* has successfully invaded the shoot. Infection by *D. pinea* can be initiated in a number of ways but it appears that the ability of a tree to limit the entry and development of the fungus is strongly influenced by its S status. Sulfur deficiency produces a metabolic malfunction that leads to apical bud resinosis (and hence facilitates the entry of the fungus into the shoot) and also leads to an accumulation of particular amino acids (for example, arginine) that have been shown to be favorable for the rapid growth of the fungus (D. W. Edwards, pers. comm.). In eucalypts, the symptoms of S deficiency are the same as for nitrogen in the early stages: interveinal areas of older leaves pale green, major veins broadly flanked with green. In the advanced stages, symptoms spread to old leaves and all leaves turn uniformly yellow. Symptoms of mineral nutrient deficiencies and the nutrient concentration ranges in seedlings of *E. maculata* at defined growth stages have been described by Dell and Robinson (1993).

Although the ratio of organic S to total N is constant, total S concentrations vary between species, and between provenances and families within species, providing a basis for genetic selection. Variation in foliage sulfate-S levels between families was demonstrated for *P. radiata* (Table 3) from a field trial in which the families were randomized within a planting area and hence were grown under the same conditions.

Table 3 Sulfate-S in Foliage of Families of *P. radiata* in NSW

	Foliar Sulfate-S (ppm)	
Family No.	Mean	Range
E80	10	0–35
E64	55	0–150
E70	65	0–180
E83	135	110–175
E87	155	65–235
E62	175	120–260
P(0.05)	45	—

Source: Lambert and Turner, 1977.

III. SOURCES OF S IN FORESTS

Three main sources of S for meeting plant requirements are the soil, atmospheric inputs, and fertilizers. Sulfur can be taken into a plant from the soil generally in the sulfate form (Epstein, 1972) and from the atmosphere as SO_2. Plants growing under normal conditions obtain most of their plant nutrient S requirements by root absorption from soluble sulfate compounds in the soil. Only a small proportion of the sulfate is reduced in the roots. Instead, the sulfate is translocated to the leaves and reduced in the chloroplasts. The sulfate compounds derive either naturally from soil minerals and the breakdown of soil organic matter, or from S added via fertilizers, soil amendments, pesticides, and so forth. Mudd (1975) reported that atmospheric SO_2 may enter the plant system through the leaves, resulting in the accumulation of sulfate-S as the major product. Little or no information is available on the concentration of SO_2 in the air that is most favorable to plant growth, nor on the relationship of this concentration to the supply of S available to plant roots. It is well known that atmospheric SO_2 can cause injury to plants but the mechanism has not been fully established.

The various chemical forms of S in soils include both organic and inorganic compounds and their amounts vary widely between soil types and within the soil profiles themselves (Williams, 1975b). Soils vary widely in their capacity to adsorb sulfate, and this factor is critical in retention of S within the profile and its subsequent distribution.

Foliar sulfate-S concentrations for *P. radiata* on soils derived from various parent materials (Lambert and Turner, 1977) are presented in Table

4. The range of foliar sulfate-S levels in most forests is very large, varying from deficiency to sufficiency and in a few instances near industrial activity, such as coal-burning power stations, is quite high. On soils derived from sedimentary materials, the mean foliar sulfate-S levels are adequate with few deficient areas, whereas the levels from granitic soils depend more on the state of weathering and the closeness of unweathered (and thus relatively S-rich) material to the surface. The foliar S levels in trees located on soils derived from extrusive materials are generally low with a few areas in the adequate range. The relationship between S availability and soil parent material has been indicated by a range of authors (for example, Williams, 1975a; Rehm and Caldwell, 1968).

Soil parent material leads to a major variation in soil properties, and it is also a significant factor in determining soil S status. The main difference between these two soil types arise from their weathering rates and physical properties. The basic extrusives (especially the basalts) are the most widespread of the extrusive rocks encountered in Australian pine plantations and represent about 15% of the planted area. The sites generally have high P and N status; high but possibly unbalanced cation levels; and deficient or marginal boron and S status (Lambert and Turner, 1977). The

Table 4 Mean and Range of Foliar Sulfate-S Concentrations in *P. radiata* Located on Soils in NSW Australia Derived from Different Parent Materials: In Order of Increasing Mean Foliar Sulfate-sulfur

Forest	Parent material	*P. radiata* foliage: Mean Sulfate-S (ppm)	*P. radiata* foliage: Range Sulfate-S (ppm)
Wee Jasper	Basic extrusives	16	0–95
Clouds Creek	Tertiary basalt	20	0–60
Mount Topper	Tertiary basalt	26	0–60
Green Hills	Diorite	29	0–110
Nundle	Tertiary basalt	43	0–120
Mullions Range	Rhyolite	83	0–370
Carabost	Ordovician sediments	91	0–365
Penrose	Hawkesbury sandstone	119	10–285
Jenolan	Silurian sandstone	135	25–320
Newnes[a]	Triassic sandstone	252[a]	100–525
Lidsdale[a]	Permian sediments	356[a]	120–740

[a]Adjacent to coal-burning power station.
Source: Lambert and Turner, 1977.

soils developing from these parent materials are usually kraznozems or chocolate soils. It is on these soils in areas on lower slopes or hollows that trees are most prone to dieback and disease infection. Productivity is generally high on these soils, but despite the high total S status of these soils, marginal or gross boron and/or S deficiency is produced because of the low availability of S and high N and this leads to deformities and high disease susceptibility.

It is generally accepted that plants take up their S in the sulfate form and thus the availability of soil organic S depends on mineralization of the S. Apart from fully available water-soluble sulfate, adsorbed sulfate has been shown to be available and to make up a major source of plant-available S in many soils (Freney and Spencer, 1960; Barrow, 1961; Williams and Steinbergs, 1962). The levels of total S in soils in areas where there are minimal inputs via precipitation are primarily a factor of soil parent material together with an effect of land use (Table 5). Although the surface soil is used as an indicator, there is considerable variation in the depletion down the profile also related to parent material, partly as a leaching effect. For example, Williams (1975a) reported on a sediment with 666 and 350 ppm total S in surface and subsoil respectively; "grit" had 613 and 68 ppm total S in surface and subsoil, respectively. While these types of differences between surface and subsoil S concentrations were attributed to drainage classes, such differences have also been related to drainage and pH (Virmani and Kanwar, 1971).

The forms of S in soil also vary, in particular the relative quantities of organic and inorganic forms, and this affects availability of S to the vegetation. A high proportion of S in soils is organically related and only small concentrations of inorganic sulfate are present (Melville, 1971). For example, for three forest-related soils, the percent organic S was 72–89% (Virmani and Kanwar, 1971) although on drier, more alkaline areas, it was 33–64%. Acquaye and Kang (1987) found that 94% of the total S of Ghanaian soils was in the organic form. The small fraction of inorganic S occurred primarily as sulphate in the water-soluble or adsorbed form. The estimation of net mineralization of S has to take the rapid re-incorporation of S by microorganisms into account. Melville (1971) found that the amounts of S taken up by *E. blakelyi* were closely correlated with the concentrations of phosphate-soluble sulfate in the soil. Mineralization appeared to be independent of the rate of incorporation and hence, unless sufficient sulfate is available, could limit the S available to the plant. Using data from a range of published studies, it is apparent that a reasonably consistent relationship between organic S and total N exists within soil. There is some variability due to the interpretation of what constitutes organic S; however, the relationship is sufficiently consistent to use as a basis

Table 5 Concentrations of Total C, N, and S in a Selection of Southern Hemisphere Surface Soils

		%				
Location	Geology	Total C	Total N	Total S	N:S ratio	Reference
Australia	Basalt	2.24	0.16	0.021	7.6	Barrow (1961)
Australia	Basalt	—	0.252	0.027	9.3	Freney et al. (1969)
Australia	Basalt	—	0.53	0.075	7.1	Williams (1975b)
Australia	Granite	0.63	0.06	0.008	7.5	Barrow (1961)
Australia	Granite	—	0.154	0.017	9.1	Freney et al. (1971)
Australia	Granite	—	0.200	0.022	9.1	Hurditch (1981)
Australia	Granite	—	0.18	0.022	8.2	Hurditch (1981)
Ghana	Granite	2.60	0.22	0.032	7.9	Acquaye and Beringer (1989)
Ghana	Granite	1.54	0.10	0.012	8.2	Acquaye and Beringer (1989)
Ghana	Granite	3.94	0.31	0.030	12.7	Acquaye and Beringer (1989)
Ghana	Granite	4.04	0.31	0.035	8.9	Acquaye and Beringer (1989)
Ghana	Granite	1.57	0.11	0.014	7.8	Acquaye and Beringer (1989)
Ghana	Granite	2.26	0.18	0.015	12.4	Acquaye and Beringer (1989)
Ghana	Granite	0.87	0.04	0.006	7.0	Acquaye and Beringer (1989)
Ghana	Granite	0.83	0.05	0.009	5.6	Acquaye and Beringer (1989)
Ghana	Granite	0.85	0.05	0.011	4.7	Acquaye and Beringer (1989)

Australia	Granodiorite	—	0.137	0.014	9.8	Freney et al. (1971)
Australia	Granodiorite	—	0.270	0.026	10.4	Freney et al. (1969)
Australia	Slate/chert	1.11	0.09	0.009	10.0	Barrow (1961)
Australia	Shale/slate	1.71	0.14	0.016	8.8	Barrow (1961)
Australia	Shale	—	0.32	0.043	7.4	Hurditch (1981)
Australia	Shale	—	0.18	0.190	9.5	Hurditch (1981)
Australia	Shale	0.23	0.23	0.021	10.9	Hurditch (1981)
India	Shale	1.62	—	0.026	—	Virmani and Kanwar (1971)
India	Shale	2.04	—	0.025	—	Virmani and Kanwar (1971)
India	Shale	1.40	—	0.027	—	Virmani and Kanwar (1971)
Australia	Siliceous shale	—	0.156	0.017	9.2	Freney et al. (1969)
Australia	Calcareous sand	—	—	0.082	—	Williams & Steinbergs (1962)
Australia	Sand	—	0.045	0.002	22.5	Hurditch (1981)
Australia	Sand	—	0.020	0.004	5.0	Hurditch (1981)
Australia	Sand	—	0.029	0.004	7.3	Hurditch (1981)
India	Alluvium	0.35	—	0.011	—	Virmani and Kanwar (1971)
India	Alluvium	0.28	—	0.014	—	Virmani and Kanwar (1971)
India	Alluvium	0.77	—	0.028	—	Virmani and Kanwar (1971)

for further analysis. Using a linear model (Table 6), the ratio of total N to organic S is 0.074 and this is reasonably similar to that in plant material; however, analysis of the data indicates this model to be biased, with the ratio being higher at lower concentrations. A polynomial model described the relationship more appropriately, and this variation is important in analyses of S availability to forest soils because these soils have generally higher organic matter than agricultural soils.

As a demonstration of the value of such a model, N utilization in a reasonably productive 27-year-old *E. grandis* plantation (Turner and Lambert, 1983) can be used to calculate organic S requirements. Uptake of N from the soil was 100.4 kg N ha^{-1} yr^{-1} and this requires 6.83 kg S ha^{-1} yr^{-1}, plus a minimum 10% additional in sulfate form (as described earlier) for optimum utilization—that is, 7.5 kg S ha^{-1} yr^{-1} uptake. Analysis of soils indicates that most S is in organic form so that availability is dependent on the net effect of mineralization and re-incorporation (Melville, 1971). The soil organic mineralization required to fulfill N uptake (100.4 kg N ha^{-1}), based on the linear model of Table 6, would theoretically release 7.47 kg S ha^{-1}, which would mean there would be sufficient S assuming no re-incorporation of S in organic matter. The alternative polynomial model would indicate that at lower soil N concentrations, the mineralization of the required N would theoretically release substantially more S (approximately 11 kg S ha^{-1}). This would be adequate even allowing for significant re-incorporation (required to maintain the ratio of N to S). However, at N concentrations equivalent to those in many forest soils, the release of S would be equivalent to or less than the 7.5 kg S ha^{-1} even without any impact of re-incorporation. The result of this for plantations, particularly those on more fertile granites or basalts, is that there will be a dependence on atmospheric inputs—and where these inputs are low (for example, away from oceanic effects), S deficiency would probably occur. Based on Table 8 (see later), this would be more than 100 km for S sources with low ecosystem sorption effectiveness (for example, low soil S sorp-

Table 6 Calculated Relationships Between Organic S and Total N in Soil (Based on Published Information)

Model	No. of soils	R^2
Organic-S (%) = 0.0744 (Total N[a]) + 0.0051	116	0.882
Organic-S (%) = −0.0583 (Total N^2) + 0.1161 (N) + 0.0018	116	0.903

[a]Total N (%).

tion) that S deficiency becomes apparent. Such a situation occurs for plantations on the tablelands of Australia, in Africa, and in parts of South America. The application of high analysis N fertilizers (low in S) exacerbates the problem of S deficiency.

Acquaye and Beringer (1989) indicated how climate and vegetation affected S availability and form in three toposequences in three vegetation forms in Ghana; across evergreen rainforest (2200 mm rainfall), semideciduous rainforest (1270–2159 mm rainfall), and savanna (762–1143 mm rainfall). The area had indications of S deficiency. Considering surface horizons only, the changes due to these factors showed that S concentrations in soils associated with savanna (lowest rainfall) were very much lower than in soils from the two rainforests. The percentage organic S and percent C-bonded S were higher under this former vegetation form, and the ratios of total N to organic S were lower than in soil in the other two. For the two rainforest communities, S concentrations in the soil were very much higher at the upper slope than the mid or lower slope (Table 7). The differences in the S content at the deepest horizons indicate high variation in the S content of the granite. Bromfield (1972) observed that the distribution of S in profiles is affected by the rooting pattern of the vegetation, and Kang et al. (1981) noted the better ability of deep-rooted trees and shrubs in pumping up and recycling S than the shallow-rooted vegetation of savanna regions.

The high proportion of organic S in many soils would mean that with a cycle of soil disturbance (clearing forest for agriculture), mineralization, and low soil S sorption capacity, S is one nutrient at high risk of loss or removal as a result of forest management.

IV. IMPACTS OF ATMOSPHERIC NUTRIENT INPUTS

Nutrient inputs to ecosystems are mainly the result of dryfall, wetfall, and rock weathering. In the first two cases, S is derived from various stages of atmospheric oxidation of SO_2 and from sulfate-containing aerosols from the sea. Hydrogen sulfide is introduced into the atmosphere from both sea and land but is rapidly converted into SO_2, SO_3, and sulfate by a variety of pathways within the atmosphere. Industrial processes and the burning of fossil fuels also introduce large amounts of S into the atmosphere each year. The SO_2 concentration in the air ranges from 0.3 $\mu g\ m^{-3}$ for unpolluted air to more than 4000 $\mu g\ m^{-3}$ for polluted air, but there is no evidence of a total atmospheric buildup of S (Ross, 1971; Junge, 1972). The balance is apparently maintained by adsorption of SO_2 by oceans and lakes, by direct uptake of SO_2 by plants and soil, and by removal in rain. Weathering

Table 7 Changes in S Forms in Surface Soil (0–5 cm) as a Result of Climate, Vegetation, and Topographical Position on Soils Derived from Granite Parent Material

Topographical position	Evergreen Rainforest (2200 mm RF[a])	Semi-deciduous Rainforest (1270–2159 mm RF)	Savanna (762–1143 mm RF)
		Upper slope	
Total S (ppm)	318	347	57
Organic-S (ppm)	280	336	52
C-bonded S (ppm)	163	223	29
Sulfate-S (ppm)	38	11	5
Total N:Organic S	7.85	9.22	7.69
% Organic-S	88	97	91
% C-bonded S	51	64	51
		Mid slope	
Total S (ppm)	122	141	89
Organic-S (ppm)	105	130	84
C-bonded S (ppm)	55	109	70
Sulfate-S (ppm)	17	11	5
Total N:Organic S	9.52	8.46	5.95
% Organic-S	86	92	94
% C-bonded S	45	77	78
		Lower slope	
Total S (ppm)	295	145	106
Organic-S (ppm)	244	144	102
C-bonded S (ppm)	129	103	81
Sulfate-S (ppm)	51	1	4
Total Nitrogen: Organic S	12.7	12.5	4.90
% Organic-S	83	99	96
% C-bonded S	44	71	76

[a]RF = Annual rainfall.
Source: Acquaye and Beringer, 1989.

is difficult to measure and is generally considered to be of the same order as for runoff waters at base flow (Likens et al. 1977).

Sulfur availability to trees from all sources is reflected in foliar sulfate-S concentrations, accepting that this index integrates the effects of S inputs and N availability. Sulfur inputs can be assessed in terms of distance from source of S inputs (Table 8). In addition there is an interaction be-

Table 8 Foliage Sulfate-S Concentrations in Australia and New Zealand in Relation to Soil Parent Material and Distance from Source of Atmospheric Input

Form of S	Distance from S source (km)	Soil parent material	Foliar sulfate-S (ppm) Mean	Range
H_2S[a]	0.1	Pumices	755	205–985
SO_2	2	Quaternary sands	1115	380–1840
SO_4	5–40	Quaternary sands	520	270–850
SO_2	5	Permian sediments	355	120–740
SO_2	18	Triassic sandstone	250	100–525
SO_2	25	Silurian sediments	210	0–445
H_2S[a]	30	Pumices	190	80–130
SO_2	>28	Quaternary sands	125	50–250
SO_2	>100	Tertiary basalts	45	0–120
SO_2	>100	Diorite	30	0–110
SO_2	>100	Basic extrusives	15	0–95

[a]New Zealand geothermal areas.
Sources: Lambert and Turner, 1978.

tween soil parent material and rainfall on sulfate adsorption by soils (Barrow et al., 1969). Above 750 mm annual rainfall, the ability of soils to adsorb sulfate increased with increasing rainfall. Adsorption was greater on soils derived from basic rocks than those derived from acidic rocks. The interactions of these factors explain differences in S cycling with changing rainfall and soils. Further, inputs on any site vary; for example, Muller (1975) reported 20 years of data for S inputs to a site in New Zealand. An average of 13 kg S ha^{-1} was received annually in the rainfall, ranging from 6.6 to 17.6 kg S ha^{-1}.

Few determinations of S inputs are available for the Southern Hemisphere. Emissions of S from the combustion of fossil fuels are relatively small. In western Australia, Hingston and Gailitis (1976) found that sulfate precipitation was generally about 15 kg S ha^{-1} yr^{-1} on the coast with decreases inland. They calculated that the excess of sulfate relative to sea-salt deposition was 3 kg S ha^{-1} yr^{-1}, which is low by European standards. In the Victorian hinterland in southeastern Australia, Hutton and Leslie (1958) found accessions generally to be below their limit of detection of 2.2 kg S ha^{-1}. They found that up to 7.8 kg S ha^{-1} was added annually near the coastline but the contribution fell very sharply with increasing distance from the coast. At six collection stations in New Zealand (Anonymous, 1961), none of which are near the sea or industrial activity, annual accessions ranged from 0.5 to 6.8 kg S ha^{-1}. The mean annual S accession in the northeast of Tasmania in Australia was 13.91 kg S ha^{-1} and this was in those coastal areas directly exposed to prevailing sea winds (Anonymous, 1972). The rapid fall-off in S in precipitation with increasing distances from the coast in various parts of the world (Hutton and Leslie, 1958; Ericksson, 1960; Anonymous, 1972) emphasizes the restricted contribution made by sea spray to S accessions.

Estimations from New South Wales in Australia show a pattern of S inputs in relation to distance from the coast (Fig. 1) and this in part is reflected in rainfall concentration (Fig. 2) rather than changes in precipitation itself. This pattern reflects the concentrations found in foliage (Table 8).

V. USE OF FERTILIZERS

Sulfur deficiency has been of minor importance in most parts of the world because of the wide use of fertilizers containing sulfates and the accession of large amounts of S in rainfall arising from the combustion of fuels. The extent of S deficiency in Australia was realized only as recently as the 1950s and 1960s (Freney et al., 1962). One of the reasons it escaped attention for so long was the widespread agricultural use of superphosphate, which contains S as well as Ca and P. It is now known that many soils in southern Australia are low in S. These cover large areas of the tablelands in New South Wales, southern Victoria, and western Australia. The current swing to high-N and high-P fertilizers, in an effort to lessen freight and handling costs, means that the amount of S being applied indirectly is much less than previously. High-analysis fertilizers such as triple-superphosphate contain very little S, and antipollution environmental protection legislation is aiming for a reduction in the S available from the atmosphere. Although this source of S exerts a major effect in Europe and North America, it is less important in Australia because of smaller emissions from industry, the drier climate, and the remoteness of much of the agricultural and forested land from industrial emissions.

In Australia, forest soils requiring S additions are usually those with the poorest sulfate sorption capacities (Humphreys et al., 1975). Soluble fertilizers such as gypsum are rapidly leached from these soils and responses are short-lived. Because of the difficulties encountered with these types of soils, slow release fertilizers such as iron pyrites or various S forms together with the selection of tolerant clones by tree breeding are being further investigated as possibilities to overcome soil S deficiency problems.

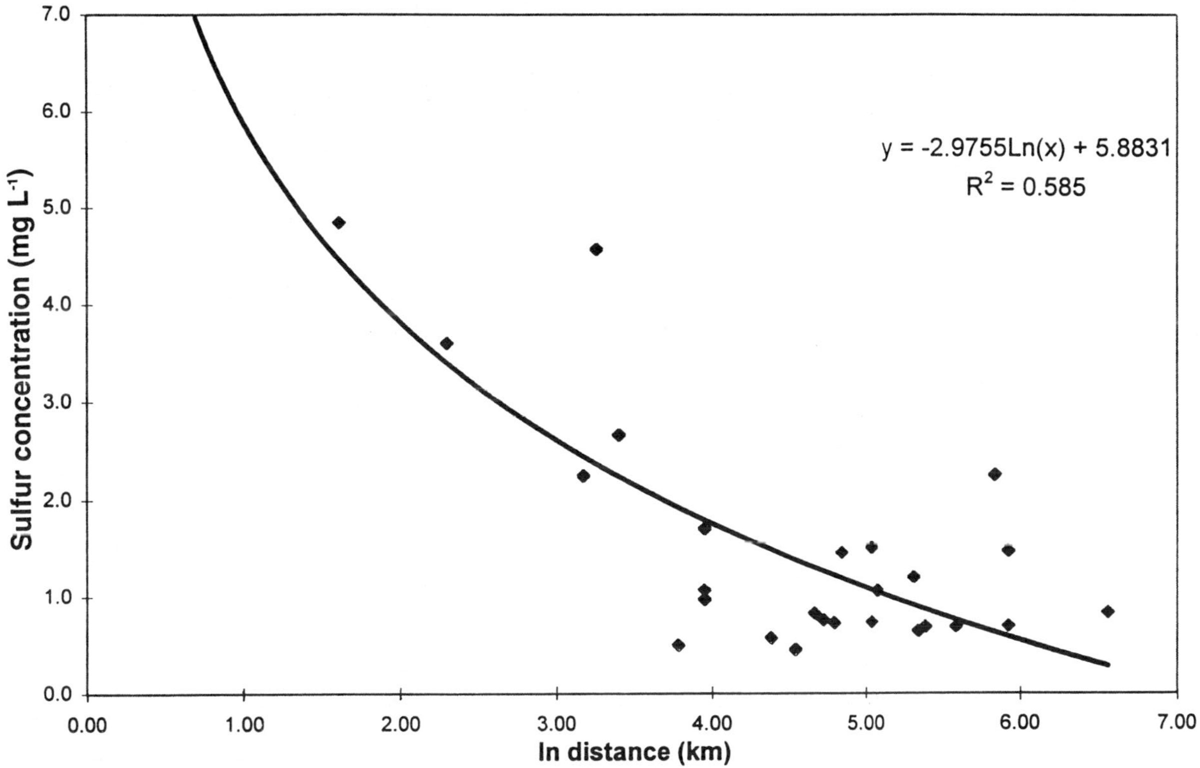

Figure 1 Sulfur inputs (ka ha^{-1}) in rainfall in relation to distance (km) from the coast in New South Wales, Australia.

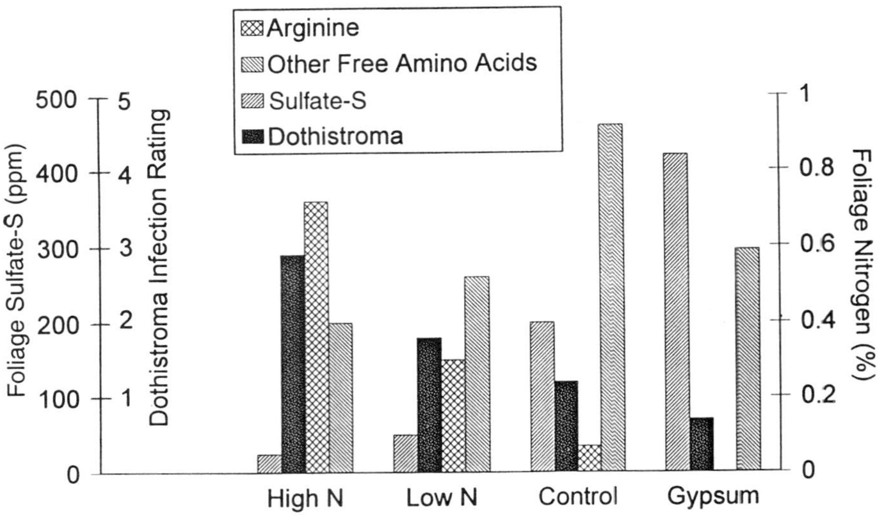

Figure 2 Sulfate S, N, and arginine concentrations in foliage from a fertilizer trial in *P. radiata*. Associated *Dothistroma* infection in the stand is also shown.

Fertilizers applied for amendment of S deficiency, either induced or existing, need to address the interaction between S nutrition and other applied nutrients such as N. Few instances exist of applications of S only being applied and a response being obtained. This is because few instances of S deficiency in the absence of other nutrient deficiencies have been diagnosed. Where areas of S deficiency have been diagnosed, most commonly P deficiency has also been found in conjunction with S. The more common situation is where S applications in addition to other nutrient applications result in a growth response. In a fertilizer experiment established in a 45-year-old stand of ponderosa pine (*P. ponderosa*) on a pumice soil in central Oregon, 112 kg S ha^{-1} added as gypsum resulted in a small increase in tree growth in the first 5-year period. The combination of S with 224 kg N ha^{-1} gave a greater response (Table 9). *P. radiata* was grown in pot trials in soil collected from the rooting zone of the treatments in the field trial, and in all cases, the growth of radiata pine was improved by the addition of S. Increases in plant weight of between 25% and 210% due to N/P/S compared with N/P alone, indicated a supplementary S response (Will and Youngberg, 1978).

It has been shown that the form of S stored in trees can be greatly influenced by management procedures such as fertilization (Turner et al., 1980; Lambert 1986). In a N-deficient Douglas fir ecosystem, a large pro-

Table 9 Basal Area Increase (%) Compared with Control (No Treatment) 10 Years After Treatments Were Applied to a 45-Year-Old *P. ponderosa* Stand

	Nil P		Phosphorus applied ($57\ kg\ P\ ha^{-1}$)	
Sulfur	Nil N	Nitrogen applied ($224\ kg\ N\ ha^{-1}$)	Nil N	Nitrogen applied ($224\ kg\ N\ ha^{-1}$)
Nil S	0	15.4	1.3	6.0
Sulfur applied ($112\ kg\ S\ ha^{-1}$)	4.6	18.9	2.4	16.6

Source: Will and Youngberg, 1978.

poration (14%) of the total S in the aboveground components was found to be sulfate-S and nearly all the S transferred to the forest floor (81%) was also in this form. The site, in the Pacific Northwest of the United States, is subject to acid rain and elevated S inputs, and because of the biochemical relationship between organic S and organic N, excess S was found to be stored and cycled as sulfate-S. Nitrogen fertilization resulted in the incorporation of foliar sulfate-S into organic-S and a reduction in litterfall sulfate-S concentrations. Nitrogen stress induced by carbohydrate additions to the forest floor resulted in greater sulfate-S return via litterfall.

VI. IMPACTS ON DISEASES

In Australian *P. radiata* plantations, S deficiency does not appear to cause a direct reduction in primary productivity. However, there appears to be a high correlation between S deficiency and foliar pathogen infection, in particular *Diplodia* (Lambert et al., 1976; Lambert and Turner, 1977) and associated with this are stem malformations with subsequent losses in merchantable productivity and volume. Sulfur deficiency produces a metabolic malfunction that leads to apical bud resinosis (hence facilitating the entry of the fungus into the shoot) and also leads to an accumulation of particular high N-containing amino acids (for example, arginine), which would be favorable for the rapid growth of the fungus. This has been demonstrated from glasshouse work and in a field fertilizer trial with *P. radiata* that was established in a N-rich, S-deficient environment to determine the effects of N and S fertilizers on the concentrations of these nutrients in the foliage and to relate any differences to tree growth and health (Lambert, 1986; Turner and Lambert, 1979). The results for the foliar S concentrations indicated that the most striking changes occurred in the plots where N fertilizer had been applied (Fig. 2). Within three months after treatment, S deficiency was induced and has persisted to the present. The foliar sulfate-S in the N-treated plots had been utilized with a corresponding increase in the total concentration of free amino acids, particularly arginine. It was further noted that *Dothistroma septospora* (Dorog.) Morelet infection, a severe needle cast fungus, was highest in the N-treated plots, with the degree of infection depending upon the intensity of S deficiency (Eldridge et al., 1981).

VII. CYCLING OF S IN FORESTS

Cycling of S in forest ecosystems involves uptake, utilization and return of S to the soil, together with accumulation, inputs, and losses. The difference in the analysis of cycling of S compared with that of most other nutrients is the functional relationship between S and N and hence the cycling needs to be considered in terms of the organic-related S and the free sulfate. While the aspect of free sulfate has been significant in the Northern Hemisphere due to elevated and often excessive atmospheric S inputs, this is not the case for forests in the Southern Hemisphere, where S inputs are generally very close to the natural levels.

Differences exist between species in their S requirements and this has an effect on S cycling. Lambert (1979) compared S uptake among coniferous species using a series of arboreta and trial plantings in New South Wales, Australia. The sources of S available to these trees were from soil and atmospheric origins (specifically maritime inputs). In the case of *P. radiata*, for which most data were available, foliar total-S concentrations decreased with increasing distance from the coast (Table 8, Fig. 1). There was an effect from soil fertility in that trees located on the most fertile soils, usually basalt derived, had the highest S accumulations, whereas trees on the nutritionally poor sedimentary soils (Triassic, Permian, and Devonian sediments) had the lowest S accumulations.

Species other than *P. radiata* had different accumulation patterns, with the data suggesting that when species native to coastal situations were planted near the coast, higher concentrations of S were accumulated than when planted inland where the concentrations of total-S were lower. Species native to inland locations (for example, *P. ponderosa*) had a pattern of foliar total-S accumulation that was independent of their geographical location. This tended to show that inland species were adjusted to low S uptake from the soil, but in areas of higher atmospheric S inputs they could not efficiently utilize this additional source of S.

To further compare strategies of S uptake and utilization, planted and natural conifer stands from a range of sites were compared—a planted Douglas fir stand with moderate S availability at Bago, New South Wales (Lambert, 1979); a natural Douglas fir stand located on a site at Cedar River, Washington, which is subject to elevated inputs of S from a nearby copper smelter (Turner et al. 1980); and a nutritionally poor *P. radiata* site at Lidsdale, New South Wales, with comparatively recently increased S inputs following the installation of a coal-burning power station nearby (Lambert, 1979).

At the Bago site, which was located inland, the Douglas fir had low accumulations of sulfate-S in the various tree components (Table 10) (Fig. 3). The S inputs to the site were extremely low, but the soil had relatively high S concentrations and tree S uptake fulfilled about 90% of the requirement for growth. The data from the Douglas fir stand at Cedar River clearly showed that S was in excess at that site and that excess S was cycled in sulfate form (Table 10, Fig. 3). The stand was N deficient and there was a high proportion of N redistribution within the tree to fulfill current growth requirements. However, S in the tree followed a different pattern, with the accent on accumulation in old foliage rather than on redistribution. The excess of S in the system accumulated as sulfate in both current and older tissue. Although N uptake was less than the growth requirement, the S uptake was very much higher than the amount required for growth.

In contrast to the Bago site, two forest types were assessed at Lidsdale where S was naturally low but had been elevated by atmospheric inputs from a coal burning power station. The native eucalypt species (*E. rossii*, E. maculata) were adapted to the low S environment and did not appear to be affected by the elevated inputs as indicated by the relatively small increases in accumulation of sulfate-sulfur. By contrast, planted *P. radiata* (a species native to coastal environments and hence adapted to high S inputs) accumulated high levels of S (as sulfate-S) in the foliage and maintained relatively high growth rates (Lambert, 1979). This was shown in the estimates per unit area rather than concentrations (Table 11).

These accumulations were further compared with adjacent native eucalypt woodland species, and it was found that the S content of the *P. radiata* standing crop was one-third greater than that in the native woodlands and in addition an amount equivalent to the standing biomass had been removed in logging. The adjacent native eucalypts had developed in a low S, P, and Ca environment, and these species appeared to be inelastic in their utilization of the elevated S (mainly as sulfate inputs) and redistributed S very efficiently—that is, to fulfill 54–70% of their S requirements. Generally, tree species that have adapted to low S growth are efficient at uptake from the soil and redistribute S efficiently within the tree.

Table 10 Nutrient Contents and Transfers in Douglas Fir Forests with Low (Lambert, 1979) and High (Turner et al., 1980) S Inputs

	Low S Inputs Bago, NSW, Australia				High S Inputs Cedar River, Washington, USA			
Component	O.M.[a]	Total-N	Total-S	SO_4-S	O.M.	Total-N	Total-S	SO_4-S
	Nutrient content (kg ha^{-1})							
Tree—foliage	16,720	186	16	1.3	9,440	98	18	11.2
—total	404,290	737	63	11.4	229,400	316	41	18.6
Understorey	0	0	0	0	3,390	29	4	1.1
Forest floor—woody	2,990	19	1	0	3,480	15	2	0.2
—total	35,680	369	27	1	26,670	223	26	3.0
Soil (0–15 cm)	1,680	80	5	—	1,300	220	40	—
Total ecosystem	439,970	2,786	170	19	259,460	1,868	291	62.7
	Nutrient transfers (kg ha^{-1} yr^{-1})							
Litterfall	3,920	29.5	3.1	1.2	3,745	25.5	7.0	5.2
Total return to forest floor	3,920	35.7	6.1	4.1	3,745	27.7	12.7	10.8
Requirement for growth[b]	8,730[c]	51	3.8	—	7,460	45.8	5.6	—
Uptake[d]	—	35	3.5	—	—	25.1	6.8	—
% Uptake/Requirement	—	69	92	—	—	55	120	—
% Contained in tree component	92	26	37	—	88	17	14	—

[a]Organic matter.
[b]Nutrient content of current issue.
[c]Aboveground primary productivity.
[d]Difference between requirement, internal redistribution, and leaching. Includes foliar intake.

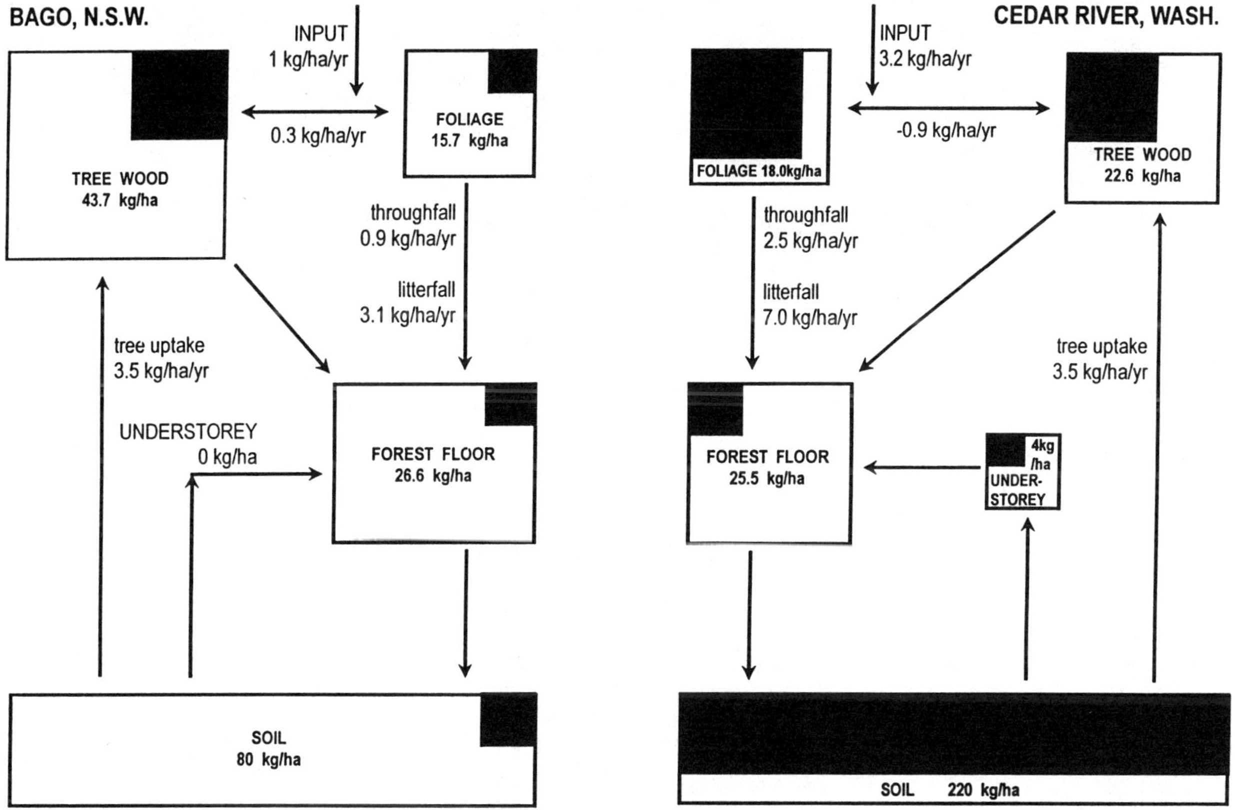

Figure 3 Nutrient contents and transfers in Douglas fir forests with low (Bago, New South Wales, Australia) and high (Cedar River, Washington, U.S.) sulfur inputs.

Table 11 Biomass, S Content, and Transfers in a *P. radiata* Plantation and an Adjacent Eucalypt Woodland (*E. rossii*/*E. maculosa*) at Lidsdale in the Vicinity of a Coal-burning Power Station in New South Wales, Australia

	Eucalypt			Pine		
Component	Biomass	Total-S	Sulfate-S	Biomass	Total-S	Sulfate-S
	Nutrient distribution (kg ha^{-1})					
Tree foliage	2,340	2.3	0.6	7,240	8.8	3.1
Total vegetation	62,355	11.6	3.3	160,500	22.3	9.1
Forest floor						
wood	5,205	1.5	0.5	9,880	2.4	1.0
total	10,480	4.5	1.6	13,900	6.0	2.1
Soil (0–10 cm)	—	5.9	16	—	55	14
	Nutrient transfers (kg ha^{-1} yr^{-1})					
Input (open area precipitation)	—	9	9	—	9	9
Throughfall	—	11	9	—	19	16
Litterfall	2,910	3.1	—	3,790	1.3	—
Logging removal	—	—	—	5,590	0.50	—

Source: Turner and Lambert, 1979.

Communities with higher productivity and organic matter turnover appear to be buffered against environmental changes. However, in some situations in which S and other nutrients are limiting to growth, productivity is generally low and responses of species to changes in nutrient status vary from virtually nil (as in the case of S in the eucalypts at Lidsdale where elevated S inputs were not absorbed) to species that accumulate and often utilize the added nutrients (for example, for S in *P. radiata* at Lidsdale).

Where there are changes in S inputs to forest management areas, there are several options to maintain healthy stands.

1. Leave the stand unaltered when it is healthy and productive.
2. Change either the species or the provenance of a species to one that can more efficiently utilize this changed resource.
3. Amend the site with other nutrients so that the stand can more fully utilize the added inputs.

This last option has been studied in a western Washington Douglas fir forest in which a nitrogen manipulation experiment was carried out in a 42-year-old, poor-quality stand limited in growth by a deficiency of nitrogen (Table 12) (Turner et al., 1980). Sulfur was in excess. Treatments consisted of control, a high nitrogen treatment (880 kg N ha^{-1}), a treatment of carbohydrate (8,000 kg sucrose ha^{-1} plus 10,000 kg sawdust ha^{-1}) to widen the C/N ratio and reduce N availability, and a carbohydrate treatment as above plus P, K, Ca, and S additions. The primary effects on the N cycle have been described elsewhere (Turner and Olson, 1976; Turner, 1977). The S cycle has also been described (Turner et al., 1980) and showed that excess S can cycle as sulfate in forest ecosystems and in that sense operate somewhat independently of the N or C cycles. The N additions resulted in the incorporation of sulfate-S into organic forms and even though the stand was overstocked, a stimulus to growth was obtained. The changes (kg ha^{-1}) also indicated that in the foliage, and presumably in the wood because of the growth stimulus, there were increases in N and S uptake. In the first year, the N application reduced leaf litterfall because older needles were retained, but this was probably only a temporary phase.

A further extension of this N fertilization indicated that unless in excess of 400 ppm sulfate-S is accumulated in tree foliage, no response to N application of 220 kg N ha^{-1} could be expected (Turner and Lambert, 1978; Turner et al., 1979), and that many of the soil types involved were naturally low in S and atmospheric sources provided the necessary forms of S inputs.

Table 12 Nitrogen and S Contents of Specific Stand Components in the Various Fertilizer Treatments of a Pacific Northwest Douglas Fir Field Experiment

Component	Treatments														
	880 kg N ha^{-1}			220 kg N ha^{-1}			Control			Carbohydrate			Carbohydrate +P,K,S,Ca		
	Total N	Total S	Sulfate S	Total N	Total S	Sulfate S	Total N	Total S	Sulfate S	Total N	Total S	Sulfate S	Total N	Total S	Sulfate S
Current foliage (kg ha^{-1})	50	3.38	0.06	39	2.97	0.27	20.7	3.26	1.86	20.2	2.71	1.28	22.6	2.86	1.41
Total foliage (kg ha^{-1})	160	18.82	3.67	138	15.89	6.17	76.6	17.40	12.60	86.2	15.29	9.41	92.9	17.68	11.33
Leaf litter (kg ha^{-1} yr^{-1})	11	1.03	0.27	13	1.69	1.00	14	3.9	2.8	14	5.0	3.6	16	5.0	3.6
Foliage redistribution (kg ha^{-1} yr^{-1})	−20[a]	0.78		6	1.58		10	−0.80		16	−1.59		15.4	−2.06	
Foliage uptake (kg ha^{-1} yr^{-1})	70	2.60		33	1.97		12	4.11		4	4.37		7.2	4.29	

[a]Negative sign indicates accumulation.
Source: Turner et al., 1980.

VIII. CONCLUSIONS

1. Sulfur is an essential major tree nutrient and the biochemical relationship between S and N in plant protein dictates that one element cannot be adequately studied without reference to the other.
2. The availability of S to forests, where anthropogenic inputs are low, is related to soil parent material, mineralization rates, and distance from natural sources (for example, maritime inputs).
3. Because trees accumulate any excess S beyond that required to balance the nitrogen available as sulfate-S, the sulfate-S status of the foliage provides an indicator of the S status of both the tree and the site.
4. Sulfur deficiency is associated with some forms of crown dieback in *P. radiata* and also leads to the accumulation of particular amino acids, such as arginine, which is associated with high foliar disease susceptibility.
5. When high quantities of S have been applied to stands in the sulfate-S form rather than as SO_2, no damage has been noted.
6. Tree species vary intrinsically in their ability to utilize S and particularly so when the S status of the site is altered.
7. A comparison of various conifer species has indicated that species native to coastal situations with high maritime S inputs are more elastic in their utilization of S than species native to inland locations.
8. Additions of N to forest stands result in the incorporation of sulfate-S into organic forms, and thus the sulfate cycle in forest ecosystems is regulated to a large extent by N status and N cycling.
9. Application of high analysis fertilizers (N and P) that increase growth will futher stress available S. This will be more apparent in areas with low S inputs.
10. A high proportion of S in soils is organically related and only small concentrations of inorganic sulfate are present. The high proportion of organic S in many soils means that a cycle of soil disturbance (clearing forest for agriculture), mineralization, and low soil S sorption capacity would indicate that S is a nutrient at high risk of loss or removal as a result of forest management.
11. Establishment of rapidly growing plantations distant from elevated S inputs and on soils with low S supplying power will result in a high risk of S deficiency.

REFERENCES

Acquaye, D. K., and H. Beringer. (1989). Sulphur in Ghanaian soils. I. Status and distribution of different forms of sulfur in some typical profiles. *Plant and Soil* 113: 197–203.

Acquaye, D. K., and B. T. Kang. (1987). Sulphur status and forms present in some surface soils of Ghana. *Soil Sci.* 144: 43–52.

Adamson, R. S. (1956). South Africa. *In* A World Geography of Forest Resources. p. 385–392. (Eds. S. Haden-Guest, J. K. Wright & E. M. Teclaff). The Ronald Press Company, New York. 736 pp.

Afzal, M. (1972). Boron and sulphur nutritional studies in *P. radiata* D. Don. B.Sc. (Honours) thesis, Australian National University, Canberra, Australia.

Anderson, J. W. (1975). The function of sulphur in plant growth and metabolism. *In* Sulphur in Australian Agriculture. pp 87–97. (Ed. K. D. McLachlan) Sydney Univ. Press, Sydney. 261 pp.

Anonymous (1961). Sulphur in air and rainfall. *New Zealand J. Agric.* 101: 328.

Anonymous (1972). Environmental pollution in Tasmania. Department of Environment, Hobart, Tasmania, Australia. Unnumbered paper.

Attiwill, P. M. (1991). Productivity of *Eucalyptus regnans* forest regenerating after bushfire. IUFRO Symposium in Intensive Forestry: The role of eucalypts. Duban, South Africa.

Barkow, W. R. (1956). South America. *In* A World Geography of Forest Resources. p. 201–230. (Eds. S. Haden-Guest, J. K. Wright & E. M. Teclaff). The Ronald Press Company, New York. 736 pp.

Barrow, N. J. (1961). Studies on mineralization of sulfur from soil organic matter. *Aust. J. Agr. Res.* 12: 306–319.

Barrow, N. J., I. K. Spencer, and W. M. McArthur. (1969). Effects of rainfall and parent material on the ability of soils to adsorb sulphate. *Soc. Sci.* 108: 120–126.

Beaton, J. D. (1966). Sulfur requirements of cereals, tree fruits, vegetables and other crops. *Soil Sci.* 101: 267–282.

Boland D. J., M. I. H. Brooker, G. M. Chippendale, N. Hall, B. P. M. Hyland, R. D. Johnston, D. A. Kleinig, and J. D. Turner. (1992). Forest trees of Australia. CSIRO, Melbourne. 687 pp.

Bolton, J., and B. Benzian. (1970). Sulphur as a nutrient for Sitka spruce (*Picea sitchensis*) seedlings and radish (*Raphanus sativus*) grown on a sandy podzol in England *J. Agric. Sci.* 74: 501–504.

Bromfield, A. R. (1972). Sulphur in Northern Nigeria soils. 1. Effects of cultivation and fertilizers on total sulphur and sulphate patterns in soil profiles. *J. Agric. Sci.* 78: 465–470.

Dell, B., and J. M. Robinson. (1993). Symptoms of mineral nutrient deficiencies and the nutrient concentration ranges in seedlings of *Eucalyptus maculata* Hook. *Plant and Soil* 155/156: 255–261.

Duffy, P. J. B., and R. F. Shepherd. (1969). Australian forestry highlights. *For. Chron.* 45:180–206.

Eldridge, R. H., J. Turner, and M. J. Lambert. (1981). *Dothistroma* needle blight in a New South Wales *Pinus radiata* plantation in relation to soil types. *Aust. For.* 44: 42–45.

Epstein, E. (1972). Mineral nutrition of plants: Principles and Perspectives. John Wiley and Sons, Inc., New York. 412 pp.

Ericksson, E. (1960). The yearly circulation of chlorine and sulphur in nature; meteorological, geochemical and pedological implications. *Tellus* 12: 63–109.

Freney, J. R., and K. Spencer. (1960). Soil sulphate changes in the presence and absence of growing plants. *Aust. J. Agric. Res.* 11: 339–345.

Freney, J. R., N. J. Barrow, and K. Spencer. (1962). A review of certain aspects of sulfur as a soil constituent and plant nutrient. *Plant and Soil* 17: 295–308.

Freney, J. R., G. E. Melville, and C. H. Williams. (1969). Extraction, chemical nature and properties of soil organic sulphur. *J. Sci. Fd. Agric.* 20: 440–445.

Freney, J. R., G. E. Melville, and C. H. Williams. (1971). Organic sulphur fractions labelled by addition of ^{35}S-sulphate to soil. *Soil Biol. Biochem.* 3: 133–141.

Herbert, M. A. (1992). Nutrition of eucalypts in South Africa. ICFR Bull. 27/92. 25 pp.

Hingston, F. J., and V. Gailitis (1976). The geographic variation of salt precipitated over Western Australia. *Aust. J. Soil Res.* 14: 319–335.

Hingston, F. J., A. G. Turton, and G. M. Dimmock. (1979). Nutrient distribution in Karri (*Eucalyptus diversicolor* F. Muell.) ecosystems in southwest Western Australia. *For. Ecol. Manage.* 2: 133–158.

Hingston, F. J., G. M. Dimmock, and A. G. Turton. (1980). Nutrient distribution in a Jarrah (*Eucalyptus marginata*) Donn ex. Sm.) ecosystem in southwest Western Australia. *For. Ecol. Mange.* 3: 183–207.

Humphreys, F. R., M. J. Lambert, and J. Kelly. (1975). The occurrence of sulphur deficiency in forests. *In* Sulphur in Australasian Agriculture. pp. 154–162. (Ed. K. D. McLachlan). *Sydney Univ. Press.* 261 pp.

Hurditch, W. J. (1981). The biogeochemistry of sulphur in coastal forest ecosystems. PhD Thesis. University of New England. 410 pp.

Hutton, J. T., and T. I. Leslie. (1958). Accession of non-nitrogenous ions dissolved in rainwater to soils in Victoria, Australia. *Aust. J. Agr. Res.* 9: 492–507.

Ingestad, T. (1962). Macroelement nutrition of pine, spruce and birch seedlings in nutrient solutions. *Medd. Skogsforskn. Inst. Stockholm* 51(7).

Junge, C. (1972). International Symposium on Sulphur in Agriculture, Versailles, 1970. *Annales agronomiques, Numéro hors Série* 72-1, 305 pp.

Kang, B. T., E. Okoro, D. K. Acquaye, and O. A. Osiname. (1981). Sulfur status of some Nigerian soils from the savanna and forest zones. *Soil Sci.* 132: 220–227.

Kelly, J., and M. J. Lambert. (1972). The relationship between sulphur and nitrogen in the foliage of *Pinus radiata. Plant and Soil* 37: 395–407.

Lambert, M. J. (1979). Sulphur relationships of native and exotic tree species. M.Sc (Hons). Macquarie University, Sydney, Australia. 181 pp.

Lambert, M. J. (1986). Sulphur and nitrogen nutrition and their interactive effects on *Dothistroma* infection in *Pinus radiata*. *Can. J. For. Res.* 16: 1055–1062.

Lambert, M. J., J. Turner, and D. W. Edwards. (1976). Effects of sulphur deficiency in forests. *Proceedings of the 16th IUFRO World Congress, Oslo, Norway, November 1976, IUFRO, Oslo. Group 2 Paper*. 6 pp.

Lambert, M. J., and J. Turner. (1977). Dieback in high site quality *Pinus radiata* stands—the role of sulphur and boron deficiencies. *N.Z. J. For. Sci.* 7: 333–348.

Lambert, M. J., and J. Turner. (1978). Interaction of nitrogen with phosphorus, sulphur and boron in N.S.W. *Pinus radiata* plantations. pp. 255–262, Plantation Nutrition 1978, *Proceedings International Colloquium on Plant Analysis and Fertilizer Problems, Auckland, N.Z., 1978*. (Eds. A. R. Ferguson, R. L. Bieleski, I. B. Ferguson). *N.Z. D.S.I.R. Information Series* No. 134. 629 pp.

Likens, G. E., F. H. Bormann, R. S. Pierce, S. Eaton, and N. M. Johnson. (1977). Biogeochemistry of a Forested Ecosystem. Springer-Verlag, New York Inc. 146 pp.

Linzon, S. N. (1965). Sulphur dioxide injury to trees in the vicinity of petroleum refineries. *For. Chron.* 41: 245–250.

Linzon, S. N. (1966). Damage to eastern white pine by sulphur dioxide, seminature tissue needle blight, and ozone. *J. Air Pollution Control Assoc.* 16: 140–144.

Lyle, E. S. (1969). Mineral deficiency symptoms in Loblolly pine seedlings. *Agron. J.* 61: 395–397.

Melville, G. E. (1971). Chemical nature of organic-sulphur compounds present in soil and their availability to plants. PhD thesis, ANU, Canberra.

Mudd, J. B. (1975). Sulphur dioxide in response of plants to air pollution. *In* Responses of Plants to Air Pollution. pp. 9–22. (Eds. J. B. Mudd and T. T. Kozlowski) Academic Press. New York.

Muller, F. B. (1975). Sulfur received in rainfall and leached from a yellow brown loam. *N.Z. J. Sci.* 18: 243–252.

Rehm, G. W., and A. C. Caldwell. (1968). Sulfur suppling capacity of soils and the relationship to soil type. *Soil Sci.* 105: 355–361.

Ross, F. F. (1971). What sulphur dioxide problem? *Combustion* 42: 6–11.

Sidhu, S. S., and P. Singh. (1977). Foliar sulphur content and related damage to forest vegetation near a linerboard mill in Newfoundland. *Plant Disease Reporter* 61: 7–11.

Turner, J., and P. R. Olson. (1976). Nitrogen relations in a Douglas-fir plantation. *Ann. Bot.* 40: 1185–1193.

Turner, J. (1977). Effect of nitrogen availability on nitrogen cycling in a Douglas-fir stand. *For. Sci.* 23: 307–316.

Turner, J., M. J. Lambert, and S. P. Gessel. (1977). Use of foliage sulphate-sulphur concentrations to predict response to urea application by Douglas-fir. *Can. J. For. Res.* 7: 476–480.

Turner, J., and M. J. Lambert. (1978). Sulphur nutrition of conifers in relation to response to fertilizer nitrogen, to fungal infection and to soil parent materials. *In* Forest Soils and Land Use. pp. 546–564. (Ed. C. T. Youngberg). Proc.

Fifth North American Forest Soils Conference. August 6–9, 1978. Colorado State University. 623 pp.

Turner, J., and M. J. Lambert. (1979). Comparison of sulphur cycling between a conifer and a native forest in the vicinity of a coal-burning power station. p. 228–230. *In* Internat. Symp. Sulphur Emissions & the Environment, Society of Chemical Industry, London.

Turner, J., and M. J. Lambert. (1980). Sulphur nutrition of forests. *In* Atmospheric Sulphur Deposition—Enivironmental Impact and Health Effects. pp. 321–333. (Eds. D. S. Shriner, C. R. Richmond, S. E. Lindberg). Ann. Arbor Science Publishers Inc. 586 pp.

Turner, J., D. W. Johnson, and M. J. Lambert. (1980). Sulphur cycling in a Douglas-fir forest and its modification by nitrogen applications. *Acta Oecol. Plant.* 15: 27–35.

Turner, J., and M. J. Lambert. (1983). Nutrient cycling within a 27-year-old *Eucalyptus grandis* plantation in New South Wales. *For. Ecol. Manage.* 6: 156–168.

Virmani, S. M., and J. S. Kanwar. (1971). Distribution of forms of sulphur in six soil profiles of North-east India. *Indian Soc. Soil. Sci.* 19: 73–77.

Will, G. M., and C. T. Youngberg. (1978). Sulphur status of some Central Oregon pumice soils. *Soil Sci. Soc. Am. J.* 42: 132–134.

Williams, C. (1975a). The distribution of sulphur in the soils and herbage of North West Pembrokeshire. *J. Agric. Sci.* 84: 445–452.

Williams, C. H. (1975b). The chemical nature of sulphur compounds in soils. *In* Sulphur in Australasian Agriculture. pp. 21–30. (Ed. K. D. McLachlan). Sydney University Press. Sydney. 261 pp.

Williams, C. H., and A. Steinbergs. (1962). The evaluation of plant-available sulphur in soils. *Plant and Soil* 17: 279–294.

9

Modeling Effects of SO_2 on Net Photosynthesis in Forest Trees, from Twig to Canopy Level: A Synthesis

Paul A. Arp and Fan-Rui Meng
University of New Brunswick, Fredericton, New Brunswick, Canada

I. INTRODUCTION

Assessing the effects of air pollutants such as sulfur dioxide (SO_2) on forest growth is a complex undertaking and requires a synthesis of many pieces of anecdotal observations, experimental results, and an evaluation of existing concepts, principles, and theories (Winner and Atkinson, 1986). This chapter attempts to produce such a synthesis. First, it presents a list of guiding ideas and observations about the nature and scope of the problem; then, a synthesis by way of the Forest Net PhotoSynthesis model (ForNePS model). This model, as designed, can be used to simulate the effects of low rates of SO_2 exposure on net carbon dioxide (CO_2) assimilation, from the twig to the canopy level, and from the beginning to the end of a growing season. This is followed by a red/black spruce case study that is used for model performance checking (red/black spruce = hybrids of *Picea rubens* and *Picea mariana*), and other studies involving low SO_2 exposure effects on net photosynthesis in forest trees. An overview of physiological responses and related mechanisms regarding stomatal conductance and net photosynthesis to SO_2 is given in Appendix 1 and Appendix 2.

The following ideas have guided the writing of this chapter:

1. Leaf internal damage occurs when air pollutants are actually absorbed by the foliage (Meng, 1994).
2. The extent of damage is likely proportional to the amount of SO_2 absorbed (Meng et al., 1994).
3. The plants may or may not have an ability to repair SO_2-induced damage (Winner et al., 1982).
4. If the plants have repair mechanisms to deal with SO_2 damage, then it is important to determine factors that affect the recovery time from such damage.
5. There are several damage levels: some that are readily seen through foliage inspection, such as needle discoloration, premature foliage drop (premature abscission), distorted growth, or reduced growth in foliage following the year of damage. Others can be observed through microscopes, i.e., physical changes of epicuticular structures and stomatal structures.
6. Pollutant damage may not be visible to the eye, macroscopically or microscopically. Yet changes could be quite apparent by way of physiological investigations, such as changes in metabolic rates which—in total—affect net CO_2 assimilation and, therefore, productivity (Smith, 1990).

This chapter focuses on quantifying the effects of SO_2 absorption on net assimilation by considering:

1. Boundary layer conductance models that can be used to estimate atmosphere-foliage gas exchange rates
2. Stomatal conductance and stomatal control theories that can be related to SO_2 effects observations
3. CO_2 assimilation theories that can be related to SO_2 effects observations.

In any of this, many complicating factors play a role. For example,

1. Leaf-external conditions (e.g., air temperature, solar radiation, atmospheric CO_2 concentrations, relative humidity, soil water potential)
2. Leaf-internal conditions (e.g., plant water potential, amount of growth regulators; sink strength for CO_2 assimilation) and
3. Leaf type (e.g., sunlit versus shaded leaves, position in canopy, leaf age)

All these factors affect boundary conductance, stomatal conductance, and net CO_2 assimilation. In turn, these factors are affected by weather conditions, time of day, and day in season. Furthermore, SO_2 exposure needs to be quantified by in-air concentrations, duration, and timing and frequency of recurring fumigating events.

Foliage reactions differ according to exposure categories, i.e., short-time exposure to low SO_2 concentrations, short-time exposure to high SO_2 concentrations, and long-time exposure to low SO_2 concentrations. For example, exposure to high SO_2 concentrations reduces net photosynthesis in a short time (Smith, 1990). Thresholds for SO_2-caused inhibition of net photosynthesis are dependent on species and exposure time (Umbach and Davis, 1984; Farooq *et al.*, 1988). For example, net photosynthesis of sugar maple (*Acer saccharum* Marsh.) seedling is inhibited when exposed to 1 ppm SO_2 for 4 to 6 hours (Keller, 1978). Foliage exposure at low SO_2 concentration but for long periods of time may inhibit net photosynthesis. For example, 0.05 to 0.2 ppm SO_2 cause substantial reductions in net photosynthesis of Norway spruce and silver fir (*Abies alba* Mill.) at the end of a 60 day experiment (Keller, 1978).

Documented responses of stomates to SO_2 exposure vary from increasing stomatal conductance, through no response, to decreasing stomatal conductance. At low SO_2 concentration exposure lasting for a few hours, SO_2 causes stomates to open (Unsworth *et al.*, 1972; Black and Black, 1979; Black and Unsworth, 1980; Kropff, 1987; Darrall, 1989). Stomatal closure was reported when high or low SO_2 concentrations were applied for several weeks or longer (Kropff, 1987; Darrall, 1989). Water stress during and before SO_2 exposure may cause stomatal closure. Low stomatal conductance not only reduces water losses from leaves but also reduces SO_2 absorption (Atkinson and Winner, 1990), as some plants avoid damage by closing the stomates during SO_2 exposure.

Inside the leaves, the cells surrounding stomatal pores are the first to encounter atmospheric SO_2. Subsidiary cells are particularly vulnerable because guard cells are usually protected by a thin layer of wax, and dissolved SO_2 has to diffuse through subsidiary cells before entering the guard cells (Fig. 1). Other cells are also affected. This may lead to uncontrolled opening or closing of stomates.

Sometimes, separate stomate response stages are noticed with increasing SO_2 exposure. For example, Unsworth and Black (1981) reported three stages:

1. Reversible stomate opening at low dosage
2. Irreversible stomate opening at medium dosage
3. Irreversible stomate closure at high dosage.

In other research, Koch and Maier-Maercher (1986) identified three distinct SO_2 response types with Norway spruce (*Picea abies* (L.) Karst.):

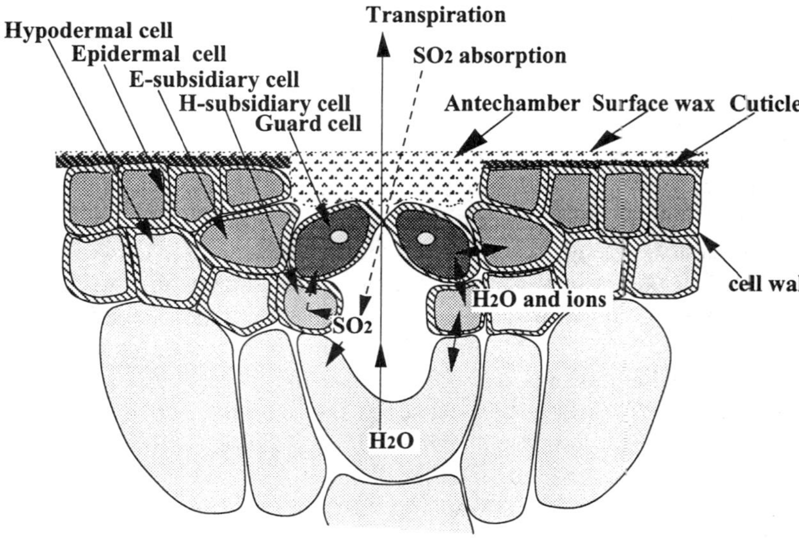

Figure 1 Stomatal complex SO_2 diffusion pathways (dashed lines) and water evaporation pathways (solid lines) of spruce foliage. The thick cuticle layer not only prevents water losses, but also reduces non-stomatal SO_2 absorption. Sulfur dioxide enters foliage through the stomatal pores and gets into guard cells through subsidiary cells. It is possible that subsidiary cells are damaged before guard cells are destroyed. (Drawing based on ideas of Black and Black, 1979; Black and Unsworth, 1980, Maier-Maercker and Koch, 1992b.)

1. An initial reduction in transpiration
2. A subsequent high but erratic transpiration
3. A final decrease in transpiration.

The length of time from phase 1 to phase 2 tends to decrease with increasing SO_2 concentrations. The final decrease tends to be accompanied by visible injuries.

To facilitate and simplify the analysis and the synthesis of these many complicating factors, we now proceed by adopting the following two principles:

1. Foliar responses are directly related to the amount of SO_2 taken up by the foliage
2. SO_2 is mainly taken up through the stomates.

From this, a general dose-response relationship at the foliage level can be formulated as follows:

$$\text{Net photosynthesis response} = f\left\{\int_{t_0}^{t} SO_2\_\text{uptake}(t)\, dt\right\} \tag{1}$$

where t is time, t_0 is the starting time of the fumigation, and $SO_2_$uptake(t) is the rate of SO_2 uptake at time t.

This chapter proceeds by presenting the ForNePS model that allows one to solve Equation (1) in principle. However, existing information about experiments done on SO_2 effects on net photosynthesis does not allow for a full evaluation of Equation (1) for these situations. Instead,

$$\text{Net photosynthesis response} = f\left[\int_{t_0}^{t} SO_2\_\text{concentration}(t)\, dt\right] \tag{2}$$

is used as a general guide, with $SO_2_$concentration(t) as the stated concentration of SO_2 at some time t. For constant SO_2 concentration experiments, this equation simplifies to:

$$\text{Net photosynthesis response} = f\,[SO_2\_\text{concentration} \times (t - t_0)] \tag{3}$$

Setting

Net photosynthesis response = relative change in net photosynthesis

and examining the literature in this way (Appendix) generates the cross-species data shown in Figure 2. Apparently, sensitivity to SO_2 exposure differs by species (Klein et al., 1978; Umbach and Davis, 1984; Darrall, 1986; Farooq et al., 1988; Smith, 1990). Figure 2 indicates that some of the plant species used in agriculture are most sensitive to SO_2 exposure. Among tree species, shade-intolerant hardwoods appear to be most sensitive to SO_2, followed by shade-tolerant hardwoods and firs. Spruce species appear to be fairly resistant to SO_2. The responses of pines to SO_2 appear to be highly variable, ranging from very sensitive (similar to shade-intolerant hardwoods) to relatively resistant. In many cases, the differences shown can be derived from differences in SO_2 absorption rate (Klein et al., 1978). Some insensitive species, however, are also shown to have a high H^+ buffer capacity and a high superoxide dismutase activity.

In summary, we suggest that a large part of the variations in Figure 2 is caused by differences in foliar SO_2 absorption, as affected by experimental conditions, and by species-specific stomate controls. Related differences would be due to differences in the following:

1. The leaf-boundary resistance to gas exchange as determined by leaf structure and rigidity
2. Rates of stomatal conductance

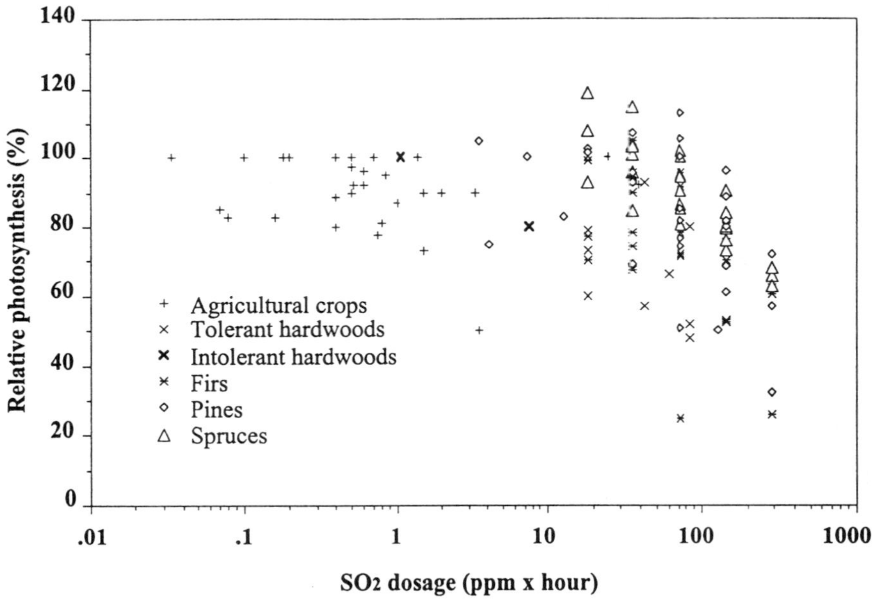

Figure 2 Relative net photosynthesis versus SO_2 exposure dosage.

3. Intercellular mechanisms by which plants detoxify SO_2 and its damage-causing by-products.

II. ForNePS MODEL

In order to monitor and model some of the many complications by which SO_2 absorption affects net photosynthesis, one must consider the following for each experimental situation:

1. Atmospheric hour-to-hour and day-to-day variations regarding air temperature, relative humidity, photosynthetically active and global solar radiation, atmospheric CO_2 and SO_2 levels
2. Soil water potential variations, to estimate the rate of leaf water replenishment in relation to evaporative losses.
3. Stomatal conductance in relation to all of the above, at the twig level.
4. Net CO_2 assimilation in relation to all of the above, at the twig level.

5. Integration of all of the above across the varying canopy conditions, from canopy top to canopy bottom.
6. Integration of the twig and canopy results from the beginning to the end of the growing season.

These considerations are presented in Figure 3 as a flowchart, which, in turn, also presents the basis and an overview of the ForNePS modeling effort.

A. Stomatal Conductance

As seen from the above and the Appendices, there are many conflicting observations and alternative theories about stomatal control mechanisms. A few guidelines, however, have some general validity for the modeling purpose. For example, stomates tend to remain closed in dry conditions and when leaf water potentials are low. In contrast, stomates tend to open in response to increasing levels of short-wave radiation and relative hu-

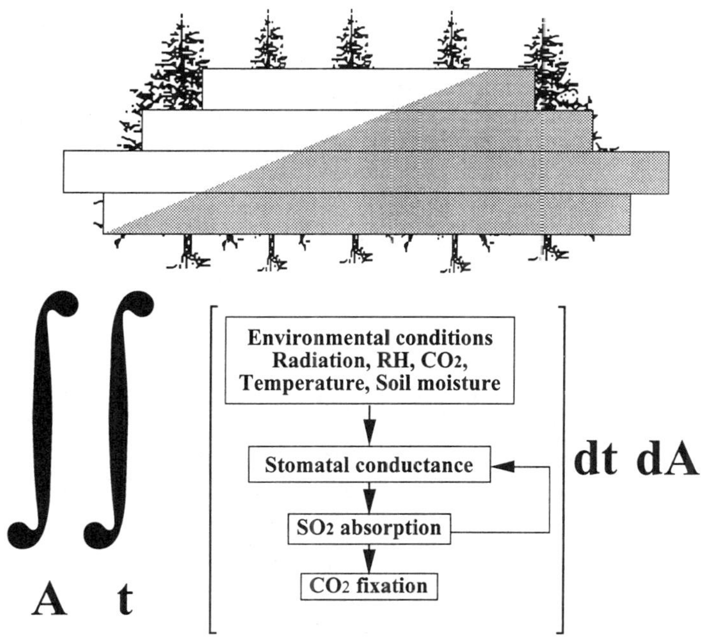

Figure 3 Structure of the integrative ForNePS model, with its flow of logic between its submodules. The diagram on top shows the model representation of the forest canopy: four leaf layers, with increasing shading from top to bottom.

midity, and decreasing intercellular CO_2 concentration (Jarvis and Morison, 1981). For our model, we favor the following theory: stomates tend to open whenever CO_2 fixation occurs and close when CO_2 fixation stops—i.e., stomatal movements are strongly synchronized with photosynthesis and likely follow the optimization principle of maximal photosynthesis at minimal water loss (Farquhar, 1979; Eamus, 1991). Formally, Farquhar and Sharkey (1982) have quantified this notion in terms of the "water use economic index," which expresses transpiration as an expense and photosynthesis as a gain. Symbolically, the water use economic index is given by

$$\lambda = \frac{\Delta E}{\Delta A} \tag{4}$$

where

ΔE = the rate of transpiration (in $\mu mol\ m^{-2}\ s^{-1}$)
ΔA = the rate of net CO_2 assimilation (also in $\mu mol\ m^{-2}\ s^{-1}$)

At the foliage level, gas movements (water vapor, CO_2, SO_2) follow Fick's diffusion law (Thornley, 1976; Thornley and Johnson 1990). Hence, net photosynthesis (defined as net CO_2 uptake through stomates) is determined by (1) the CO_2 concentration gradient that exists across guard cells and (2) stomatal conductance. The rate of net photosynthesis (A) and transpiration (E) can then be calculated from the water-vapor differences across the stomates (Farquhar and Sharkey, 1982), i.e.,

$$A = \beta g(C_a - C_i) - \frac{C_a + C_i}{2} E, \text{ and} \tag{5}$$

$$E = \frac{g \cdot \Delta w_{bs}}{P - (e_i + e_a)/2} = g \Delta w_{bs}/M, \tag{6}$$

with

g = the stomatal conductance to water vapor ($mol\ m^{-2}\ s^{-1}$)
M = $P- (e_i + e_a)/2$
C_a = the atmospheric CO_2 concentration (ppm)
C_i = the intercellular CO_2 concentration (ppm)
e_i = the intercellular water-vapor pressure (mb)
e_a = the atmospheric water-vapor pressure (mb)
β = the diffusivity ratio between CO_2 and H_2O
Δw_{bs} = the water-vapor pressure deficiency between inside and outside of stomates, i.e., $e_a - e_i$

Differentiating λ with respect to A generates a quantitative expression for stomatal conductance, namely:

$$g = \frac{\lambda \cdot [\beta M \Delta C_{ai}/\Delta w_{bs} - (C_i + C_a)/2] - 1}{\beta + \Delta w_{bs}/2M} \cdot \frac{\delta A}{\delta Ci} \tag{7}$$

where

$$\frac{\delta A}{\delta Ci} = \text{the gain in photosynthesis rate with increasing } C_i$$

According to this equation, stomatal conductance should (1) increase with increasing values of λ, and (2) decrease/increase with decreasing/increasing water vapor pressure differences across the stomatal complex. Stomatal conductance should also increase with increasing CO_2 concentration differences across the stomates between the leaf exterior (atmosphere) and the leaf interior, i.e. $\Delta C_{ai} = C_a - C_i$. In humid conditions, stomatal conductance is likely to be limited by light intensity (Dubbe et al., 1978; Gross et al., 1991; Lawlor and Mitchell, 1991).

In this chapter, it is assumed that light-limited stomatal conductance follows the photosynthesis light response pattern(A_{rel}), given by

$$g^o = A_{rel}\, g^o_{max} \tag{8}$$

where g^o_{max} = the maximum stomatal conductance observed in the field.

In general, stomatal conductance is not substantially affected by leaf water potential until this potential drops below a sensitive threshold value. After that, stomatal conductance should decrease rapidly with decreasing leaf water potential, and a linear model can be used to describe this relationship (Baldocchi et al., 1987; Bache, 1986). We propose that the water use economic index follows the same relationship, i.e.,

$$\lambda = \begin{cases} \lambda_o & \text{for } \psi_{leaf} > \psi_{sens} \\ \lambda_o \dfrac{(\psi_{leaf} - \psi_{clos})}{(\psi_{sens} - \psi_{clos})} & \text{for } \psi_{leaf} \quad \psi_{sens} \end{cases} \tag{9}$$

where

ψ_{leaf} = the leaf water potential when soil plant water movements are in steady state (= $\psi_{soil} - E/g_{sl}$; MPa)
ψ_{sens} = the sensitive water potential (= −2.1 MPa, for spruce)
ψ_{clos} = the water potential that causes total stomatal closure (= −4.5MPa)
λ_o = the species-specific water use economic index
g_{sl} = the soil-to-leaf water conductance (μmol $m^{-2}\, s^{-1}$ Mpa^{-1})

Stomates can only respond to humidity changes only within the air close to the guard cells. At this location, the humidity is affected by transpiration and, to a large extent, by the boundary-layer conductance (Jarvis and McNaughton, 1986). Actual water-vapor deficiency across the stomatal guard cells can be calculated from

$$\Delta w_{bs} = \frac{g_b}{g_s + g_b} \cdot \Delta w_{as} \tag{10}$$

where Δw_{as} is the water vapor deficiency of the atmosphere [$= e_i(1 - RH)$; mb]. This equation indicates that the actual water-vapor-pressure differences across the guard cells should increase with increasing boundary-layer conductance. In turn, stomatal conductance should decrease with increasing boundary-layer conductance.

B. SO_2 Absorption

The rate of SO_2 absorption by the foliage is calculated from

$$SO_2^{abs} = [S_{ext} - S_{int}]g_{SO2} \tag{11}$$

where

S_{ext} = the instantaneous atmospheric SO_2 concentrations
S_{int} = the mesophyll SO_2 concentration
g_{SO2} = the rate of total SO_2 conductance from atmosphere to leaves

Mesophyllic environments are generally moist. Hence, the mesophyllic resistance to the uptake of a highly water-soluble gas such as SO_2 is low. Consequently, S_{int} should be negligible (see also Baldocchi et al., 1987). For this reason, only S_{ext} and g_{SO2} are needed to calculate SO_2 absorption.

Sulfur dioxide absorption is also controlled by boundary-layer conductance and stomatal conductance. In addition, SO_2 can also be absorbed through the cuticle, and cuticle SO_2 conductance is not always negligible (Baldocchi et al., 1987; Huttunen et al., 1989). Hence,

$$g_{SO2} = \frac{g_b(\beta_{SO2} \cdot g + g_c)}{\beta_{SO2} \cdot g + g_c + g_b} \tag{12}$$

where

g_c = the cuticle conductance for SO_2 [$= LAI/(5000 \cdot f_w)$]
f_w = the fractional foliage surface that is free of water

In summary, the effect of SO_2 on maximum stomatal conductance can be calculated as follows:

$$g^o_{max}(SO_2) = \frac{(1 - K_{CSO2}{*}ASO_2)\ PAR\ [PAR_o + 2.1\ J_{max}\ Kp_{SO2}\ ASO_2/PAR_o/f]}{2.1\ J_{max} + PAR} \times g^o_{max} \tag{13}$$

where K_{CSO2} is the slope of relative stomatal conductance rate versus cumulative SO_2 absorption (= -0.0114 ha kg^{-1}; see Meng, 1993).

C. CO_2 Fixation

In the literature, there are many carbon fixation models. Some of them are strictly mechanistic, and others are strictly empirical. Empirical models tend to be quite simple and use one or two predictor variables (PAR, CO_2) as input, with two to four adjustable parameters (Thornley, 1976). In contrast, mechanistic models are complex. For example, Hahn's model (1987) uses 33 state variables, one differential equation for each state variable, all reaction rates are determined by kinetics only, and all reactions that relate to photosynthesis are driven by simulated values of available ATP. This model was found to be numerically stable; it behaved consistently with general expectations; and it was able to explain observed photosynthetic responses at the cellular level. The integration of this model into the canopy level, however, would be difficult, and would generally be inaccurate due to many cumulative errors. We proceed by adopting the CO_2 fixation model by Farquhar et al. (1978), which is widely cited and accepted by many researchers (Farquhar et al., 1978; Farquhar and von Caemmerer, 1982; Farazdaghi and Edwards, 1988a,b). In this model, net photosynthesis is given by

$$A = A_g - R_d \tag{14}$$

with

A_g as gross photosynthesis,
R_d as dark respiration.

When photosynthesis is limited by RuBP regeneration, then the rate of gross photosynthesis (A_C) is given by

$$A_C = \frac{V_{cmax}\ (C_i - \Gamma^*)}{C_i + Kc + (2\Gamma^*/k)} \tag{15}$$

When the maximum carboxylation speed is limited by electron transport capacity (i.e., limited by available radiation for the production of NADPH and ATP), then the rate of photosynthesis (A_J) is given by

$$A_J = \frac{A_{rel}(C_i - \Gamma^*)}{(C_i + 2.33\ \Gamma^*)} \tag{16}$$

with

$$A_{rel} = \sum_{j=1}^{2}\sum_{i=1}^{4} \left[\frac{LAI_{ij}\ (J_{max}/4.5)\ PARij}{PAR_{ij} + 2.1\ J_{max}/f)\ LAI}\right]$$

as the average relative photosynthesis rate of the forest canopy. In this expression,

- i = 1, 2, 3, 4 = canopy layers
- j = 1, 2 = leaf classes representing sunlit and shaded leaves respectively
- LAI_{ij} and PAR_{ij} = leaf area and PAR in layer i and leaf class j
- LAI = the leaf area index of the stand
- Γ^* = the CO_2 compensation point without day respiration = $O_i\ k\ K_c/(2K_o)]$
- k = V_{omax}/V_{cmax} (= 0.23, dimensionless)
- J_{max} = the potential rate of electron transport including pseudo cyclic electron transport to O_2(= 210 μmol m^{-2} s^{-1})
- K_c = the Michaelis constant for CO_2 (= 460 ppm)
- K_o = the Michaelis constant for O_2 (= 330 mmol mol^{-1})
- f = the fraction of light not absorbed by functional photosynthetic pigments (= 0.23, dimensionless)
- R_d = the rate of dark respiration (= 1.1 μmol m^{-2} s^{-1})
- C_i = the intercellular CO_2 concentration (ppm)
- O_i = the intercellular O_2 concentration (ppm)
- V_{cmax} = the maximum carboxylation velocity at C_i saturation (= 98 μmol m^{-2} s^{-1})
- V_{omax} = the maximum oxygen fixation velocity at O_i saturation.

Altogether, gross photosynthesis rate (A_g) is either limited by the electron transport capacity of the phtosynthesizing cells, or by the ribulose 1,5-bisphosphate regeneration within these cells, depending on environmental conditions. Hence, the actual rate of photosynthesis is given by

$$A_g = \min\ (A_J, A_C). \tag{17}$$

We further note that the gain in photosynthesis rate with increasing C_i is given by

$$\frac{\delta A}{\delta C_i} = \frac{A\text{›}15\text{rel}\ (3.33\ \Gamma^*)}{(C_i + 2.33\ \Gamma^*)^2} \qquad \text{when } A_J \geq A_C \tag{18a}$$

and

$$\frac{\delta A}{\delta C_i} = \frac{V_{cmax}\ [Kc(1 + O_i/K_o) + \Gamma^*]}{[C_i + Kc(1 + O_i/K_o)]^2} \qquad \text{when } A_J < A_C \tag{18b}$$

D. Temperature Effects

In principle, numerical values for the various parameters (V_{cmax}, Γ^*, E_{Rd}, K_o, Kc) in all of the foregoing equations depend on enzyme reactions which, in turn, are affected by leaf temperature T_{leaf} (°C). The general influence of temperature on the rate of enzyme reactions is given by the following equation (Farquhar and von Caemmerer, 1982):

$$Kt = \left[\frac{T_{leaf}}{298}\right]^{0.5} \exp\left[\frac{Ek(T_{leaf} - 25)}{298(T_{leaf} + 273)R_{gas}}\right] \tag{19}$$

where

E_K = the activation energy for enzymes
E_{Kc} = 65800; E_{Ko} = 1400; E_{Rd} = 66405; E_{Vcmax} = 6800 J mol^{-1}
R_{gas} = the gas constant (= 8.314 J mol^{-1} K^{-1}).

All of the foregoing synthesis suggests that maximum photosynthesis rates should be proportional to either the potential rate of electron transport J_{max} or the maximum carboxylation capacity V_{cmax}.

E. SO_2 Effects

We now assume that maximum rates of net photosynthesis and stomatal conductance decrease linearly with cumulative SO_2 absorption, i.e.

$$A/A_o = (1 - K_{PSO2}\ ASO_2), \tag{20}$$

where A_o is the rate of net photosynthesis and stomatal conductance at reference conditions (Meng and Arp, 1993). The corresponding effects of SO_2 absorption on electron-transport capacity and the maximum carboxylation velocity are given by:

$$J_{max}(SO_2) = PAR_o\ J_{max}(1 - K_{PSO2}\ ASO_2)/[PAR_o + 2.1\ J_{max}\ K_{PSO2}\ ASO_2/f] \tag{21}$$

and

$$V_{cmax}(SO_2) = V_{cmax}(1 - K_{PSO2}\ ASO_2) \tag{22}$$

where

K_{PSO2} = the slope of relative net photosynthesis rate versus cumulative SO_2 absorption (= −0.0348, ha kg^{-1})
ASO_2 = the cumulative foliar SO_2 absorption (kg ha^{-1})
PAR_o = the saturation photosynthetically active radiation (= 1800 μmol m^{-2} s^{-1}).

Some enzymes responsible for dark respiration are inhibited under light when foliage switches to photosynthesis. Dark respiration, however, is unlikely to be totally suppressed during the day, but estimates for foliage respiration cannot be separated from net photosynthesis and photorespiration. Following Farquhar et al. (1978), we assume that dark respiration in the light is one half of the dark respiration without light.

The effect of SO_2 on dark respiration is poorly understood (Meng, 1994). In our studies, the rate of dark respiration of both current-year needles and previous-year needles was found to be 40 to 50% higher on high SO_2 deposition sites than on low SO_2 deposition sites (Meng and Arp, 1993). Because dark respiration did not increase beyond 50% with one more year of SO_2 exposure, we propose that (1) dark respiration increases with SO_2 absorption up to 50% and (2) dark respiration is maintained at that level at higher SO_2 dosage, i.e.,

$$R_{SO2} = \left[1.5 - \frac{1}{2}\exp(-K_{RSO2}\ ASO_2)\right] R_d \tag{23}$$

where

R_{SO2} = the rate of SO_2 affected dark respiration (μmol m^{-2} s^{-1})
K_{RSO2} = an adjustable parameter (0.2 ha kg^{-1}).

F. Model Sensitivity to Model Parameters

Most of the ForNePS parameters were obtained from published literature (see Table 1). Model sensitivities regarding stomatal conductance, net photosynthesis, intercellular CO_2 concentration, and leaf water potential regarding some of the crucial and unavailable parameters such as J_{max}, λ_o, g_{sl}, and V_{cmax} are shown in Figure 4. For example, it was found that:

1. Changing the maximum electron transport capacity parameter J_{max} has a larger impact on net photosynthesis than on stomatal conductance.

Table 1 Fixed ForNePS Parameters and References

Parameter	Ref.	Description and default value
β	(5)	Diffusivity ratio between CO_2 and H_2O (= 0.63, dimensionless)
β_{SO2}	(1)	Diffusivity ratio between SO_2 and H_2O (= 0.52, dimensionless)
ψ_{sens}	(1)	Sensitive foliage water potential (= 2.1 MPa)
ψ_{clos}	(4)	Foliage water potential that causes stomatal closure (= 4.5 MPa)
E_{Kc}	(3)	Activation energy for enzymes for Kc (= 6580 J mol^{-1})
E_{Ko}	(3)	Activation energy for enzymes for Ko (= 1400 J mol^{-1})
E_{Rd}	(3)	Activation energy for enzymes for Rd (= 66405 J mol^{-1})
E_{Vcmax}	(3)	Activation energy for enzymes for Vcmax (= 6800 J mol^{-1})
f	(3)	Fraction of light not absorbed by functional photosynthetic pigments (= 0.23, dimensionless)
k	(3)	Ratio of CO_2 and O_2 fixation velocity of Rubsco (= 0.23, dimensionless)
V_{cmax}	(2)	Maximum carboxylation velocity at Ci saturation (= 98 μmol m^{-2} s^{-1})
J_{max}	(2)	Potential rate of electron transport (= 210 μmol m^{-2} s^{-1})
K_c	(3)	Michaelis constant for CO_2 (= 460 ppm)
K_o	(3)	Michaelis constant for O_2 (= 330 mmol mol^{-1})
K_{CSO2}	(6)	Slope of relative stomatal conductance rate versus cumulative SO_2 absorption (= −0.0114 ha kg^{-1})
K_{PSO2}	(6)	(slope of relative net photosynthesis rate versus cumulative SO_2 absorption = −0.0348 ha kg^{-1})
K_{RSO2}	(4)	rate of SO_2-affected dark respiration (= 0.2 ha kg^{-1})

1. Baldocchi et al., 1987.
2. Farquhar, 1979.
3. Gross et al., 1991.
4. Meng, 1994.
5. Thornley and Johnson, 1990.
6. Meng et al., 1994.

2. Changing λ_o (the water use economic index) has a large impact on stomatal conductance, but a small impact on net photosynthesis.
3. The rate of net photosynthesis is not sensitive to changing values of the soil-leaf water conductance g_{sl}, unless g_{sl} is very small (zero). In contrast, stomatal conductance is very sensitive to g_{sl}.

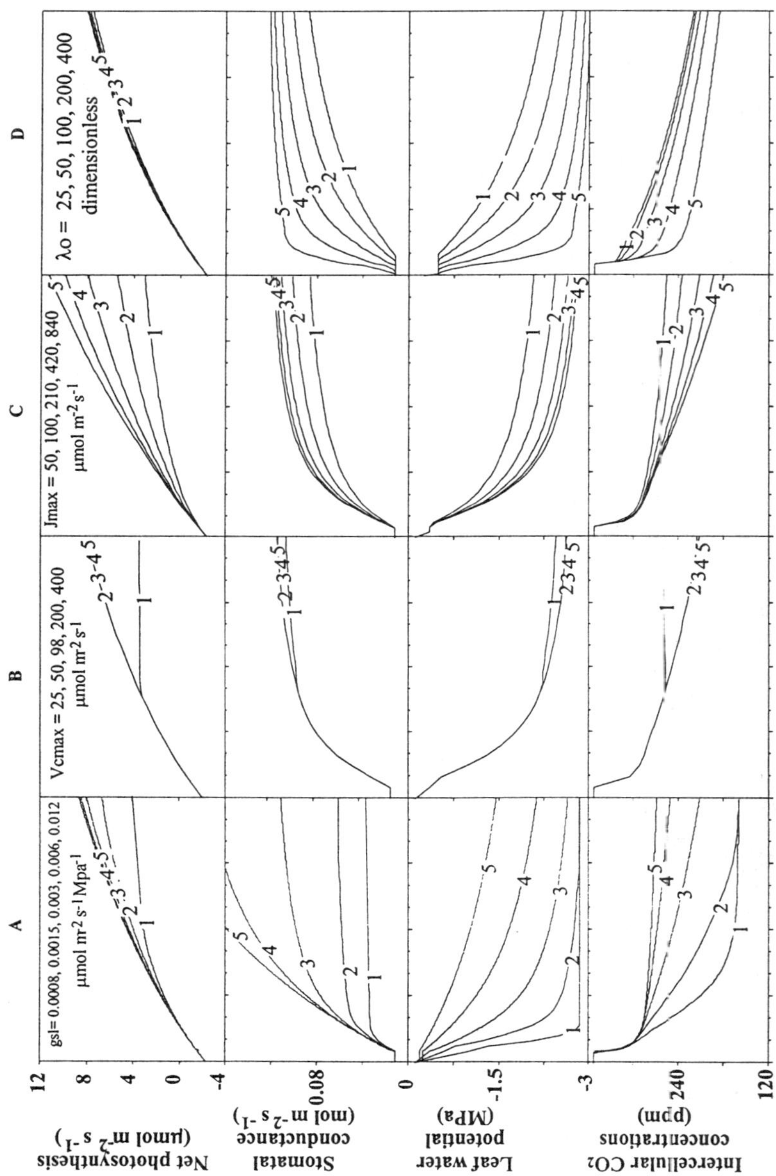

Figure 4 Sensitivity of net photosynthesis model to model parameters g_{sl} (part A), V_{cmax} (part B), J_{max} (part C) and λ_o (part D). Environmental variables were fixed at Tleaf = 20°C; RH = 50%; soil moisture content = 80 mm. Parameters other than analyzed were fixed. Line numbers correspond to parameter values from low to high.

4. The model simulations are not sensitive to V_{cmax} unless this parameter is smaller than one-quarter of the recommended value, and when the set rates for global radiation are high.

G. Stomatal Conductance and Net Photosynthesis Simulations in Relation to Changing Atmospheric Conditions and Soil Moisture

Model responses to global radiation at different settings for leaf temperature and model responses to leaf temperatures at different settings for global radiation are plotted in Figure 5, Parts A and B. In general, the model responds to global radiation by following the typical photosynthesis light-response curve (Thornley, 1976; Thornley and Johnson, 1990). As shown, the optimum temperatures for stomatal conductance and net photosynthesis increase with increasing global radiation. At constant relative humidity, simulated values for leaf water potential decrease with increasing temperature and radiation. This is because water vapor pressure differences across the stomates should increase with increasing temperature, and excessive water losses from the leaves should reduce the leaf water potential. At temperatures above 35°C, CO_2 fixation should become smaller than respiration, i.e., net photosynthesis should become negative. The maximum temperature for net photosynthesis to remain positive should increase with increasing radiation.

Model responses to radiation at different settings for relative humidity and soil moisture are shown in Figure 5, Parts C and D. Evidently, changing these variables should have a large impact on stomatal conductance. In contrast, the impact of humidity and soil moisture on net photosynthesis would be moderate.

Model responses in terms of net photosynthesis to degree of cloudiness [approximated by (day length-sunshine hour)/day length] are shown in Figure 6. In particular, rate of net photosynthesis on sunny, cloudy, and overcast days in relation to maximum rate of net photosynthesis are plotted in Figure 6A. Relative photosynthesis by eight different leaf classes (four layers with shaded and sunlit leaves) are presented in Figure 6B. As shown, the relative rate of net photosynthesis of shaded leaves should increase with increasing cloudiness. In contrast, the relative rate of net photosynthesis of sunlit leaves should decrease with increasing cloudiness. The proportion of leaves that would remain sunlit in each of the four canopy layers is plotted in Figure 6B. As shown, cloudy conditions would not change the proportion of leaves that remain sunlit, but time of the day (or zenith angle) would have a strong effect.

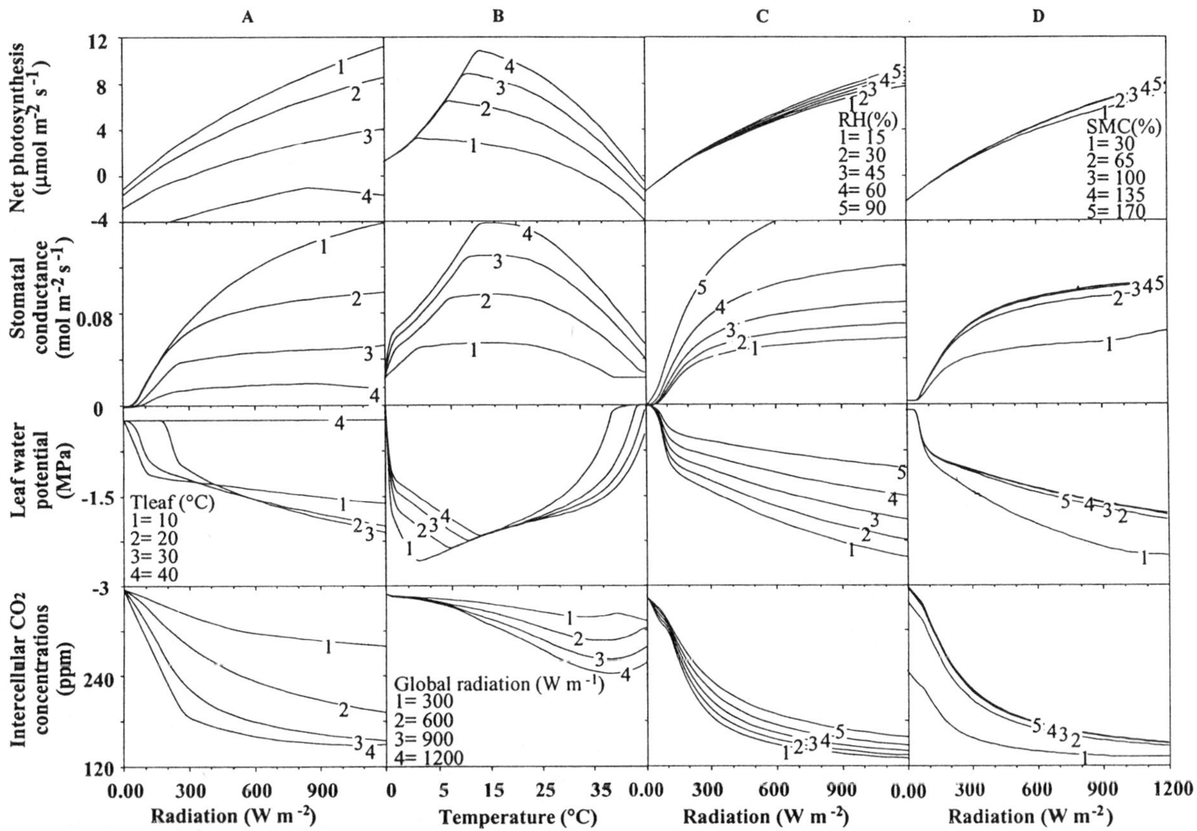

Figure 5 Simulated rates of net photosynthesis, stomatal conductance, intercellular CO_2 concentrations and leaf water potential in response to global radiation (part A), leaf temperature (part B), relative humidity (part C) and soil moisture (part D). Unless specified otherwise, variables were fixed at RH = 50% and soil moisture = 80 mm.

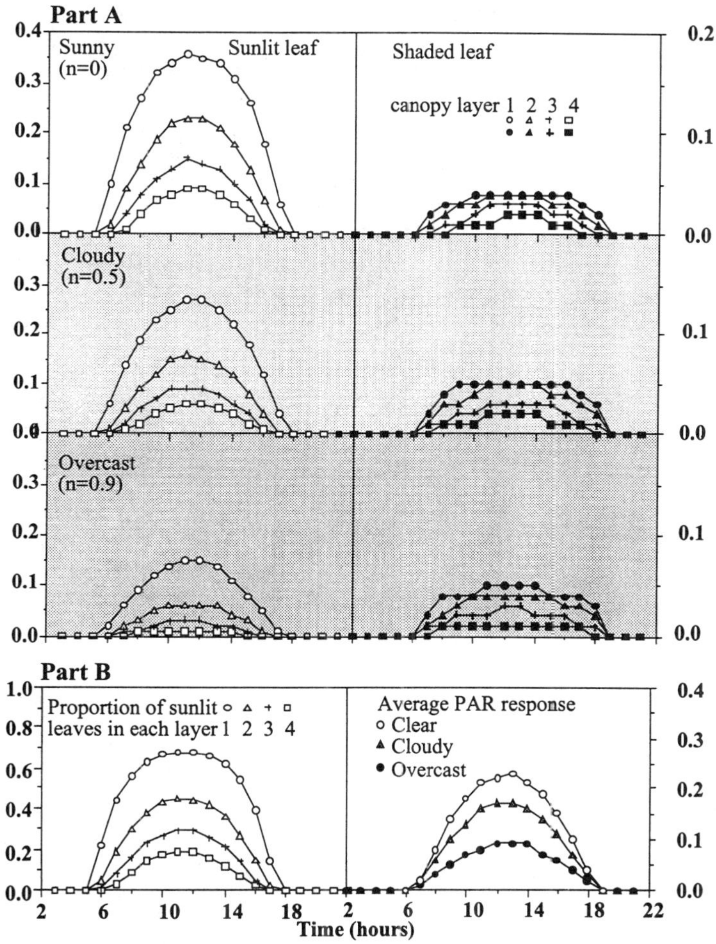

Figure 6 Effects of cloudiness on canopy photosynthesis. (a) Relative net photosynthesis rate of different layer of canopy versus time of day in response to variable cloud conditions. (b) Proportion of sunlit leaves for each canopy layer (1 to 4, left) and average relative net photosynthesis rates of the canopy for various cloudy conditions (right).

H. SO_2 Effects Calibrations

Parameters that deal specifically with the influence of SO_2 on stomatal conductance and net photosynthesis (i.e., Kp_{SO2}, K_{CSO2}, and K_{RSO2}) need to be obtained by model calibration, using data that are specifically generated for this purpose. For example, we calibrated Kp_{SO2}, K_{CSO2}, and K_{RSO2} for red/black spruce foliage with net photosynthesis and stomatal conductance data generated by SO_2-fumigating spruce foliage on mature spruce trees by way of a branch chamber experiment (Meng and Arp, 1993). These trees were about 50 years old, and 15 to 18 m tall. Four SO_2 concentration treatments (0, 0.1, 0.2, 0.4 ppm) were delivered to four branch chambers. Each treatment was replicated on three trees. Concentrations of SO_2 were doubled near the end of the experiment. Rates of net photosynthesis and stomatal conductance were measured with detached shoots with a Li-Cor 6200 photosynthesis system. Measured rates were standardized with nonlinear regression models (Meng and Arp, 1993). Relative net photosynthesis and stomatal conductance were obtained by dividing measured rates with control. Parameters Kp_{SO2}, K_{CSO2} were then obtained by regressing relative net photosynthesis and stomatal conductance against simulated SO_2 absorption. The related plots are shown in Figure 7. Parameter K_{RSO2}

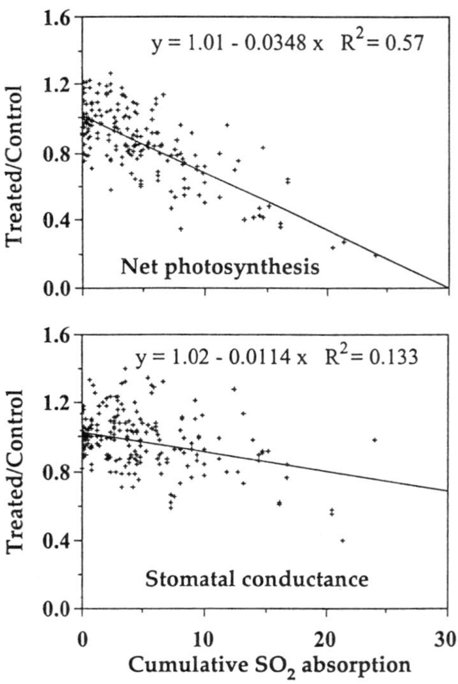

Figure 7 Relative net photosynthesis and relative stomatal conductance of red spruce foliage varies with cumulative SO_2 absorption.

was obtained by fine-tuning the model simulations towards the actual determinations. Calibrated parameter values are listed in Table 2.

To illustrate how model simulations compare with experimental results, we plotted observations and best-fitted simulations for net photosynthesis and stomatal conductance of spruce foliage in Figure 8. As shown, the simulations with the SO_2 treatment dropped gradually with time of treatment, and this drop accelerated with increased SO_2 dosage. The actual observations were much more variable than the simulations, but the average trends were in reasonable agreement, thereby confirming the model calibration and the numerical choices for the model parameters, as listed in Tables 1 and 2.

I. Model Performance Check: Case Study

The performance of the model was further checked by comparing ForNePS generated values for the SO_2-influenced net photosynthesis response of twigs on mature red/black spruce trees with actual observations. The area selected for this study is downwind from the Grand Lake power generator in central New Brunswick. This power generator has been active since 1930. Its capacity for generating electrical energy by burning locally mined coal (S content between 7 and 8%) was upgraded in 1960. Since then, SO_2 emission rates have been fairly steady at about 30,000 tons per year.

Midsummer rates of net photosynthesis and dark respiration were determined for the foliage of mature red spruce trees on eight uplands sites receiving about 2 or 50 kg ha^{-1} yr^{-1} of SO_2. Typical SO_2 deposition rates were estimated to vary from 2 (background) to 70 kg ha^{-1} yr^{-1} within a 40 km radius from the emission source (Bourque, 1992). Average SO_2 concentrations were estimated for the location of each measurement tree, and it was assumed that actual hourly SO_2 concentrations were exponentially distributed. Daytime averages were about 0.00, 0.010, 0.025, and 0.05 ppm, depending on tree location. Nighttime SO_2 concentrations were set at 0 ppm.

Hourly weather records (air temperature, global radiation, relative humidity, sunshine hours) were obtained from the nearby Fredericton area.

Table 2 Adjusted ForNePS Parameters

λ_o	Species-specific water use economic index (= 100, dimensionless)
g_{sl}	Soil to leaf water transport conductance (= 0.003 $\mu mol\ m^{-2}\ s^{-1}\ Mpa^{-1}$)

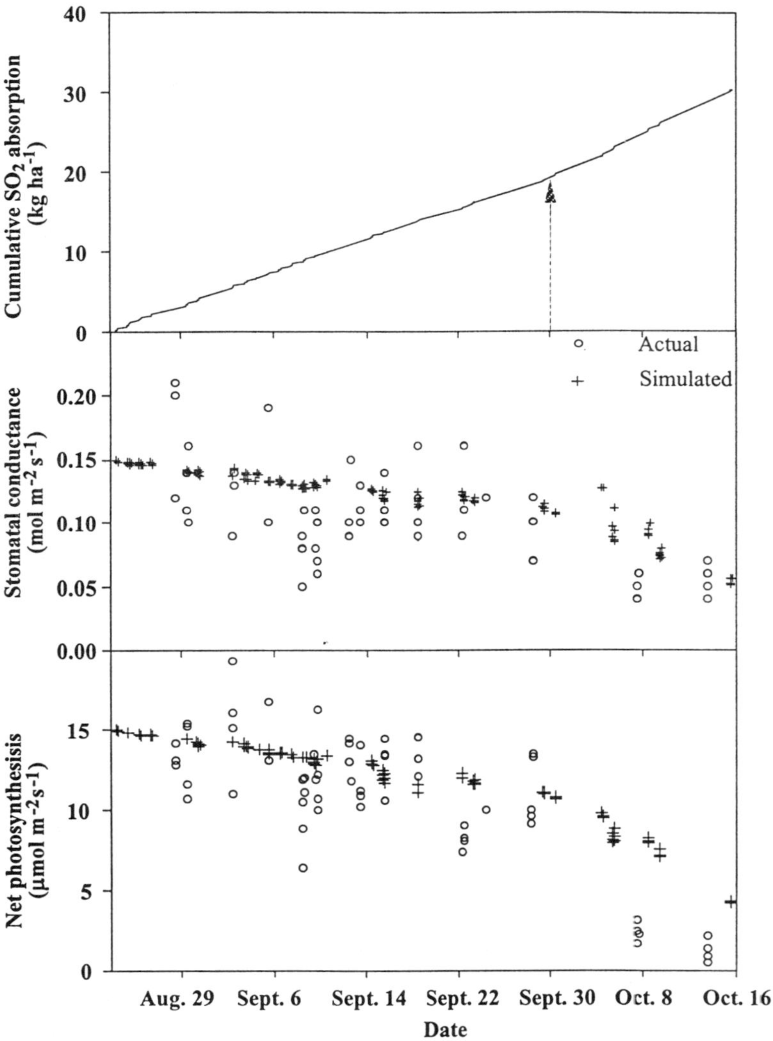

Figure 8 Simulated and actual rates of stomatal conductance and net photosynthesis of red spruce foliage and cumulative SO_2 absorption exposed to high SO_2 concentration. Data were obtained at or above saturation light intensity. Arrow indicated the time when SO_2 concentration was increased.

The clear-sky solar radiation (or global solar radiation) was calculated by specifying solar angle or zenith angle (Meng 1994). The resulting calculation for the direct beam radiation was obtained from Iqbal (1983), Spitters (1986), and Oke (1987)

Soil moisture contents in the soil below the measurement trees were simulated with the ForHyM hydrology model (Arp and Yin, 1992). This model uses air temperature, rain volume, and a basic description of soil physical properties (depth of forest floor, depth of rooting space within the mineral soil, clay content, and rooting habit of the vegetation) as input.

Model simulations for cumulative CO_2 fixation of the measurement trees are plotted versus cumulative SO_2 absorption Figure 9. In summary, these simulations take into account:

1. The local weather record and the corresponding simulations for soil moisture
2. The exponentially distributed SO_2 fumigation events (hourly basis), which range from 0 to 0.05 ppm on average
3. A spruce canopy, which is lumped into four leaf layers, and two foliage types in each layer (shaded versus sunlit); see Figure 3.

As shown in Figure 9; increasing cumulative SO_2 absorption decrease simulated rates for transpiration and cumulative CO_2 fixation. These plots also suggest that relative SO_2 effects should be stronger on CO_2 fixation than on transpiration. Consequently, stand water use efficiency (= cumulative CO_2 fixation/cumulative transpiration) should decrease with increasing SO_2 exposure. If this is true in general, then trees that are continuously exposed to SO_2 should become increasingly less productive with time, and they may become more vulnerable to drought stress (see also Maier-Maercker and Koch, 1992a,b).

Also shown in Figure 9 are cumulative SO_2 absorption by spruce foliage of the measurement trees at low and high SO_2 deposition areas. The corresponding plot in this figure suggests that the amount of SO_2 absorbed by the foliage has caused a small reduction in cumulative CO_2 fixation. This reduction, however, is likely not detectable by way of instantaneous photosynthesis measurements on account of high measurement variability (Fig. 7) and limited sample size.

Best-fitted regression curves and equations for relative cumulative CO_2 fixation, water loss, and water use efficiency with respect to cumulative SO_2 absorption are also shown in Figure 9. Knowing these regression equations, and knowing the annual SO_2 absorption map for this area (Bourque, 1992), it is possible to plot the net effect of the SO_2 emissions on cumulative CO_2 fixation in this area. The result is shown in Figure 10.

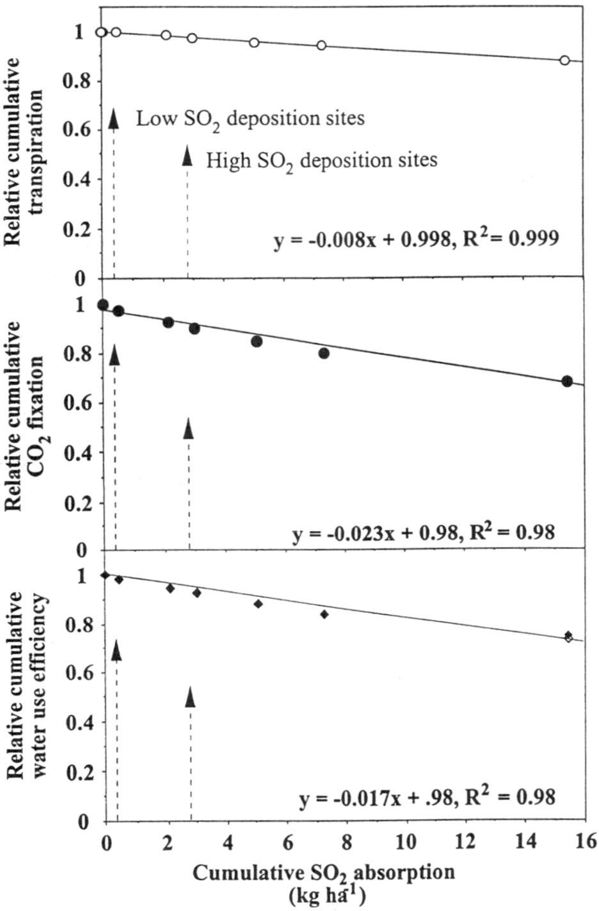

Figure 9 Relative total evapotranspiration, relative net CO_2 fixation, and relative water use efficiency versus cumulative SO_2 absorption at end of September 1991. Dashed arrows indicate the levels of foliar SO_2 absorption at high and low SO_2 absorption sites downwind from the Grand Lake power plant (see also Fig. 11).

According to this map, it is evident that current annual SO_2 deposition rate could reduce cumulative CO_2 fixation of red spruce foliage by as much as 7% from the end of April to the beginning of October. This reduction, however, would occur only in isolated spots near the power generator, and should decrease rapidly with increasing distance from the generator. Locations of high and low spots are a function of actual SO_2 emission rates as well as local meteorology and topography (Bourque, 1992).

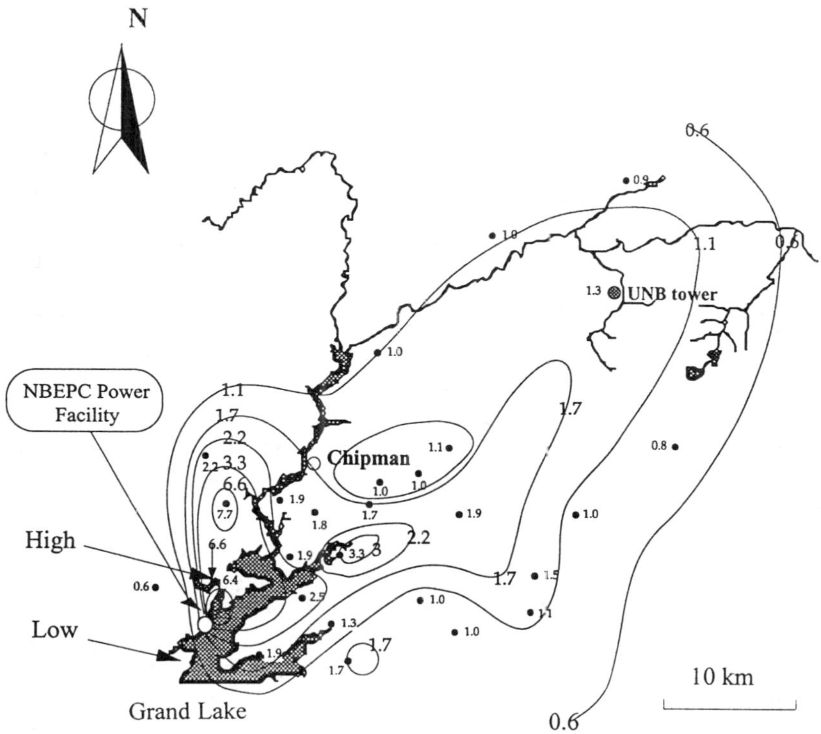

Figure 10 ForNePS model projection of relative reduction in percent of cumulative carbon fixation of red spruce foliage downwind from the Grand Lake power plant from May 1 to October 1, 1991, according to annual SO_2 absorption estimates (Bourque, 1992), and by assuming that half of annual amount of SO_2 absorbed by red spruce is taken up from May 1 to October 1.

III. OTHER STUDIES

Exposing plants to low SO_2 concentrations for a short time may stimulate net photosynthesis (Smith, 1990). For example, Boyer et al., (1986) experimented with grafted eastern white pine (*Pinus strobus* L.) clones which were classified either as sensitive or tolerant to SO_2. The grafts were cultivated in a greenhouse and were grown in clean air. Both clones were fumigated with 0.05 ppm SO_2 for 3 days. It was found that SO_2 slightly increased net photosynthesis on all three tolerant clones (Fig. 11). This effect cannot be simulated with ForNePS directly, because direct effects of SO_2 stomatal conductance and net photosynthesis are prescribed to be negative (other factors being the same). Indirectly, however, positive effects

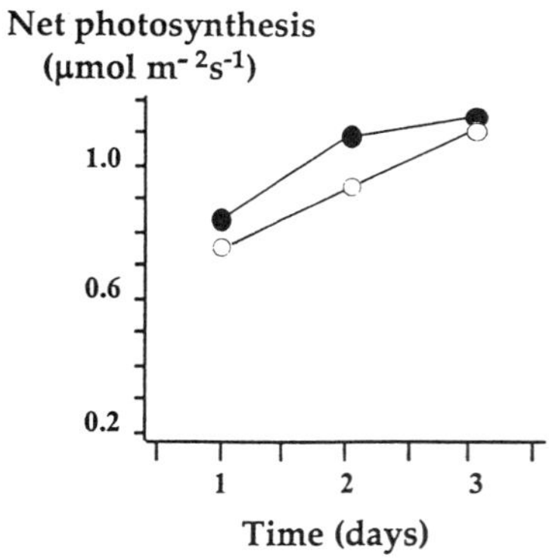

Figure 11 Net photosynthesis for sensitive and tolerant eastern white pine clones over time; in response to 0.05 ppm SO_2 for 6 h/d for 3 days. Plants were monitored in clean air on day 1 and in polluted air on days 2 and 3. [(●) sensitive, (○) tolerant] (From Boyer et al., 1986.)

can be simulated to occur by way of forcing special correlation among the environmental variations, e.g., correlated increases in CO_2, humidity, temperature, and SO_2 could generate overall increases in net photosynthesis.

Other experiments expose Norway spruce and silver fir seedlings with long-term (5 years) low concentration SO_2(0.0 to 0.12 ppm; Seufert et al., 1990). No visible injury was found on spruce and fir needles by the end of the experiment. Rates of net photosynthesis and transpiration were measured in clean air with cut branches after 5 years' exposure. No differences in the rates of net photosynthesis or transpiration were observed on SO_2-treated spruce foliage, but the amount of pigments decreased with increasing SO_2 exposure. The amount of pigments also decreased with increasing needle age, but the chlorophyll ratio was not changed by the SO_2 exposure (Siefermann-Harms, 1990). In the case of silver fir, SO_2 alone reduced the rates of net photosynthesis and transpiration. Both light and dark reactions were depressed by SO_2 (Schweizer and Arndt, 1990). Since the actual rates of SO_2 absorption by these trees cannot be calculated (insufficient data), it is difficult to compare these observations with those that could be simulated with the ForNePS model.

A. Effects of Light Intensity and Relative Humidity

We checked several studies that contain information about the effects of SO_2 on net photosynthesis under varying light intensity and humidity conditions (Darrall, 1989). The result were, at least qualitatively, in agreement with what one can simulate with ForNePS in general. Data related to SO_2 effects alone, however, appear to be limited. In one study, Carlson (1979) experimented with black oak (*Quercus velutina* Lam.), sugar maple, and white ash (*Fraxinus americana* L.) to study the interactive effects between SO_2 exposure, O_3 exposure, light intensity, and relative humidity. Three light intensity levels (low, intermediate, high) and two relative humidity levels were applied to branches on sugar maple and white ash. All trees (one oak, two sugar maple, and four white ash trees) were exposed to 0.5 ppm of SO_2 plus 0.5 ppm of O_3 for 8 h/d. The results indicated that the reduction of net photosynthesis of plants grown under intermediate light conditions was much less than that of plants grown in low and high light intensity (Fig. 12).

In Carlson's experiment, visible SO_2 damage of white ash increased with increased light intensity at high relative humidity (Fig. 13), but the damage reported may or may not have been caused by SO_2 alone. There are records that ozone causes extensive damage to leaves of some agriculture plants under low light intensity (Heck and Dunning, 1967). The possibility that SO_2 also causes increased damage at low light intensity cannot be ruled out (Mansfield and Jones, 1985).

McLeod (1995) and his colleagues exposed 4-to 5-year-old Norway spruce trees to SO_2 and O_3 with an open-air fumigation system for up to

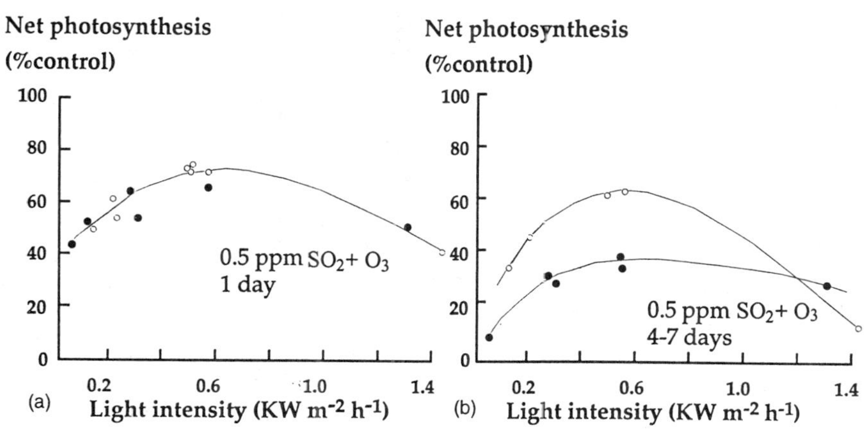

Figure 12 Effect of fumigation with 0.5 ppm of SO_2 plus 0.5 ppm O_3 on net photosynthesis of sugar maple (○) and white ash (●) at different intensities (a) after 9 h of fumigation on one day and (b) after 4 to 7 days of fumigation for 7–11 h/d. The trends are similar for the two species after one day of fumigation. (From Carlson, 1979.)

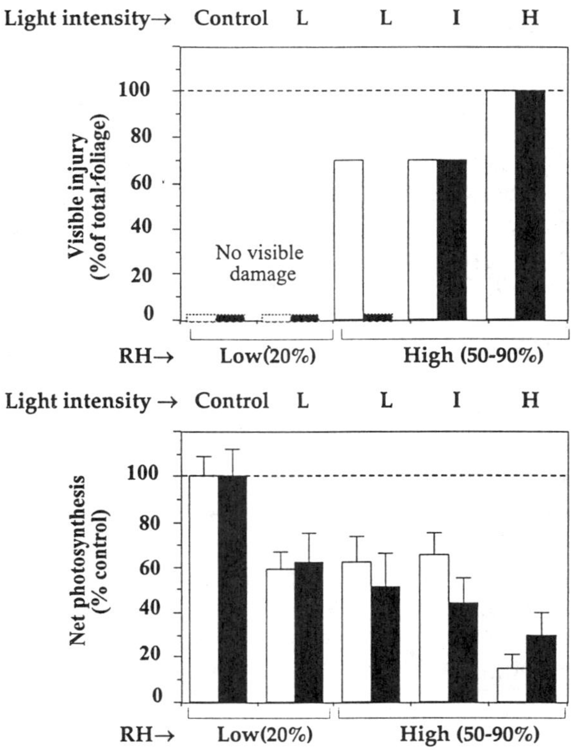

Figure 13 Visible damage and net photosynthesis represented by percentage of control for sugar maple (□) and white ash (■) at various light and humidity conditions. H = high; L = low; I = intermediate. Showing on top of each bar is the standard error of the mean. (Redrawn from Carlson, 1979.)

4 years. Wedler et al. (1995) found that long-term (4 years) low SO_2 concentration (13 and 22 ppb, long-term mean) depressed net photosynthesis under light saturation. They noticed that SO_2 exposure accelerated a nitrogen-related carboxylation efficiency.

B. Effects of CO_2 Concentrations

With ForNePS, we simulate that SO_2 affects net photosynthesis less at high than at low CO_2 concentrations (all other factors being the same). The study by Carlson and Bazzaz (1982) confirms this: these authors exposed three C_3 species and three C_4 species to 0.25 ppm of SO_2 at different CO_2 concentration. They found that C_3 plants were more sensitive to SO_2 exposure when grown with 300 ppm of CO_2 than when grown with 600 or 1200 ppm of CO_2. Leaf areas were much smaller in fumigated plants than in nonfumigated plants at 300 ppm CO_2. No leaf-area differences were found between fumigated and nonfumigated plants grown at high CO_2 concentrations. Water uptake of C_3 species during the fumigation was less with fumigated plants than with nonfumigated plants at low CO_2 concentration. Net photosynthesis was determined in clean air 5 days after completion of SO_2 exposure. At that time, net photosynthesis was slightly reduced within the fumigated plants.

C. Effects of Temperature

The effects of temperature on net photosynthesis response to SO_2 are complicated (Fig. 5), and data are few. Stomatal conductance and net photosynthesis have an optimum temperature. High stomatal conductance at optimum temperature tends to increase SO_2 absorption, as observed by Taylor et al. (1985), and as modeled with ForNePS.

D. Effects of Seasons

Conifer trees may be more sensitive to SO_2 air pollution stress during spring than summer (Smith, 1990). Keller (1977, 1978) fumigated three species with low SO_2 concentration ranging from 0.05 to 0.2 ppm for up to 60 days. The experiments were conducted in spring, summer, and fall. He found substantial reductions in net photosynthesis of silver fir and Norway spruce. The data in Figure 14 show that net photosynthesis was reduced with time and with increasing SO_2 concentrations. Short-time fumigations with low concentrations stimulated net photosynthesis. Large variations in the spring were likely caused by large environmental and leaf physiological variations, which can be partially traced back to stomatal conductance and SO_2 absorption. With ForNePS, we are able to simulate the effects of different weather conditions on SO_2 absorption and resulting net photosynthesis response. We assume that the leaf physiological conditions that affect SO_2 absorption remain the same from the beginning to the end of the growing season.

IV. CONCLUDING REMARKS

Predictions generated with ForNePS appear to be useful for evaluating the relative impact of SO_2 on cumulative carbon fixation and on net primary production in general. In this respect, lumping the canopy into four layers

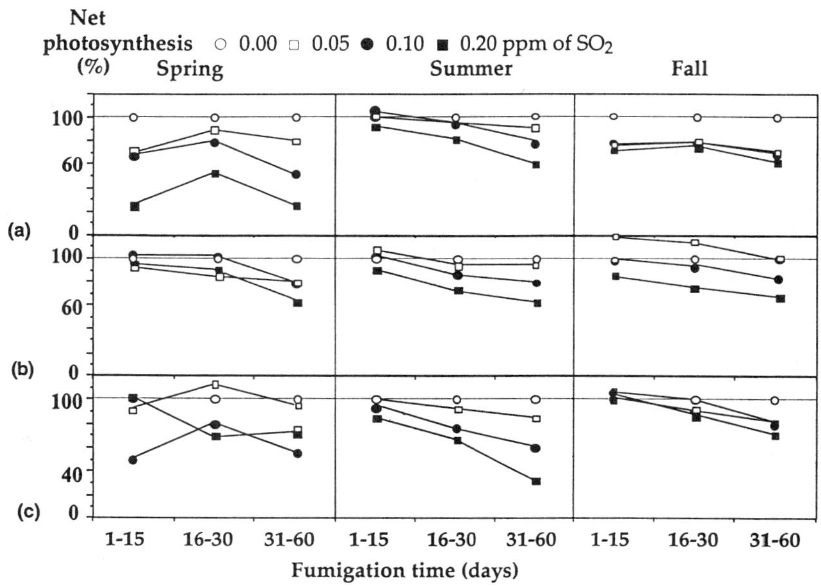

Figure 14 Relative net photosynthesis of three conifer species after 15, 30, and 60 days of SO_2 exposure (0.05, 0.10, and 0.20 ppm) in spring, summer, and fall. A, Silver fir; B, Norway spruce; and C, Scots pine. (Redrawn from Keller, 1977.)

with two classes (sunlit versus shaded leaves) is only a very modest admission of the complexities involved. On the whole, ForNePS treats the canopy as a black box but focuses on the quantification of light capture, SO_2 absorption, and net mass exchange of H_2O and CO_2 between the atmosphere and the canopy. With little modification, it should be possible to use the model and the accompanying procedures to evaluate net photosynthesis and stomatal conductance impacts on plants when subjected to stresses other than SO_2 (notably UV, ozone), or a combination of such stresses, including those related to site-specific changes regarding canopy temperature and soil moisture.

REFERENCES

Alscher, R. G., and Amthor, J. S. 1988. The physiology of free-radical scavenging: maintenance and repair processes. *In* Air pollution and plant metabolism. *Edited by* Schulte-Hostede S., Darrall, N. M., Blank, L. W., and Wellburn, A. R. Elsevier Applied Science. New York. pp. 94–115.

Alscher, R., 1985. Effects of SO_2 on light-modulated enzyme reactions. *In* Sulfur dioxide and vegetation. *Edited by* Winner, W. E., Mooney, H. A., and Goldstein, R. A. Stanford University Press, California. pp. 181–191.

Arp, P. A., and Yin, X. 1992. Predicting water fluxes through forests from monthly precipitation and mean monthly air temperature records. Can. J. For. Res. 22: 864–877.

Atkinson, C. J., and Winner, W.E. 1990. Interactions of SO_2 with other environmental leaf gas exchange. Oecologia 84:500–505.

Bache, D. H. 1986. On the theory of gaseous transport to plant canopies. Atmos. Environ. 20:1379–1388.

Baldocchi, D. D., Hicks, B. B., and Camara, P. 1987. A canopy stomatal resistance model for gaseous deposition to vegetated surfaces. Atmos. Environ. 21: 91–101.

Bender, J., Manderscheid, R., and Jäger, H. J. 1990. Analyses of enzyme activities and other metabolic criteria after five years of fumigation. Environ. Pollut. 68:331–345.

Black, C. R., and Black, V. J. 1979. The effects of low concentrations of sulphur dioxide on stomatal conductance and epidermal cell survival in field bean (*Vicia faba L.*). J. Exp. Bot. 30:291–298.

Black, V. J., and Unsworth, M. H. 1980. Stomatal responses to sulphur dioxide and vapour pressure deficit. J. Exp. Botany 31:667–677.

Black, V. J., Ormrod, D. P., and Unsworth, M. H. 1982. Effects of low concentrations of ozone, singly, and in combination with sulphur dioxide on net photosynthesis rates of *Vicia faba* L. J. Exp. Bot. 33:1302–1311.

Bourque, C. P. 1992. Sulfur dioxide absorption by spruce forests downwind from a coal-fired generator. Ph D. Thesis, University of New Brunswick, Fredericton, N. B. Canada.

Boyer, J. N., Houston, D. B., and Jensen, K. F. 1986. Impact of chronic SO_2, O_3 and SO_2 + O_3 exposures on photosynthesis of *Pinus strobus* clones. Eur. J. For. Path. 16:293–299.

Carlson, R. W. 1979. Reduction in the photosynthetic rate of *Acer*, *Quercus* and *Fraxinus* species caused by sulphur dioxide and ozone. Environ. Pollut. 18: 159–170.

Carlson, R. W. 1983. The effect of SO_2 on photosynthesis and leaf resistance at varying concentrations of CO_2. Environ. Pollut. (Ser. A) 30:309–321.

Carlson, R. W., and Bazzaz, F. A. 1982. Photosynthetic and growth response to fumigation with SO_2 at elevated CO_2 for C3 and C4 plants. Oecologia 54: 50–54.

Darrall, N. M. 1986. The sensitivity of net photosynthesis in several plant species to short term fumigation with sulphur dioxide. J. Exp. Bot. 37:1313–1322.

Darrall, N.M. 1989. The effect of air pollutants on physiological processes in plants. Plant Cell Environ. 12:1–30.

Dizengremel, P. and Citerne, A. 1988. Air pollutant effect on mitochondria and respiration. *In* Air pollution and plant metabolism. *Edited by* Schulte-Hostede

S., Darrall, N. M., Blank, L. W., and Wellburn, A. R. Elsevier Applied Science. New York. pp. 169–188.

Dubbe, D. R., Farquhar, G. D., and Raschke, K. 1978. Effect of abscisic acid on the gain of the feedback loop involving carbon dioxide and stomata. Plant. Physiol. 62:413–417.

Eamus, D. 1991. The interaction of rising CO_2 and temperature with water use efficiency. Plant Cell Environ. 14:843–852.

Farazdaghi, H., and Edwards, G. E. 1988*b*. A model for photosynthesis and photorespiration in C_3 plants based on the biochemistry and stoichiometry of the pathways. Plant Cell Environ. 11:799–809.

Farazdaghi, H., and Edwards, G. E. 1988*a*. A mechanistic model for photosynthesis based on the multisubstrate ordered reaction of ribulose 1,5 bisphosphate carboxylase. Plant Cell Environ. 11:789–798.

Farooq, M., Masood, A., and Beg, M. U. 1985. Effect of acute exposure of sulphur dioxide on the metabolism of *Holoptelea integrifolia* plants. Environ. Pollut. 39:197–205.

Farooq, M., Saxena, R. P., and Beg, M. U. 1988. Sulfur dioxide resistance of Indian trees I. Experimental evaluation of visible symptoms and SO_2 sorption. Water Air Soil Pollut. 40:307–316.

Farquhar, G. D. 1979. Models describing the kinetics of ribulose biphosphate carboxylase-oxygenase. Arch. Biochem. Biophys. 193:456–468.

Farquhar, G. D., and Sharkey, T. D. 1982. Stomatal conductance and photosynthesis. Ann. Rev. Plant. Physiol. 33:317–345.

Farquhar, G. D., and Von Caemmerer, S. 1982. Modelling of photosynthetic response to environmental conditions. *In* Physiological plant ecology II. Water relation and carbon assimilation. 12 B:550–587.

Farquhar, G. D., Dubbe, D. R., and Raschke, K. 1978. Gain of the feedback loop involving carbon dioxide and stomata. Theory and measurement. Plant Physiol. 62:406–412.

Farquhar, G. D., Von Caemmerer, S., and Berry, J. A. 1980. A biochemical model of photosynthetic CO_2 assimilation in leaves of C_3 Species. Planta, 149: 78–90.

Gross, L. J., Kirschbaum, M. U. F., and Pearcy, R. W. 1991. A dynamic model of photosynthesis in varying light taking account of stomatal conductance, C_3-cycle intermediates, photorespiration and Rubisco activation. Plant Cell Environ. 14:881–893.

Hahn, B. D. 1987. A mathematical model of photosynthesis and photorespiration. Ann. Bot. 60:157–169.

Halbwachs G. 1984. Organismal responses of higher plants to atmospheric pollutants: Sulphur dioxide and fluoride. *In* Air pollution and plant life. *Edited by* Treshow, M. John Wiley and Sons, London. pp. 175–214.

Hällgen, J. E., and Gezelius, K. 1984. Effects of SO_2 on photosynthesis and ribulose bisphosphate carboxylase in pine tree seedlings. Physiol. Plant. 54:153–161.

Heck, W. W., and Dunning, J. A. 1967. The effects of ozone on tobacco and bintc bean as conditioning by several ecological factors. J. Air Pollut. Cont. Assoc. 19:331–336.

Huttunen, S., Turunen, M., and Reinikainen, J. 1989. Studies on Scots pine (*Pinus sylvestris* L.) and Norway spruce (*Picea abies* (L.) Karst.) needle cuticles. Ann. Sci. For. 46:553–556.

Iqbal, M. 1983. An introduction to solar radiation. Academic Press. Toronto. 390 pp.

Jarvis, P. G., and McNaughton, K. G. 1986. Stomatal control of transpiration scaling up from leaf to region. Adv. Ecol. Res. 15:1–45.

Jarvis, P. G., and Morison, J. I. L. 1981. The control of transpiration and photosynthesis by the stomata, Stomatal Physiology. Cambridge University Press, London. pp. 248–279.

Keller, T. 1977. The effect of long term low SO_2 concentration upon photosynthesis of conifers. Proc. IV Intl. Clean Air Congress, Union Air Pollut. Prevent. Ass. Tokyo, Japan. pp. 81–83.

Keller, T. 1978. The influence of low SO_2 concentrations upon CO_2 uptake of fir and spruce. Photosynthetica 12:316–322.

Klein, H., H. J. Jüger, W. Domes, and C. H. Wong. 1978. Mechanisms contributing to differential sensitivities of plants to SO_2. Oecologia 33:203–208.

Koch, W., and Maier-Maercher, U. 1986. Changes in the humidity responses characteristics of *Picea abies* (L.) Karst produced by fumigation with SO_2. Eur. J. For. Path. 16:329–341.

Kropff, M. J. 1987. Physiological effects of sulphur dioxide. 1. The effect of SO_2 on photosynthesis and stomatal regulation of *Vicia faba* L. Plant Cell Environ. 10:753–760.

Krupa, S., and Kickert, R. N. 1987. An analysis of numerical models of air pollutant exposure and vegetation response. Environ. Pollut. 44:127–158.

L'Hirondelle, S. J., and Addison, P. A. 1985. Effects of SO_2 on leaf conductance, xylem tension, fructose and sulphur levels of jack pine seedlings. Environ. Pollut. 39(A):373–386.

Lawlor, D. W., and Mitchell, A. C. 1991. The effects of increasing CO_2 on crop photosynthesis and productivity: a review of field studies. Plant Cell Environ. 14:807–818.

Leininger, T. D., Miller, P. R., Schilling, S. L., and Dunn, P. H. 1991. Seedling responses of five species of western conifers to simulated ambient sulfur dioxide exposures. For. Sci. 37:1538–1549.

Macrez, V., and Hubac, C. 1988. Exposure to SO_2 and water stress in *Picea abies* (L.) Karst. Water Air Soil Pollut. 40:251–259.

Maier-Maercker, U., and Koch, W. 1992a. Histological examination of spruce needles from a long-term gas exchange experiment in pure and polluted air in the field. Trees 6:186–194.

Maier-Maercker, U., and Koch, W. 1992b. The effect of air pollution on the mechanism of stomatal control. Trees 7:12–25.

Malhotra, S. S. 1976. Effects of sulphur dioxide on biochemical activity and ultrastructural organization of pine needle chloroplasts. New Phytol. 76: 239–245.

Malhotra, S. S. 1977. Effects of aqueous sulphur dioxide on chlorophyll destruction in *Pinus contorta*. New Phytol. 78:101–109.

Malhotra, S. S., and Khan, A. A. 1984. Biochemical and physiological impact of major pollutants. *In* Air pollution and plant life. *Edited by* Treshow, M. Wiley and Sons, London. pp. 113–158.

Mansfield, T. A., and Jones, T. 1985. Growth/environment interactions in SO_2 responses of grasses. *In* Sulphur dioxide and vegetation. *Edited by* Winner, W. E., Mooney, H. A., Goldstein, R. A. Stanford University Press, California. pp. 332–345.

McLeod, A. R. 1995. An open-air system for exposure of young forest trees to sulphur dioxide and ozone. Plant Cell Environ. 18:215–225.

Meng, F. R., and Arp, P. A., 1993. Net photosynthesis and stomatal conductance of red spruce twigs before and after twig detachment. Can. J. For. Res. 23: 716–721.

Meng, F. R. 1994. Sulfur dioxide effects on net photosynthesis and cumulative CO_2 fixation in red spruce stand. Ph.D. Thesis, University of New Brunswick. 186 pp.

Meng, F.-R., Cox, R. M., and Arp, P. A. 1994. Fumigating spruce branches with SO_2: effects on net photosynthesis and stomatal conductance. Can. J. For. Res. 24:1467–1471.

Muller, R. N., Miller, J. E., and Sprugel, D. G. 1979. Photosynthetic response of field-grown soybeans to fumigations with sulfur dioxide. J. Appl. Ecol. 16: 567–576.

Oke, T. R. 1987. Boundary layer climates. Methuen, New York. pp. 339–356.

Oleksyn, J., and Bialobok. S. 1986. Net photosynthesis, dark respiration and susceptibility to air pollution of 20 European provenances of Scots pine *Pinus sylvestris* L. Environ. Pollut. 40(A):287–302.

Reynolds, J. F., Chen, Jalin, Harley, P. Hilbert, D. W., Dougherty, R. L., and Tenhunen, J. D. 1992. Modeling the effects of elevated CO_2 on plants: extrapolating leaf response to a canopy. Agri. For. Meteor. 61:69–94.

Schweizer, B., and Arndt, U. 1990. CO_2/H_2O parameters of one- and two-year-old needles of spruce and fir. Environ. Pollut. 68:275–292.

Seufert, G., Hoyer, V., Wöllmer, H. and Arndt, U. 1990. General methods and materials. Environ. Pollut. 68:231–252.

Shanklin, J., and Kozlowski, T. T. 1984. Effect of temperature preconditioning on responses of *Fraxinus pennsylvanica* seedlings to SO_2. Environ. Pollut. 36(A): 311–326.

Shimazaki, K., and Sugahara, K. 1980. Specific inhibition of photosystem II activity in chloroplasts by fumigation of spinach leaves with SO_2. Studies on the effects of air pollutants on plants and mechanisms of phytotoxicity. Res. Rep. Natl Inst. Environ. Stud., Japan, 11:79–89.

Siefermann-Harms, D. 1990. Chlorophyll, carotenoids and the activity of the xanthophyll cycle, Enviro. Pollut. 68:293–303.

Smith, F. A., and Raven, J. A. 1987. Intracellular pH and its regulation. Annu. Rev. Plant Physiol. 30:286–311.

Smith, W. H. 1990. Air pollution and forests, interaction between air contaminants and forest ecosystems. Springer-Verlag, New York. pp. 313–345.

Spitters, C. J. T. 1986. Separating the diffuse and direct component of global radiation and its implications for modeling canopy photosynthesis Part II. Calculation of canopy photosynthesis. Agri. For. Meteor. 38:231–242.

Taylor, G. E., Selvidge, W. J., and Crumbly, I. J. 1985. Temperature effects on plant response to sulfur dioxide in *Zea mays*, *Liriodendron tulipifera*, and *Fraxinus pennsylvanica*. Water Air Soil Pollut. 24:405–418.

Thornley, J. H. 1976. Mathematical models in plant physiology. A quantitative approach to problems in plant and crop physiology. Academic Press, London. pp. 93–110.

Thornley, J. H., and Johnson, I. R. 1990. Plant and crop modelling, a mathematical approach to plant and crop physiology. Oxford Science Publications, Clarendon Press, Oxford. pp. 224–237.

Umbach, D. M., and Davis, D. D. 1984. Severity and frequency of SO_2-induced leaf necrosis on seedlings of 57 tree species. Forest Sci. 30:587–596.

Unsworth, M. H., and Black, V. J. 1981. Stomatal responses to pollutants. *In* Stomatal physiology. *Edited by* Jarvis, P. G., and Mansfield, T. A. Cambridge Univ. Press, London. pp. 187–203.

Unsworth, M. H., Biscoe, P.V., and Pinckney, H. R. 1972. Stomatal responses to sulphur dioxide. Nature 239:458–459.

Warrington, S., and Whittaker, J. B. 1990. Interactions between Sitka spruce, the green spruce aphid, sulphur dioxide pollution and drought. Environ. Pollut. 65:363–370.

Wedler, M. Weikert, R. M., and Lippert, M. 1995. Photosynthetic performace, chloroplast pigments and mineral content of Norway spruce (Picea abies L. Karst.) exposed to SO_2 and O_3 in an open-air fumigation experiment. Plant Cell Environ. 18:263–276.

Winner, W. E., and Atkinson, C. J. 1986. Absorption of air pollution by plants and consequences for growth. Trees 1:21–26.

Winner, W. E., and Mooney, H. A. 1980. Ecology of SO_2 resistance: I. Effects of fumigations on gas exchange of deciduous and evergreen shrubs. Oecologia 44:290–295.

Winner, W. E., Koch, G. W., and Mooney, H. A. 1982. Ecology of SO_2 resistance. IV. Predicting metabolic responses of fumigated shrubs and trees. Oecologia 44:296–302.

Winner, W.E., and Atkinson, C. J. 1986. Absorption of air pollution by plants and consequences for growth. Trees 1:21–26.

Ziegler, I. 1974. Action of sulphite on plant malate dehydrogenase. Photochemistry 13:2411–2416.

Ziegler, I., Marewa, A., and Schoepe, E. 1976. Action of sulphite on the substrate kinetics of chloroplastic NADP-dependent glyceraldehyde 3-phosphate dehydrogenase. Phytochemistry 15:1627–2632.

APPENDIX 1

Documented Photosynthetic and Stomatal Response to SO_2 Exposure (pp. 331–333)

Species	SO_2(ppb)	Duration	Response	Reference
Holoptela integrifolia (chilbil tree)	100–800	48 h	Increase free sugar, increase acid phosphate, decrease in starch, no change in chlorophyll, no change in photosynthesis	Farooq et al., 1985
Geranium carolinianum (resistant)	300	5 h	−10%	Taylor et al. 1985
Geranium carolinianum (sensitive)	300	5 h	−27%	Taylor et al., 1985
Diplacus aurantiacus (Curtis Jeps)	100	7 h	No change in photosynthesis; increase in stomatal conductance	Atkinson and Winner, 1990
	250	8h	−10%	Winner and Mooney, 1980
Acer saccharum (sugar maple)	500	1–2 d	−22%	Carlson, 1979
	500	5	−27%	Carlson, 1979
	500	1–2 d	−22%	Carlson, 1979
Acer saccharum (sugar maple)	500	1–2d	−21%	Carlson, 1979
	500	5 d	−54%	Carlson, 1979
	1000	4–6 h	+ ?	Smith, 1990
	6000	4–6 h	−photosynthesis; foliar necrosis	Smith, 1990
Fraxinus americana white ash)	500	1–2 d	No change	Carlson, 1979
	500	5	No change	Carlson, 1979
	500	1–2 d	−27%	Carlson, 1979
	500	5 d	−20%	Carlson, 1979
Fraxinus pennsylvanica (red ash)	500–1500	30 h	No change in stem and leaves dry weight, reduce root dry weight, decrease R/S ratio	Shanklin and Kozlowski, 1984

Species	SO_2(ppb)	Duration	Response	Reference
Quercus velutina (black oak)	500	1–2 d	−30%	Carlson, 1979
	500	5	−46%	Carlson, 1979
	500	1–2 d	−40%	Carlson, 1979
	500	5 d	−48%	Carlson, 1979
	500+500	5 d	−44%	Carlson, 1979
Pinus banksiana (jack pine)	100–1000	0–96 h	−stomatal conductance, xylem tension and fructose level decrease, S increase after 10 h	L'Hirondelle and Addison, 1985
Pinus sylvestris (Scots pine)	45 mg m^{-3}	>1 yr	Growth reduction	Halbwachs, 1984
	750	5.5 h	−25%	Hällgren and Gezelius, 1984
Pinus sylvestris (Scots pine)	50–100	70 d	−50%	Keller, 1977
	50–100	70 d	−50%	Keller, 1977
	750	5 d;6 h/d	Visible needle damage, net photosynthesis is a function of needle damage	Oleksyn and Bialobok, 1986
Pinus ponderosa (ponderosa pine)	0–2500 event	30 d	−in diameter growth and needle wt- Leininger et al., 1991	
Pinus strobus (white pine)	50	3 d	Increase net photosynthesis	Boyer et al., 1986
	45 mg m^{-3}	7 m	Growth reduction	Halbwachs, 1984

Abies alba (silver fir) spring	100	14 d	−48%	Keller, 1977
summer	100	70 d	−28%	Keller, 1977
fall	100	14 d	−48%	Keller, 1977
Abies alba fall	20–120 $\mu g\ m^{-3}$	5 yr	Decreased net photosynthesis	Schweizer and Arndt, 1990
	50–200	70 d	−80%	Keller, 1977
	50–200	70 d	−80%	Keller, 1977
Abies lasiocarpa (subalpine fir)	0–2500 Event	30 d	Decrease in diameter growth and needle weight	Leininger et al., 1991
Abies concolor (white fir)	0–2500 Event	30 d	Decrease in diameter growth and needle weight	Leininger et al., 1991
Picea engelmannii (Engelmann spruce)	0–2500 Event	30 d	Decrease in diameter growth and needle weight	Leininger et al., 1991
Picea sitchensis (Sitka spruce)	25 + Drought	2 m	Increased green spruce aphid by 2–3 times, reduce root wt 24%	Warrington and Whittaker, 1990
Picea abies (Norway spruce)	50–200	70 d	−40%	Keller, 1978
	Events	5 yr	Decreased pigments, especially chlorophyll a content	Siefermann-Harms, 1990
	25–40 $mg\ m^{-3}$	>1 yr	Visible needle injury	Halbwachs, 1984
	?	?	Could not recover from draught	Macrez and Hubac, 1988
	?	?	Reduced buffering capacity index; N content stimulated peroxidase activities	Bender et al., 1990
	20–120 $\mu g\ m^{-3}$	5 yr	No change	Schweizer and Arndt, 1990
	50–200	70 d	−40%	Keller, 1977
	13–22 ppb	4 year	−in photosynthesis	Wedler et al. 1995

APPENDIX 2 SO_2 RESPONSE MECHANISMS

Stomatal Responses

Two hypotheses are proposed:

1. SO_2 at low concentrations diffuses through the stomates, attacks subsidiary cells next to the guard cells, and weakens the membranes of the subsidiary cells (Fig. 1). As a consequence, guard cells lose mechanical support; this process is reversible; at high SO_2 concentrations, damage to the cell membranes is beyond repair, and stomates open irreversibly (Black and Black, 1979; Black and Unsworth, 1980);

2. SO_2 may induce stomatal opening indirectly; at low concentrations, SO_2 stimulates photosynthesis; increased photosynthesis depletes intercellular CO_2 concentrations and stimulates stomate opening (Kropff, 1987).

Once SO_2 gets into the guard cells, there are many ways that lead to stomatal closure.

1. Abscisic acid (ABA) is either released or synthesized in response to SO_2 and causes stomatal closure (Malhotra and Khan, 1984). Because closed stomates can reduce SO_2 absorption, Malhotra (1976) suggests that the release or production of ABA is a SO_2 defense mechanism of SO_2-tolerant plants.

2. Within the guard cells, high H^+ levels due to SO_2 uptake cause cross-membrane leakage of ions such as K^+, Cl^-, and malate (Smith and Raven, 1987). Leaking of these ions from the guard cells (1) decreases the osmotic potential and (2) decreases turgor pressure. Decreased turgor pressure leads to stomatal closure.

3. At high concentrations, absorbed SO_2 is transferred through subsidiary cells to guard cells. High levels of sulfite ions within the guard cells cause swelling of chloroplast membranes. This kills the guard cells. Stomates with collapsed guard cells are closed (Black and Black, 1979).

4. If SO_2 exposure causes stomatal closure, then resistance to the diffusion of CO_2 through the stomates would increase. Consequently, CO_2 concentrations within the chloroplasts should decrease, and net photosynthesis should be reduced. There is, however, little evidence that stomatal closure actually reduces net photosynthesis. For example, soybeans show decreased stomatal conductance with increased fumigation from 0.12 to 0.79 ppm SO_2, but the stomatal response lags behind the net photosynthesis response (Muller et al., 979; Kropff, 1987). This means that SO_2 tends to inhibit net photosynthesis either by inhibiting carbon fixation or by stimulating respiration.

Photosynthetic Responses

Sulfur Dioxide Effects on the Light Reaction

Low concentrations of SO_2 (or sulfite ions) stimulate the Hill reaction (Malhotra, 1976). Exposure to high SO_2 concentrations or to low concentrations of a longer duration inhibits the Hill reaction (Malhotra, 1976; Shimazaki and Sugahara, 1980).

Sulfur Dioxide Effects on the Dark Reaction

The dark reaction is inhibited by SO_2 (or by sulfite ion) through interfering with several enzymes such as *triphosphate dehydrogenase* and *phosphatase*. Sulfite ions also compete with CO_2 for enzyme binding sites (Ziegler, 1974; Hällgen and Gezelius, 1984). This mechanism contributes to short-term inhibitions of net photosynthesis. It is also known that the scavenging of free radicals requires NADPH (Alscher and Amthor, 1988). This means that there will be less NADPH available for carbon fixation whenever free radicals are present.

Sulfur Dioxide Effects on Photorespiration

Sulfur dioxide may stimulate net photosynthesis by reducing photorespiration. For example, SO_2 exposure was found to inhibit glycine and serine synthesis. Glycolate oxidase, which catalyzes one reaction in the photorespiration pathway, is inhibited at low sulfite concentrations but is stimulated at high sulfite concentrations (Ziegler et al., 1976; Alscher, 1985).

Sulfur Dioxide Effects on Dark Respiration

Laboratory experiment suggests that dark respiration is not increased by short-time SO_2 exposure unless high SO_2 concentrations are applied. For example, Oleksyn and Bialobok (1986) found no change in dark respiration in Scots pine (*Pinus sylvestris* L.) after 2 days of SO_2 exposure (0.5 ppm $\times$ 6 h/d) although net photosynthesis was reduced. Higher concentrations ($<$1 ppm $\times$ 6 h/d) stimulated dark respiration and inhibited net photosynthesis.

Dizengremel and Citerne (1988) found that dark respiration of 1-year-old spruce needles collected from visibly damaged spruce stands close to a SO_2 source was substantially higher than dark respiration of needles in stands away from the source.

Normally, plant responses to SO_2 pollution are assessed with fumigation systems that expose twig, branches or small trees to known concentration of SO_2 (Meng et al., 1994; Mcleod, 1995). In order to predict stand responses to SO_2, a model is needed to combine existing data and knowledge about leaf reactions to elevated SO_2 exposure during the fumigation experiments (Krupa and Kickert, 1987). Problems related to the temporal and spatial scaling have been discussed (Reynolds et al., 1992).

10

Assessing Atmospheric Sulfur and Nitrogen Loads Critical to the Maintenance of Upland Forest Soils

Tõnu Oja
Institute of Geography, University of Tartu, Tartu, Estonia

Paul A. Arp
University of New Brunswick, Fredericton, New Brunswick, Canada

I. INTRODUCTION

The maintenance and sustainability of upland forest soils, forest type, and forest growth is, in part, dependent on the atmospheric deposition of plant-growth affecting elements such as nitrogen (N), phosphorus (P), calcium (Ca), magnesium (Mg), potassium (K), hydrogen (H), and sulfur (S). This chapter addresses the levels of atmospheric S and N deposition rates that can be considered critical for the long-term maintenance and sustainability of forest growth on upland forest soils in the context of increased soil acidification. Atmospheric deposition rates are considered *critical*, or *above critical* when:

1. The soil of the ecosystem is subjected to indefinite increases in rates of acidification beyond the natural acid buffering potential of the soil (Ulrich et al., 1979).
2. There is an intolerable increase in the mobilization of water-soluble Al, and of heavy metals, as a result of increased soil acidification (Cronan et al., 1986; Klein and Perkins, 1988).

3. The soil and the ecosystem are subject to an unrecoverable loss of base cations (Ca, Mg, K), thereby affecting forest growth (Tomlinson, 1987, 1990).
4. A combined but imbalanced deposition of S, N, P, Ca, Mg, K, and H may lead to growth-reducing nutrition deficiencies within trees (Roelofs et al., 1985; Roberts et al., 1989).
5. Increased N inputs may lead to eutrophication and hence species shifts within ecosystems that need protection against increased N deposition (Bobbink et al., 1992).

This chapter focuses on quantifying ecosystem-specific levels of atmospheric S and N deposition (which are critical for preventing soil acidification beyond tolerable levels), by way of formulating critical S and N loads. To that end, a brief outline is provided for a steady-state mass balance approach (SMB) that can be used to evaluate the combined effects of atmospheric S and N deposition on forest soil acidification, and on forest biomass growth in general, at varying levels of atmospheric deposition rates, from pre- to postindustrial rates. This approach is demonstrated by:

1. Tabulating and illustrating critical S and N loads and related exceedance calculations for the locations of the IFS study (Johnson and Lindberg, 1992).
2. Comparing SMB and the corresponding dynamic model calculations for two of the IFS sites, namely the Turkey Lakes site north of Lake Superior in Canada, and the Huntington Forest site in the Adirondacks of New York State in the United States.

II. BACKGROUND

Numerous attempts have been made to quantify the results of S and N deposition on forest soils and forest growth through monitoring, experimenting, and modeling. Quantification, however, is difficult because of the many complicating physical, chemical, geological, and biological factors through which S and N affect soils and the forest ecosystem on top (Johnson and Ball, 1990; Joslin et al., 1992; Mitchell and Lindberg, 1992). Uncertainties relate to

1. Monitoring and measuring atmospheric inputs and soil/system outputs
2. Monitoring and measuring soil, plant, and faunal responses to S and N deposition, and the resulting effects on soil acidification in particular

3. Realizing that soil responses are coupled to ecosystem response through soil weathering, nutrient uptake, growth, decomposition, and cycling
4. Accounting for lateral and vertical variabilities in soils and in the vegetative strata
5. Accounting for net outcomes of individual S, N, P, Ca, Mg, and K consuming processes across scales, which range, for example, from the microscale (e.g., the rhizosphere) to cross-landscape scales.

To assess forest reactions to atmospheric S and N deposition, and the resulting effect on soil acidification in particular, much work has centered on understanding and capturing H input/output balances at the stand level. At this level, various contributions of atmospheric deposition to soil acidification have been summarized as follows (Binkley and Richter, 1987):

net H production in soil =
(atm H deposition − H loss)
+ (NH4 deposition − NH4 loss)
+ (Al deposition − Al loss)
− (sulfate deposition − sulfate loss)
− (nitrate deposition − nitrate loss)
− (chloride deposition − chloride loss)
− (bicarbonate deposition − bicarbonate loss)
− (organic anion deposition − organic anion loss)
+ net base cation uptake.

To this net gain (or loss) of H, one should also add net H contributions that arise from the net gain (or loss) of Fe, Mn, Zn, P, and so on. These gains and losses, however, are generally small and are therefore neglected. Also negligible in many cases is the net H contribution of Na and Cl deposition: these ions are generally not retained by vegetation and soils, such that inputs are equal to outputs. Exceptions occur in areas with soils that have sodic minerals (e.g., Na-containing feldspars) in the soil parent material.

For wetland soils, further complications arise as the result of H, S, N, and C losses through the formation of volatile, H-rich gases (e.g., NH_3, H_2S and CH_4) and through loss of N gases such as NO_x and N_2.

In reference to atmospheric deposition on uplands soils, the following can be stated:

1. Essentially all atmospheric inputs (= precipitation + dry deposition + fog deposition) need to be considered. Lack of account-

ing for all atmospheric inputs biases the evaluation of this equation (Posch et al., 1995).

2. Each of the factors causing acidification can be analyzed separately. There is no net effect on soil acidification when atmospheric inputs for strong acid anions (e.g., sulfate, nitrate) equal their outputs, due to soil leaching. However, net effects may occur from unbalanced complementary base cation input and leaching.
3. Excess leaching of strong acid anions leads to soil acidification; in contrast, absorption of strong acid anions, either by the soil or biota, is acid-neutralizing. Similarly, net absorption of base cations is acid producing, and the net production of base cations—through soil weathering, for example—is acid neutralizing.
4. Soils with high levels of aluminum (Al) and iron (Fe) oxides/hydroxides have appreciable levels of strong acid anion absorption. Soils and soil layers for which organic matter and silicate minerals dominate have low to zero potential for strong acid anion absorption (Singh, 1980; Karltuna and Gustafsson, 1993).
5. Forest vegetation tends to respond to increased N and S availability through increased uptake. Vegetational S and N storage, however, is temporary because of litterfall and subsequent mineralization. Removal of vegetation—through harvesting, for example—turns this uptake into export.
6. For forested areas with zero net growth, and for S- or N-saturated soils, S or N inputs equal S or N outputs, respectively.

It follows from the above that long-term effects of atmospheric S, N, P, Ca, Mg, K, and H deposition on upland forest soil acidification are mainly due to net input/output effects. With respect to S deposition, direct effects of sulfate-S on vegetation tend to be small, and many forested ecosystems appear to be S saturated, or near S saturation. This is not to say, however, that an increased input of S into soils has a negligible effect on soil chemistry. The following points are noteworthy:

1. Sulfate ions become the dominant co-ion for transporting cations (Ca, Mg, K, Al, Fe, Mn, etc.) through the soil (Hendershot et al., 1991).
2. Sulfate movement through soils is largely unhindered unless sulfate ions encounter soil particles with a high sulfate ion affinity (such as Al and Fe oxides/hydroxides) (Singh, 1984).
3. Sulfate ion affinity in soils that do absorb sulfate ions is pH dependent (Nodvin et al., 1986).

With respect to N, the situation differs: most forested ecosystems are not N saturated (Ågren and Bosatta, 1988; Aber et al., 1989), and in such systems, there tend to be notable growth responses to increased N inputs and increased N uptake rates. In addition, there can be strong biotic responses as well, including those that could lead to species shifts and changes in biodiversity throughout the entire floral and faunal ecosystem community (Skeffington and Wilson, 1988). In general, increased N loads may shift the existing competitive balance within N-affected plant communities in favor of nitrophilic species. For example, formerly N-poor heathlands can be transformed into grassy meadows. Also, blue-green algae and related lichen formations in forest canopies are threatened to extinction by high N loads because of the prolific N growth response of green algae (Bobbink et al., 1992).

III. STEADY-STATE MASS BALANCE MODEL (SMB)

According to de Vries (1991), Hettelingh et al. (1992), Posch et al. (1993a, 1995), and de Vries et al. (1995b), critical S and N loads with respect to soil acidification can be obtained from:

$$\begin{aligned} CL(S + N) = CL(S) + CL(N) = \\ BC_{dep} + BC_{we} - BC_{up} + N_{up} \\ + N_{dnit} + N_{imm} + Ac_{leach,crit} \end{aligned} \quad (1)$$

where BC stands for base cations (Ca, Mg, K, Na), and the subscripts dep, we, up, dnit, imm, and leach refer to total deposition, weathering, uptake, denitrification, immobilization, and leaching, respectively. In this equation, soil acidification is the net effect of all acid-producing processes (mainly net base cation uptake), and acid-neutralizing processes (all the other terms) at critical rates for S and N deposition. Specifically, acceptable levels of incoming acidity (due to critical S and N deposition) can be determined from the net input (or net output) of base cations (BC_{dep} + BC_{we} − BC_{up}), the removal of N from the soil solution (N_{up} + N_{dnit} + N_{imm}), and an acceptable acid leaching rate from the soil ($Ac_{leach,crit}$). The "effects criterion" for $Ac_{leach,crit}$ would have to be acceptable for ecosystem protection within soils, and within surface waters as well.

From the net retention balance of N, a minimum critical load for N (determined by the tolerance of the system to N eutrophication) can be obtained as follows:

$$CL(N) = N_{up} + N_{dnit} + N_{imm} + N_{leach,crit} \quad (2)$$

where $N_{leach,crit}$ specifies the effects criterion that would be considered "tolerable" with regard to N eutrophication within the soil and within surface waters. From Eqs. (1) and (2), it follows that the corresponding maximum critical load for S should be given by

$$CL(S) = BC_{dep} \pm BC_{we} - BC_{up} + Ac_{leach,crit} - N_{leach,crit}. \quad (3)$$

Hence, the ability of the soil to tolerate S deposition is therefore conceptually linked to

(1) the ability of the soil to neutralize acidity ($BC_{dep} \pm BC_{we} - BC_{up}$)
(2) the ability of the ecosystem to tolerate acidity in the percolating soil solution.

Note that $Ac_{leach,crit} - N_{leach,crit}$ is that part of tolerable acidity flux from the soil that is not due to N leaching.

It follows, from the above, that calculations regarding the ability of soils to tolerate S deposition must include an assessment of

1. the extent of atmospheric co-deposition of base cations (BC_{dep})
2. the extent of soil weathering rate (release of base cations from the soil substrate, $= BC_{we}$)
3. the vegetation uptake rate of base cations BC_{up}. The quantity refers to net uptake rate, to account for the difference between base cation uptake and base cation mineralization from forest litter and soil organic matter.

Base cation uptake rates (BC_{up}) can be estimated from considering the net increment of vegetational biomass. Typically, this requires the assessment of woody biomass increment per hectare per year (growth − mortality − woody litterfall), with wood referring to twigs, branches, stem, stump, coarse roots, and bark. The emphasis here is on the gradual storage of base cations in the aggrading part of the forest biomass. Foliage and fine root nutrient content changes also occur over the course of a forest stand rotation, but these plant components cycle faster than the wood component, and therefore tend to have a small net contribution to the long-term average of net base cation uptake.

For steady-state conditions, all nutrient uptake rates in the above expressions also refer to nutrient removal or export from each site, as averaged over the years between removal or export events (e.g., forest harvesting). Nutrients that are being taken up from the soil, and are then recycled, are not part of the steady-state uptake calculations. Furthermore, nutrients that are not taken up, and are therefore subject to leaching, are not part of the nutrient export calculations. Doing so has a practical advantage: it allows one to calculate critical loads and exceedances as they may vary with varying forest harvesting regimes, and natural disturbance mechanisms such as forest fires and blow-downs. In general, increased removal of base nutrients should decrease the critical loads. Removing both S and N, and base nutrients, at the same time may increase or decrease the critical loads, depending on the net differences of equivalents between S + N and base cation removal: positive when S + N > base cations, and negative otherwise.

Much effort has been made to calculate soil weathering rates (BC_{we}), and several approaches are available. The most detailed procedure and model yet described is that by Sverdrup (1990). For regional assessments, and stand-level assessments within regions, de Vries (1991, 1994) has adopted a simplified approach by stipulating soil weathering rates according to substrate type, fine mineral particle content (% clay content), and the prevailing soil temperature.

A conceptual overview of critical S and N loads and their exceedances is obtained by plotting the expected region of zero soil acidification and zero N eutrophication on a generalized S-versus-N deposition quadrant, as shown in Figure 1 (modified from Posch et al., 1993a). This zero-exceedance region is bounded by a simple polygon with the following (x,y) corner points:

A. Lower left: (0,0)
B. Upper left: $(0, CL(S)_{max,\ 0\ N\ leach})$
D. Point of zero N leaching: $(CL(N)_{min}, CL(S)_{max,\ 0\ N\ leach})$
E. Point of zero N eutrophication: $(CL(N)_{max,0\ N\ eutr}, CL(S)_{min,\ 0\ N\ eutr})$
C. Bottom right: $(CL(N)_{max,0\ N\ eutr}, 0)$

where

$$CL(S)_{max,\ 0\ N\ leach} = BC_{dep} - Cl_{dep} - BC_{up} + BC_{we} + Ac_{leach,crit} \tag{4}$$

$$CL(S)_{min,\ 0\ N\ eutr} = CL(S)_{max,0\ N\ leach} - N_{leach,crit} \tag{5}$$

$$CL(N)_{min,\ 0\ N\ leach} = N_{up} + N_{imm} \tag{6}$$

$$CL(N)_{max,\ 0\ N\ eutr} = CL(N)_{min,\ 0\ N\ leach} + N_{leach,crit}/(1 - fdnit) \tag{7}$$

with

$$f_{dnit} = N_{dnit} / (N_{dep} - N_{up} - N_{imm}) \tag{8}$$

as the denitrification fraction of N, and $N_{dep} = CL(N)$. The borders of this polygon change depending on a variety of situations and/or considerations. For instance:

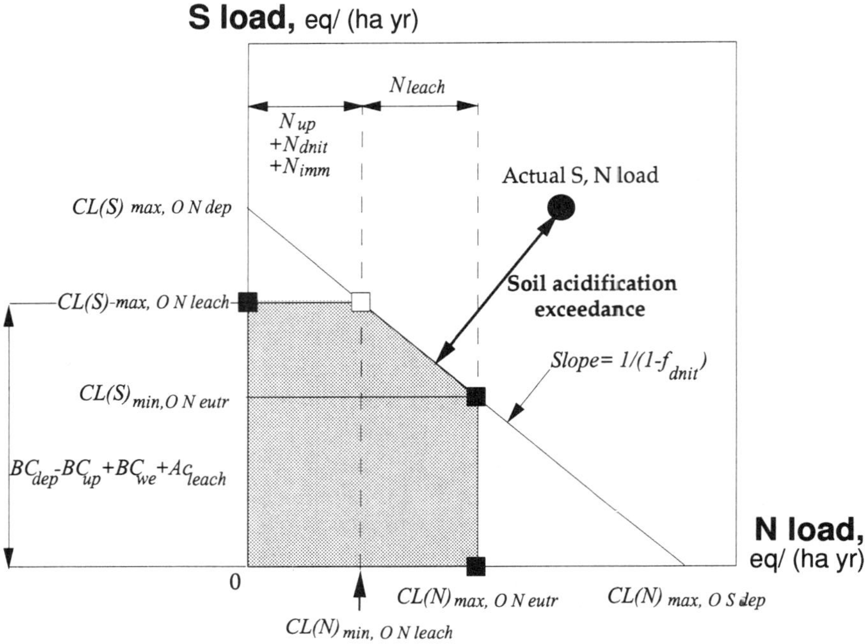

Figure 1 Region of zero exceedance (shaded portion of graph). Both S and N deposition add equally to soil acidification stress; cutoff to the right of the shaded region is due to exceedance of N eutrophication criterion. Increasing the allowable acid-leaching criterion increases the zero exceedance range by shifting the sloped line upward. Allowing for denitrification ($f_{denit}>0$) is equivalent to pivoting the sloped line about the open square corner point toward a more horizontal orientation. Allowing for increased NO_3_N leaching (N_{leach}) shifts the rightmost vertical line (slashed) to the right. Exceedances of CL(N) and CL(S) are related to actual S and N deposition rates (filled circle) and the zero exceedance region (equations 9 to 12). The critical S and N deposition loads (open circle) are symbolized as CDS and CDN in Tables 4 and 5.

1. Decreasing $N_{leach,crit}$ moves the vertical line associated with $CL(N)_{max,\ 0\ N\ eutr}$ to the left. Doing so implies that the ecosystem under consideration requires greater protection against N eutrophication. In contrast, increasing $N_{leach,crit}$ moves $CL(N)_{max,\ 0\ N\ eutr}$ to the right, possibly beyond $CL(N)_{max,\ 0\ S\ dep}$. In this case, the ecosystem under consideration is thought to require more protection from soil acidification than N eutrophication. This situa-

tion tends to occur when N is not the growth-limiting factor. For surface waters in forest lands, the growth-limiting factor is often P (Schindler, 1977).

2. Increasing the net base cation input ($BC_{dep} + BC_{we} - BC_{up}$), or increasing the amount of acceptable acid leaching, expands the zero-exceedance region upward away from the origin. The opposite occurs with decreased rates of soil weathering, decreased amounts of tolerable acid leaching, decreased atmospheric base cation deposition, or increased base cation uptake.
3. Lowering N_{up} and N_{imm} (i.e., going to ecosystems with low capabilities of N uptake and N immobilization) reduces critical N loads (all other rates kept constant). Because N_{up} and N_{imm} can be expected to depend on forest type (forest types likely differ by amounts of N stored in the wood, N availability, balance with other nutrients, and soil nitrification), one will obtain different critical loads for deciduous vegetation and coniferous vegetation, for example. The differences are further accentuated by noting that the occurrence of a specific vegetation type depends on soil substrate as well, with deciduous species tending to dominate on soil substrates with a high base content.
4. Allowing for denitrification changes the slope of the slanted side of the polygon—i.e., makes that side form an angle that is $< 45°$ with the bottom (x) axis. For the case of complete denitrification, one would obtain an unbounded rectangle with

$$CL(N)_{max,0\ N\ eutr} = \infty.$$

With respect to calculating soil acidification and N eutrophication exceedances, it follows from Figure 1 and the above that:

$$Ex(Ac) = S_{dep} + N_{dep-}\ [CL(S) + CL(N)]$$
$$\text{when } S_{dep} > CL(S)_{min,\ 0\ N\ eutr}$$
$$\text{and } N_{dep} > CL(N)_{min,\ 0\ N\ leach} \tag{9}$$

$$Ex(Ac) = S_{dep} - CL(S)_{max,\ 0\ N\ eutr}$$
$$\text{when } N_{dep} < CL(N)_{min,\ 0\ N\ leach} \tag{10}$$

$$Ex(Ac) = N_{dep} - CL(N)_{max,\ 0\ N\ eutr}$$
$$\text{when } S_{dep} < CL(S)_{min,\ 0\ N\ eutr} \tag{11}$$

and

$$\mathrm{Ex(N)} = \mathrm{N_{dep}} - \mathrm{CL(N)_{max,\ 0\ N\ eutr}} \quad \text{when } \mathrm{N_{dep}} > \mathrm{CL(N)_{max,\ 0\ N\ eutr}}. \tag{12}$$

IV. EFFECTS CRITERIA

The $Ac_{leach,crit}$ and $N_{leach,crit}$ quantities are effects criteria below which no adverse effects on ecosystem functioning are expected to occur, based on field observations. These effects criteria deal with (1) the tolerable acidity flux (exceedances above this level lead, for example, to toxic Al effects on root growth and soil organisms) and (2) the tolerable N flux (exceedance above this level leads to N eutrophication). At present, these quantities are not known precisely. Instead, information has been gleaned from experimentation, process studies, and observations about ecosystem responses to acidification and N stresses in general. Choosing appropriate levels has been the subject of much discussion (e.g., Grennfelt and Thörnelöf, 1992; Hornung et al., 1995). It is proposed that these levels should vary with forest cover type. For example, some coniferous stands will likely have a higher tolerance for water-soluble Al in soil solutions than some tolerant hardwood stands (Sverdrup et al., 1992, 1993). In this chapter, we set

$$Ac_{leach,crit} = [Al]leach,crit + [H]leach,crit\ Q \tag{13}$$

with

$$[H]_{leach,crit} = [Al]_{leach,crit} / K_{gibb}{}^{1/3} \tag{14}$$

Q as the flux of water through the rooting space (= precipitation −evapotranspiration), and K_{gibb} as the equilibrium constant for gibbsite (= 200 m^6 eq^{-2}).

There are a number of reports suggesting that [Al] = 0.2 mEq/L is critical for inducing irreversible damage to the normal functioning of tree roots and related ectomycorrhizal associations (Ulrich and Matzner, 1983; de Vries, 1991). A general acceptance of this criterion, however, would mean that all forest soils and subsoils would be allowed to approach a pH level of about 4, regardless of current soil pH and stand composition. At the same time, one would have to tolerate similar Al concentrations in most upland streams, especially for periods of the year when pH 4 water may enter freely into streams without incurring the acid-neutralizing effect that may arise from deep soil percolation and related soil weathering. In fact, high Al levels and low surface water pH levels are generally observed during periods of high runoff—i.e., during spring melt and periods of heavy precipitation. Critical Al concentrations for surface water acidification are usually much lower than 0.2 mEq/L (Hultberg, 1988). A conservative and probably acceptable critical Al concentration would be 0.02 mEq/L (Arp et al., 1996).

An alternative soil acidification criterion is obtained by considering the Al to base cation ratio in soil solutions, i.e., $[Al]/[BC]_{leach,crit}$. This criterion is based on the concept that ion uptake by plants depends on competitive ion interactions at the root surface: high [Al] tends to block the uptake of nutrient elements such as Ca, Mg, K, and NH_4-N. The advantage of this criterion is that it is dimensionless, and it expresses the competitive reality of ion uptake and related ion uptake interferences quite well (Ulrich and Matzner, 1983; Roelofs et al., 1985; Boxman and van Dijk, 1988 as quoted by de Vries, 1991). In this chapter, model calculations based on this criterion proceed:

1. by giving the ratio a particular value (i.e., 1.5 eq/eq; see, for example, Ulrich and Matzner, 1983, or, more conservatively, 0.15 eq/eq, see Arp et al., 1996)
2. by calculating the expected soil base cation loss from

$$BC_{leach,crit} = BC_{dep} + BC_{we} - BC_{up} \quad (15)$$

and

3. by setting

$$Al_{leach,crit} = [Al]/\ [BC]_{leach,crit}\ BC_{leach} \quad (16)$$

Corresponding leaching losses for H^+ ions are then calculated via Equation (14).

The rate of critical N leaching is calculated from

$$N_{leach,crit} = [NO_3]_{leach,crit} \cdot Q \quad (17)$$

with $[NO_3]_{leach,crit}$ as the suggested tolerance criterion for NO_3-N in the percolating soil solution. This tolerance criterion is generally set at 0.1 mEq/L (de Vries, 1991).

V. DYNAMIC APPROACH

In contrast to the steady-state approach, dynamic models allow quantitative assessments of short-, mid-, and long-term impacts of actual and projected atmospheric S, N, P, Ca, Mg, K, Na, Cl, and H deposition scenarios on forest soils and forest biomass growth (Hornung et al., 1995; Wright et al., 1991). Dynamic models also offer an opportunity to evaluate critical S and N loads from a variety of perspectives (e.g., sustainable biomass growth, sustainable soil acidification, base cation loss, and N eutrophication). Models have been created at different levels of complexity, from simple SMB derivates with some time-dependent parameters, to fully integrated nutrient cycling and vegetation growth models. Here are a few examples:

1. Single-layer soil models such as SAFE (Warfvinge et al., 1993), and SMART (Posch et al., 1993b). These models are, in principle, steady-state mass balance calculators, but with time-dependant input and output.
2. Soil leachate models such as MAGIC (Cosby et al., 1985). Such models can be used to simulate chemical changes in soilwater and streamwater.
3. Multiple-layer soil models such as RESAM (de Vries et al., 1994, 1995a) and NUCSAM (Groeneberg et al., 1995; Kros et al., 1996). These models can be used to calculate chemical changes in soilwater and streamwater at annual and daily scales, respectively.
4. Coupled soil and vegetation models such as NUCSAM/FORGRO (Mohren et al., 1993). Such models can be used to explore the influence of forest vegetation on chemical changes in the soil solution.
5. Fully integrated soil-vegetation models such as ForSVA (Arp and Oja, 1997; Oja and Arp, 1997; annual inputs/outputs), SOILVEG (Berdowski et al., 1991; daily inputs/outputs) and NuCM (Liu et al., 1992; daily inputs/outputs). These models can be used to keep track of atmospheric inputs, nutrient cycling, nutrient losses, and changing levels of nutrients in the soil and vegetation in response to forest growth and changing atmospheric input/output scenarios.

A similar classification of dynamic models has been presented elsewhere (Werner and Spranger, 1996).

Increasing the model complexity requires an increased effort for model calibration and model performance checking. Ideally, calibrated models should be applicable for a wide variety of forest conditions; i.e., the models should be able to take into account differences by forest site and type. To this effect, simplicity generates model portability. In fact, SMB models lend themselves to mapping of critical loads and related exceedances at various spatial resolution scales. In contrast, fully integrated models are not useful for mapping but can be used to test and widen the logical applicability of the critical loads concept, and the choices that are made in deriving the associated effects criteria. With ForSVA, for example, it is possible to evaluate S and N critical loads vis-à-vis SMB effects criteria, by comparing simulations with actual field observations/assessments, be they related to Al and base cation concentrations in soil solutions, expected and observed levels of forest biomass production, or rates of nutrient uptake and removal (Oja and Arp, 1995). Subsequent agreement between SMB and dynamic model calculations (e.g., NUCSAM, ForSVA), adds justification to the mapping of SMB critical load calculations at regional scales (van Grinsven et al., 1995 ; de Vries et al., 1995b).

VI. APPLICATIONS

The SMB model was applied to the 17 well-documented forest sites described in the Integrated Forest Study (IFS) (Johnson and Lindberg, 1992). The location and main characteristics of these sites are compiled in Figure 2 and Tables 1 and 2. These sites provide important benchmarks to assess critical loads and related exceedances as they would differ by climate, vegetation type, and soil type. All sites have been subjected to essentially

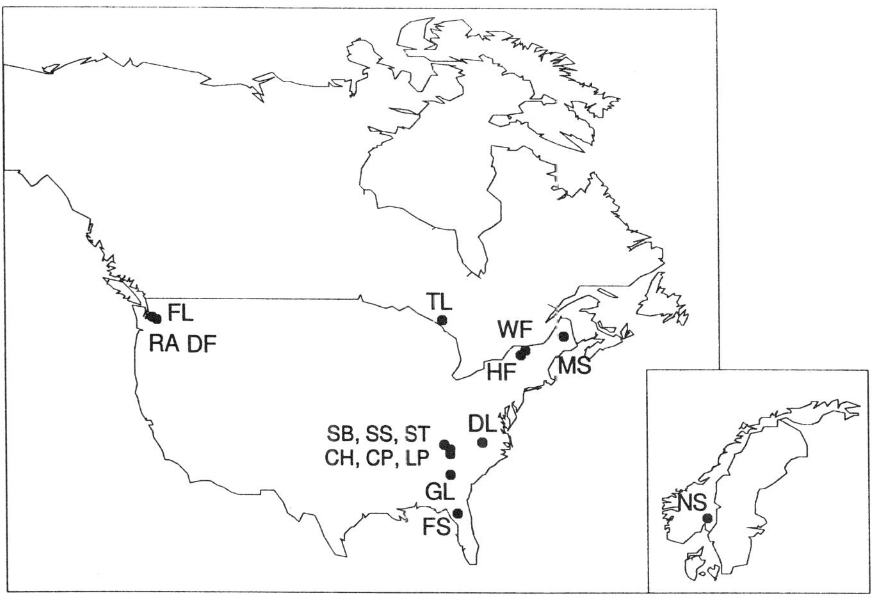

Figure 2 Locator map for IFS sites.

Table 1 Integrated Forest Study Sites: Location, Forest Type, Landform, Local Features (Seepage, Drainage, Farming History), Annual Climate, Soil Depth, and Texture

	Location					Annual				Soil		
	Latitude °	Longitude °	Elevation m	Forest Type	Landform	Precip. mm	AET mm	Temp. °C	Depth cm	Clay %	Silt %	Sand %
CH	35.05	83.45	725	Hardwoods	Colluvium[a]	1380	710	15.3	0.67	20	24	56
CP	35.05	83.45	725	White pine	Colluvium	1440	910	15.3	0.67	19	16	66
DL	36.20	79.28	213	Loblolly pine	Residual[b]	1130	730	15.6	0.60	41	19	41
GL	33.37	83.45	145	Loblolly pine	Residual[c]	870	630	18.9	0.80	30	21	50
LP	35.90	84.33	300	Loblolly pine	River terrace[a]	1140	710	15.3	0.41	26	39	35
FS	29.73	82.50	38	Slash pine	Marine[c]	1120	750	20.6	0.60	8	4	88
SB	35.62	83.43	1600	Beech	Residual	1510	700	15.3	0.40	6	45	50
SS	35.57	83.48	1800	Red spruce	Residual	1510	610	15.3	0.46	4	38	58
ST	35.57	83.47	1740	Red spruce	Residual	2440	470	15.3	0.49	11	29	61
TL	47.05	84.42	350	Hardwoods	Till	1210	420	3.9	0.57	3	42	56
HF	43.98	74.23	530	Hardwoods	Till	970	530	10.0	0.58	4	22	74
WF	44.37	73.90	1000	Red spruce	Till	1150	390	10.0	0.41	1	20	79
MS	45.60	68.67	65	Red spruce	Till[c]	990	530	6.4	0.52	4	26	71
NS	60.27	−11.10	200	Norway spruce	Outwash sand	1070	690	7.2	0.41	2	9	89
FL	47.07	121.40	1130	Fir, hemlock	Volc. ash, till	1740	770	10.0	0.71	5	36	60
DF	47.38	121.90	220	Fir	Till[c]	1140	560	10.0	0.45	5	40	55
RA	47.38	121.90	220	Red alder	Till[c]	1140	470	10.0	0.45	6	39	55

AET = evapotranspiration
[a]Site likely influenced by seepage.
[b]Forest with farming history.
[c]Restricted drainage.

Table 2 Integrated Forest Study Sites: Atmospheric Deposition Rates (Ca, Mg, K, Na, N, S, H, Cl), and Extractable Ca, Mg, and K in Soils (Rooting Space)

	Atmospheric Deposition									Soil Availability		
Site	Ca	Mg	K	Na eq/(ha yr)	NH_4	NO_3	SO_4	H	Cl	Ca	Mg eq/ha	K
CH	85	32	27	82	130	209	493	518	84	82,800	62,583	26,803
CP	143	43	34	62	180	330	552	728	74	41,300	14,833	8,338
DL	344	75	122	242	408	589	1050	335	252	65,800	23,417	7,494
GL	100	41	33	124	181	466	776	1392	415	94,250	92,917	28,389
LP	284	53	20	55	197	502	941	219	569	352,850	110,667	74,194
FS	402	205	26	608	126	305	514	315	417	15,150	5,583	691
SB	139	40	67	40	10	113	771	1007	141	23,700	9,083	8,568
SS	240	60	45	65	12	396	1609	622	44	11,300	5,333	4,143
ST	800	192	169	327	698	1241	2516	1143	103	21,550	9,917	9,003
TL	186	52	18	64	211	311	542	754	99	47,650	8,750	4,604
HF	102	21	40	33	150	417	488	628	231	45,350	7,167	5,217
WF	129	42	25	37	453	683	1096	342	251	27,600	4,333	5,320
MS	92	31	23	95	204	341	585	1007	99	33,100	12,250	7,596
NS1	109	51	63	181	356	419	663	45	115	18,200	3,333	6,573
FL	249	171	58	648	63	84	319	1981	401	17,950	7,417	8,542
DF	176	69	64	358	108	237	320	411	65	66,000	10,917	7,673
RA	176	68	62	355	104	235	325	1270	52	86,750	14,250	6,777

the same measurement and monitoring protocol for atmospheric deposition, leaching, vegetative uptake and biomass growth, and nutrient contents in soils and in vegetation.

Accurate estimates for soil weathering rates are important for a realistic assessment of critical S loads. In the model, soil weathering rates were based on the PROFILE model (Sverdrup and Warfvinge, 1990). This model requires total mineral content of soils as input. We used a mineral-oxide to mineral conversion (UPPSALA model, as presented in Posch et al. 1995, p 157) to specify mineral content for each IFS site. The resulting PROFILE weathering rates were then compared with the soil weathering rate estimates and related IFS site ranking by April and Newton (1992). The results are shown in Table 3. Uncertainties regarding PROFILE soil weathering rates have been discussed, and amount to ± 25–35% (Kolka et al., 1996; Jonsson et al., 1995).

Critical loads for sulfur, total acidity and nitrogen, and related exceedances are presented for the IFS sites in Table 4. These results were obtained for the "conventional harvesting" scenario, i.e., stem-only harvesting. Critical load exceedances below zero correspond to unused buffering potential. Positive exceedances imply that the sites so identified are subject to accelerated soil acidification because of atmospheric deposition and a general inability of the site to buffer the acid exceedance. Exceedances near zero (above or below zero) appear to be at critical loads under current S and N deposition rates.

The exceedance values, as listed in Table 4, are directly related to the choice of soil acidification criterion, which in this case was the Al to base cation ratio in the soil solution, set at 0.15 eq/eq. Increasing this criterion decreases the exceedances (Table 5). Doing so, however, is consistent with allowing for increased levels of soil acidity, Al leaching, and base cation leaching.

Conditions for nitrogen immobilization, fixation, mineralization and denitrification likely vary widely and also temporally for the IFS sites. For simplicity, net long-term accumulations for the combination of these processes were set to zero.

The impact of harvesting scenarios on critical loads is shown in Table 5. For example, the no-harvesting option is consistent with decreased critical S and N deposition loads, and therefore elevated soil acidification rates. Removing foliage and branches as well as wood and bark ("whole-tree harvesting") changes the proportions between base cation removal and nitrogen/sulfur removal by favoring N and S export, thereby increasing critical S and N deposition loads somewhat, as calculated.

Table 3 Comparing Standardized Soil Weathering Rates for the IFS Sites, by Rank: PROFILE Calculations Versus IFS Estimates by April and Newton (1992)

	IFS Ranking	PROFILE Value	PROFILE Ranking
HF	1	1.09	6
WF	2	1.32	1
TL	3	1.19	2
FL	5	1.18	3
RA	5	1.1	4
NS	5	0.66	7
DF	7	1.1	5
CP	8	0.29	12
MS	9	0.4	8
SS	10	0.34	9
CH	11	0.25	13
ST	12	0.3	11
SB	13	0.32	10
DL	14	0.04	14
FS	15	0.03	15
LP	nd	0.17	nd
GL	nd	0.08	nd
r2		0.817	0.759

PROFILE simulations (standard conditions): soil moisture = 25% (volume); soil depth = 1 m; soil surface area 106 m^2/m^3, CO_2 (soil) = 10 * CO_2 (atm), mean annual soil temperature = 4°C. SMB weathering estimates: standard PROFILE rates, adjusted by soil depth, and mean annual soil temperature.

VII. COMPARISON OF SMB TO DYNAMIC MODEL CALCULATED CRITICAL LOADS

For comparison, the dynamic soil-vegetation-atmosphere model called ForSVA (Arp and Oja, 1997) was applied to assess the atmospheric N and S deposition loads critical for the maintenance of site quality and forest biomass productivity on two of the IFS sites—Turkey Lakes (Ontario) and Huntington Forest (New York State). The application of this model to these sites has been reported before (Oja and Arp, 1997).

Table 4 Atmospheric Deposition Rates (Ndep, Sdep), Critical Depositions (CDN, CDS), Critical Soil Acidification Load (CLAc), Exceedances (ExN, ExS, ExAc) for N, S, and Total Acidity (Ac), and Critical Load Cornerpoint Values, as Shown in Figure 1

	CLminN	CLnutN	CLmaxN	CLmaxS	CDN	Ndep	ExN	CDS	Sdep	ExS	CLAc	ExAc
		eq /(ha yr)				eq /(ha yr)			eq /(ha yr)		eq /(ha yr)	
CH	108	242	590	482	262	339	77	328	493	165	522	242
CP	274	380	830	556	440	510	70	390	552	162	542	232
DL	283	363	796	513	491	997	506	305	1050	745	243	1251
GL	294	338	462	168	346	647	301	115	776	661	204	961
LP	82	168	756	674	349	699	350	407	941	534	403	884
FS	304	330	837	533	410	431	21	427	514	87	162	108
SB	16	178	835	819	45	37	−8	790	573	−217	593	−225
SS	57	237	947	890	146	134	−12	802	697	−105	642	−116
ST	302	614	1886	1584	975	1939	964	911	2214	1303	922	2267
TL	120	278	1105	985	539	522	−17	566	542	−24	967	−41
HF	139	227	833	695	463	567	104	370	488	118	814	222
WF	345	497	994	649	617	1136	519	377	1096	719	906	1238
MS	92	134	448	356	247	545	298	201	585	384	341	682
NS	105	181	604	499	356	775	419	248	663	415	431	834
FL	87	281	1493	1406	309	147	−162	1183	319	−864	1094	−1027
DF	203	319	1359	1156	559	345	−214	800	320	−480	965	−694
RA	199	333	1346	1147	544	339	−205	802	325	−477	984	−682

CDN = (1 − Sf) * [CLAc + BCdep]; ExN = Ndep − CDN
CDS = Sf * [CLAc + BCdep]; ExS = Sdep − CDN
ExAc = Ndep + Sdep − CDN − CDS; BCdep = Cadep + Mgdep + Kdep
CLAc = CL(S + N); Sf (sulfur fraction of S and net N deposition) = Sdep/(Ndep − Nup + Sdep) if Ndep > Nup, else Sf = 1.

Table 5 Cornerpoints and Critical N and S Deposition Values for Three Harvesting Scenarios and Three Critical Al/base Cation Ratio Values for the Turkey Lakes (TL) and Huntington Forest (HF) Sites

		No Harvesting					Conventional					Whole Tree				
Site	Al/BC Ratio	CL min N	CL max N eq/(ha yr)	CL max S	CDN	CDS	CL min N	CL max N eq/(ha yr)	CL max S	CDN	CDS	CL min N	CL max N eq/(ha yr)	CL max S	CDN	CDS
TL	0.15	3	1298	1295	636	661	120	1105	985	539	566	111	1174	1063	569	605
	0.4	3	1637	1634	802	835	120	1370	1250	652	718	111	1458	1347	692	766
	0.6	3	1871	1868	917	954	120	1550	1429	729	821	111	1651	1540	775	876
HF	0.15	3	1056	1053	568	488	139	833	695	463	370	179	871	692	485	385
	0.4	3	1322	1319	710	612	139	1016	877	463	370	179	1052	873	566	487
	0.6	3	1511	1508	811	700	139	1142	1003	608	534	179	1178	999	621	556

CDN, CDS: see Table 4, footnotes.

A. Derivation of Critical Loads with ForSVA

The "no decrease in biomass production" was considered as the main effects criterion. Cumulative wood biomass was checked within three consequent natural dominant tree/gap cycles for different deposition scenarios. Critical loads were determined as the maximum S and N deposition rates that would not cause a decrease in calculated forest biomass (Table 6, Oja and Arp, 1995). Also checked with ForSVA were (1) the Al concentrations and (2) the Al to Ca and Al to base cation ratios in the soil percolate below the rooting zone (60 cm).

As shown, the critical loads calculated with ForSVA (Table 6) were in reasonable agreement with the critical depositions calculated with the SMB model for the "no harvest" scenario (Table 5). It should be noted that the "no decrease in biomass production" is consistent with Al/base cation ratios somewhat higher than the SMB value of 0.15. Hence, the 0.15 value is somewhat conservative in terms of ensuring sustainable forest biomass production under the no-harvesting scenario.

Complete agreement between the ForSVA and SMB calculations is not to be expected: both models use differing means to assess soil weathering, with ForSVA requiring special calibration considerations (Oja and Arp, 1997). It would be advisable, however, to admit PROFILE type calculations in dynamic forest biomass models. This would ensure consistency (if not increased reliability) and a reduced effort regarding model calibration.

VIII. DISCUSSION

Calculating critical loads with the SMB and ForSVA models yields similar results. Hence, calculations of critical loads based on simple steady-state mass balance assumptions are as least as robust as what one can expect from dynamic calculations regarding forest biomass growth and nutrient cycling. Simplifying the SMB calculations by ignoring possible long-term effects on net N accumulation (or loss) due to N immobilization, fixation, nitrification, and denitrification produces results that are comparable with the results obtained with ForSVA, where these processes are part of the N cycling formulation. These simplifications, however, should only be made for upland soil conditions. For wetland conditions, denitrification and N immobilization could become important factors that need to be evaluated by wetland type.

Table 6 Present Deposition Rates (Pr) and Critical Loads (CL) at Turkey Lakes (TL) and Huntington Forest (HF) Sites for SO_4^{2-} and NO_3^- Deposition, Assuming that the Current Ratio of Nitrate to Sulfate Deposition is not Changing. Also Provided are Corresponding Deposition Rates for NH_4^+, H^+, Bivalent Base Cations, Al_x^+ Concentrations, Al to Ca, and Al to Total Base Cation (BC = Ca + Mg + K + Na) Ratios in the Leachates

Site		SO^{2-}	NO_3^-	NH_4^+	H^+	Ca^{2+}	Mg^{2+}	Al^{x+}	Al/Ca	Al/Bc
		Deposition eq/(ha yr)						eq/m³	in Soil Percolate eq/eq	
TL	Pr	760	460	211	479	186	52			
	CL	280	190	340	186	52		0.12	0.6	0.4
HF	Pr	488	417	150	622	102	21			
	CL	440	370	130	550	102	21	0.1	1.4	0.8

The IFS sites span soil types ranging from well drained to imperfectly drained, with some locations likely receiving nutrient-containing seepage from uplands, and with some sites losing N on account of denitrification. These differences are not yet accounted for because of lack of information. Accounting for these differences, however, should improve the model fit in terms of expected forest productivity calculations (as expressed in terms of mean annual growth), N leaching, and base cation leaching (details not shown here).

Nitrogen fixation is a definite factor at the IFS red alder site (RA). At this site, N fixation was observed to accompany high rates of nitrification, with a corresponding loss of soil base cations. Nitrification is also a factor for the Turkey Lakes site, where N leaching, although highly variable spatially and temporally, is accompanied with accelerated base cation leaching losses.

In terms of the critical soil acidification calculations, one can conclude that some of the IFS sites (ST, WF, DL, GL, LP, NS, MS) are currently subjected to S and N deposition rates that exceed the natural capacity of these sites to buffer the incoming acidity. The consequence of this is an enhanced loss of base cations and Al ions from the soils. Enhanced base cation losses would eventually lead to a decrease in forest biomass productivity, which could possibly also result in a change of vegetation type, likely shifting towards ericaceous heathlands in some cases. It is therefore advocated that ways and means be found to reduce the exceedance gap at these locations. To some extent, this can be done by controlling S emissions from industrial centers upwind. In some cases, it may be required to control N emissions as well. Reducing the exceedance gap is somewhat complicated by the fact that atmospheric S and N deposition is coupled with atmospheric deposition of base cations as well. Hence, reducing N and S emissions will reduce the exceedance gap, but the target is moving because the input of acid-buffering base cations is reduced as well.

Changing the harvesting scenario from no harvesting to whole tree harvesting benefits the forest sites by slightly increasing the critical loads for the same sites. This, however, must not be seen as a way to decrease the exceedance gap. Further calculations with SMB show that the rate of sustainable biomass production decreases with increased nutrient removal per site (details not shown here). Harvesting to decrease exceedances, however, can be used in cases where the conservation of a threatened low-nutrient ecosystem is a factor.

ACKNOWLEDGEMENTS

Financial support for the above work was provided, in part, by several organizations, namely the Canadian Forest Service (CFS), Environment Canada, the Canadian Electrical Association (CEA), the Canadian Pulp and Paper Association (CPPA), and the Ministry of Environment and Energy of Ontario, and in-kind support from Forêt Quebec. Additional support was received from the National Engineering and Science Council of Canada (NSERC). The basic SMB concepts and principles adopted in this chapter were essentially articulated within the framework of the "Convention on Long-Range Transboundary Air Pollution of the United Nations Economic Commission for Europe."

REFERENCES

Aber, J. D., Nadelhoffer, K. J., Steudler, P., and Melillo, J. M. 1989. Nitrogen saturation in northern forest ecosystems. BioScience, 39: 378–386.

Ågren, G. I., and Bosatta, E. 1988. Nitrogen saturation of terrestrial ecosystems. Environmental Pollution, 54: 185–197.

April, R., and Newton, R. 1992. Mineralogy and mineral weathering. *In* Atmospheric Deposition and Nutrient Cycling. A synthesis of the Integrated Forest Study, *Edited by* D. W. Johnson and S. E. Lindberg, Springer, New York pp. 378–425.

Arp, P. A., and Oja, T. 1997. A Forest Soil Vegetation Atmosphere Model (ForSVA), I: Concepts. Ecol. Modelling 95: 211–224.

Arp, P. A., Oja, T., and Marsh, M. 1996. Calculating critical S and N loads and current exceedances for upland forests in southern Ontario, Canada. Can. J. For. Res., 26: 696–709.

Berdowski, J. J. M., van Heerden, C., van Grisnven, J. J. M., van Minnen, J. G., and de Vries, W. 1991. SOILVEG: A model to evaluate effects of acid atmospheric deposition on soil and forest. Vol. 1: Model principles and application procedures. Dutch priority programme on acidification, Rep. no. 114.1-02. RIVM, Bilthoven, 93 pp.

Binkley, D., and Richter, D. 1987. Nutrient cycles and H+ budgets in forest ecosystems. Advances in Ecological Research, 16: 1–51.

Bobbink, R., Boxman, D., Fremstad, E., Heil, G., Houdijk, A., and Roelofs, J. 1992. Nitrogen eutrophication and critical load for nitrogen based upon changes in flora and fauna in (semi-) natural terrestrial ecosystems. *In* Critical Loads for Nitrogen, *Edited by* P. Grennfelt and E. Thörnelöf, Nord 1992: 41, pp. 111–159.

Boxman, A. W., and van Dijk, H. F. G. 1988. Het effect van landbouw ammonium depposities op bos- en heidevegetaties. Vakgroep Aquatische Oecologie en Biogeologie, Katholieke Universiteit, Nijmegen, 96 pp.

Cosby, B. J., Hornberger, G. M., and Galloway, J. N. 1985. Modeling the effects of acid deposition: Assessment of a lumped parameter model of soil and water acidification. Water Resour. Res., 21: 51–63.

Cronan, C. S., Walker, W. J., and Bloom, P. R. 1986. Predicting aqueous aluminum concentrations in natural waters. Nature (London), 324: 140–143.

de Vries, W. 1991. Methodologies for the assessment and mapping of critical loads and of the impact of abatement strategies on forest soils. DLO Winand Staring Centre, Wageningen (the Netherlands). Report 46, 109 pp.

de Vries, W. 1994. Soil acidification on a regional scale: Critical loads and long-term impacts of acidic deposition. Doctoral thesis, Agricultural University, Wageningen, the Netherlands, 487 pp.

de Vries, W., Kros, J., and van der Salm, C. 1994. The long-term impact of three emission scenarios on Dutch forest soils. Water, Air and Soil Pollution, 75: 1–35.

de Vries, W., Kros, J., and van der Salm, C. 1995a. Modelling the impact of acid deposition and nutrient cycling on forest soils. Ecol. Modelling, 79: 231–254.

de Vries, W., Posch, M., Oja, T., van Oene, H., Kros, J., and Arp, P. A. 1995b. Modelling critical loads for the Solling spruce site. Ecol. Modelling, 83: 283–293.

Grennfelt, P., and Thörnelöf, E. (*Editors*) 1992. Critical Loads for Nitrogen. Nord 1992:41.

Groenenberg, B.-J., Kros, H., van der Salm, C., and de Vries, W. 1995. Application of the model NUCSAM to the Solling spruce site. Ecol. Modelling, 83: 97–107.

Hendershot, W. H., Warfvinge, P., Courchesne, F., and Sverdrup, H. U. 1991. The mobile anion concept—time for reappraisal. J. Environ. Qual., 20: 505–509.

Hettelingh, J. P., Posch, M., de Vries, W., Bull, K., and Sverdrup H. U. 1992. Guidelines for the computation and mapping of nitrogen critical loads and exceedances in Europe. *In* Critical Loads for Nitrogen, *Edited by* P. Grennfelt and E. Thörnelöf, Nord 1992: 41, pp. 287–303.

Hornung, M., Sutton, M. A., and Wilson, R. B. (*Editors*) 1995. Mapping and Modelling of Critical Loads for Nitrogen: a Workshop Report. Proceedings of the Grange-over -Sands Workshop, 24–26. October 1994, Institute of Terrestrial Ecology, Merlewood, 207 pp.

Hultberg, H. 1988. Critical loads for sulphur to lakes and streams. *In* Critical Loads for Sulphur and Nitrogen. Report from a workshop held at Skokloster, Sweden, 19–24. March 1988, *Edited by* J. Nilsson and P. Grennfelt, NORD 1988: 15, pp. 185–200.

Johnson, D. W., and Ball, J. T. 1990. Environmental pollution and impacts on soils and forests nutrition in North America. Water, Air, and Soil Pollution, 54: 3–20.

Johnson, D. W., and Lindberg, S. E. (*Editors*) 1992. Atmospheric Deposition and Nutrient Cycling. A Synthesis of the Integrated Forest Study, Springer, New York, 707+xiv pp.

Jonsson, C., Warfvinge, P., and Sverdrup, H. 1995. Uncertainty in predicting weathering rate and environmental stress factors with PROFILE model. Water, Air and Soil Pollution, 81:1–23.

Joslin, J. D., Kelly, J. M., and Van Miegrot, H. 1992. Soil chemistry and nutrition of North American spruce-fir stands: evidence for recent change. J. Environ. Qual., 21: 12–30.

Karltuna, E., and Gustafsson, J. P. 1993. Interference by organic complexation of Fe and Al on the SO_4^{2-} adsorption in Spodic B horizons in Sweden. J. Soil Sci. 44: 625–632.

Kolka, R. K., Grigal, D. F., and Nater, E. A. 1996. Forest soil mineral weathering rates: use of multiple approaches. Geoderma, 73: 1–21.

Klein, R. M., and Perkins, T. D. 1988. Primary and secondary causes and consequences of contemporary forest decline. The Botanical Review, 54: 1–43.

Kros, J., Groenenberg, J. E., van der Salm, C., de Vries, W., and Reinds, G. J. 1996. Validation and application of soil acidification models at local, national and European scale. DLO Winand Staring Centre, Wageningen (the Netherlands). Report 98, 158 pp.

Liu, S., Munson, R., Johnson, D. W., Gherini, S., Summers, K., Hudson, R., Wilkinson, K., and Pitelka, L. F. 1992. The nutrient cycling model (NuCM): Overview and Application. *In* Atmospheric Deposition and Nutrient Cycling. A synthesis of the Integrated Forest Study, *Edited by* D. W. Johnson and S. E. Lindberg, Springer, New York pp. 583–609.

Mitchell, M. J, and Lindberg, S. E., 1992. Sulfur chemistry, deposition, and cycling in forests. *In* Atmospheric Deposition and Nutrient Cycling. A Synthesis of the Integrated Forest Study, *Edited by* D. W. Johnson and S. E. Lindberg, Springer, New York, pp. 72–149.

Mohren, G. M. J., Bartelink, H. H., Jorritsma, I. T. M., and Kramer, K. 1993. A process based growth model (FORGRO) for analysis of forest dynamics in relation to environmental factors. *In* European forest reserves. Proceedings of the European forest reserves workshop, 6–8 May 1992, the Netherlands, *Edited by* M.E.A. Broekmeyer, W. Vos and H. Koop, PUDOC, Wageningen, pp. 273–280.

Nodvin, S. C., Driscoll, C. T., and Likens, G. E. 1986. The effect of pH on sulfate adsorption by a forest soil. Soil Science 142(2): 69–75.

Oja, T., and Arp, P. A. 1995. Dynamic modelling and the analysis of critical S and N loads. *In* Mapping and Modelling of Critical Loads for Nitrogen: a Workshop Report. Proceedings of the Grange-over-Sands Workshop, 24–26. October 1994, Institute of Terrestrial Ecology, Merlewood, pp. 154–157.

Oja, T., and Arp, P. A. 1997. A Forest Soil Vegetation Atmosphere Model (ForSVA), II: Application to Northern Tolerant Hardwoods. Ecol. Modell 95: 225–247.

Posch, M., Hettelingh, J. P., Sverdrup, H., Bull, K., and de Vries, W. 1993a. Guidelines for the computation and mapping of critical loads and exceedances of sulphur and nitrogen in Europe. *In* Calculation and Mapping of Critical Loads in Europe. CCE Status Report 1993, *Edited by* R. J. Downing, J.-P. Hettelingh, and P. A. M. de Smet, RIVM Report No. 259191003. Bilthoven, the Netherlands, pp. 25–38.

Posch, M., Reinds, G. J., and de Vries, W. 1993b. A simulation model for acidification's regional trends. Model description and users manual. Mimeograph Series of the National Board of Waters and the Environment, Helsinki, Finland, Report 477, 43 pp.

Posch, M., de Smet, P. A. M., Hettelingh, J.-P., and Downing, R. (*Editors*) 1995. Calculating and Mapping of Critical Thresholds in Europe. Status Report 1995. RIVM Rep. No. 259101004. 198+iv pp.

Roberts, T. M., Skeffington, R. A., and Blank, L. W. 1989. Causes of type 1 spruce decline. Forestry, 62(3): 179–222.

Roelofs, J. G. M., Kempers, A. J., Houdijk, A. L. F. M., and Jansen, J. 1985. The effect of air-borne ammonium sulphate on *Pinus nigra* var. maritima in the Netherlands. Plant and Soil, 84: 45–56.

Schindler, D. W. 1977. Evolution of phosphorus limitation in lakes. Science, 195: 260–262.

Singh, B. R. 1980. Distribution of total and extractable S and adsorbed $^{35}SO_4^{-2}$ in some acid forest soil profiles of southern Norway. Acta Agric. Scand, 30: 357–363.

Singh, B. R. 1984. Sulfate sorption by acid forest soils: 1. Sulfate adsorption isotherms and comparison of different adsorption equations in describing sulfate adsorption. Soil Sci, 138(3): 189–197.

Skeffington, R. A., and Wilson, E. J. 1988. Excess nitrogen deposition: issues for consideration. Environmental Pollution, 54: 159–184.

Sverdrup, H. U. 1990. The kinetics of base cation release due to chemical weathering. Lund University Press, Lund, 246 pp.

Sverdrup, H., and Warfvinge, P. 1990. The role of weathering and forestry in determining the acidity of lakes in Sweden. Water, Air and Soil Pollutution, 52: 71–78.

Sverdrup, H., Warfvinge, P., and Rosen, K. 1992. A model for the impact of soil solution Ca:Al ratio on tree base cation uptake. Water, Air and Soil Pollution 61: 193–221.

Sverdrup, H., Warfvinge, P., Kuylenstierna, and Chadwick, M. 1993. Soil acidification effect on growth of trees, grasses and herbs, expressed by the (Ca+Mg)/Al ratio. SNV Report No. 1541. Statens Naturvårdsverk. 171 85 Solna, Sweden.

Tomlinson, G. H. 1987. Nutrient deficiencies and forest decline. Pulp Pap. Can., 88: T43–T48.

Tomlinson, G. H. 1990. Nutrient disturbances in forest trees and the nature of the forest decline in Quebec and Germany. Water, Air, and Soil Pollution, 54: 61–74.

Ulrich, B., and Matzner, E. 1983. Abiotische Folgewirkungen der weiträumigen Ausbreitung von Luftverunreinigung. Forschungsbericht 104 02 615, UBA Berlin.

Ulrich, B., Mayer, R., and Khanna, P. K. 1979. Die Depositionen von Luftverunreinigungen und ihre Auswirkungen in Waldökosystemen im Solling. Schriften aus der Forstl. Fak. D. Univ. Gottingen und der Nierders. Verst. Anst. Bd. 58, 291 pp.

van Grinsven, H. J. M., Driscoll, C. T., and Tiktak, A. 1995. Workshop on comparison of forest-soil-atmosphere models: preface. Ecol. Modelling, 83: 1–6.

Warfvinge, P., Falkengren-Grerup, U., Sverdrup, H., and Andersen, B. 1993. Modelling long-term cation supply in acidified forest stands. Environ. Pollut, 80: 209–221.

Werner, B., and Spranger, T. (*Editors*) 1996. Manual on methodologies for mapping critical loads/levels and geographical areas where they are exceeded. UN ECE Convention on Long-Range Transboundary Air Pollution. Texte 71/96. Umwelt BundesAmt, 144 pp.

Wright, R., Holmberg, M., Posch, M., and Warfvinge, P. 1991. Dynamic models for predicting soil and water acidification: Application to three catchments in Fenno Scandia. NIVA-Report 25, Oslo, 40 pp.

Index